M000214334

VADOSE ZONE HYDROLOGY

Daniel B. Stephens

Cover illustration by:
Andrea J. Kron
cARTography by Andrea Kron
Los Alamos, New Mexico

LEWIS PUBLISHERS

Boca Raton   New York   London   Tokyo

Library of Congress Cataloging-in-Publication Data

Stephens, Daniel B.
    Vadose zone hydrology / Daniel B. Stephens.
      p. cm.
    Includes bibliographical references and index.
    ISBN 0-87371-432-6
    1. Groundwater flow. 2. Zone of aeration. 3. Unsaturated flow. I. Title.
GB1197.7.S74   1995
551.49--dc20
                                       95-13527
                                       CIP

No claim to original U.S. Government works
International Standard Book Number 0-87371-432-6
Library of Congress Card Number 95-13527
Printed in the United States of America 1 2 3 4 5 6 7 8 9 0
Printed on acid-free paper

PREFACE

This book on vadose zone hydrology is unique in both its title and scope. Its intent is to present elements of physical processes that are most often encountered by hydrogeologists and groundwater engineers in their projects undertaking characterization and monitoring of the vadose zone.

The contents of the book generally parallel a short course of the same title I have taught since 1990. This short course, in turn, is an abbreviation of a formal, graduate-level course in vadose zone hydrology developed during my academic career in the Geoscience Department at New Mexico Tech in Socorro from 1979 to 1989. Credit for developing that course, the first with such a title in a hydrogeological curriculum to the best of my knowledge, goes to Professor Lynn Gelhar, who provided encouragement and seed money for teaching and research materials that fostered my career and that of many of my graduate students as well.

To understand more about my point of view in writing this book, it is perhaps best for me to highlight my training and the professors who most influenced me. With a bachelor's degree in geology from Penn State and after a couple of years in the field, I entered a master's program at Stanford intent on obtaining formal training as a hydrogeologist. There my interest in unsaturated flow was stimulated by Professor Irwin Remson of the Applied Earth Science Department during his inspirational lectures on capillary barriers created by gravel layers and tales about Ike Winograd's work at the Nevada Test Site. After several years of consulting as a geologist, most of which was devoted to nuclear power plant and oil shale sites, I began a doctoral program at the University of Arizona the same year Professor Shlomo Neuman, a geologist and engineer, joined the hydrology faculty. Meetings with Professor Neuman lead to discovery of mutual interest in field permeability tests in boreholes and how capillary effects could influence the results of tests above the water table. To prepare for research on the subject, I took graduate courses in traditional soil physics from Professors Dan Evans and Art Warrick, which gave me a different perspective on applications of the science of unsaturated flow to agricultural problems. During my four years at Arizona, I became friends with Dr. L. Grey Wilson, a soil physicist, with whom I shared many interests for applying soil physics to problems of a hydrogeological nature, such as waste disposal and groundwater recharge. To these gentlemen, I remain forever in debt.

The nature of this book is a unique blend of soil physics and hydrogeology, written from the perspective of a hydrogeologist. The book is intended to be used primarily as a tool for those who already have received training in geology, hydrology, or engineering and who are dealing with problems involving the vadose zone. The audience in mind includes primarily consultants and regulators without formal training in soil physics; however, the book could be a reference or supplementary text for undergraduate courses in hydrology and soil science, as well as civil, geological, and environmental engineering. The book does not delve into derivations or complex mathematics, but it does display equations for the primary purpose of recognizing the data needs for predictive models in design or regulatory decision making. The level of presentation is significantly less intense and less broad in coverage than most introductory soil physics texts. On the other hand, the text is much more detailed and extensive in coverage of this topic than any current hydrogeology texts. The book includes an introduction to physical processes, including basic theory of flow, along with discussions and examples of some of the processes at the field scale that are essential for hydrogeologists to recognize. Considerable attention is devoted to recharge, inasmuch as this process often is the most crucial and difficult to evaluate in vadose zone problems of concern to hydrogeologists. The book includes a chapter devoted to a review of field and laboratory methods to characterize the hydraulic properties in the vadose zone, with case studies. The final two chapters deal with the timely subject of vadose zone monitoring. In this area more than any other, the field is rapidly evolving toward new and more sophisticated methods to detect contaminants above the water table. Case studies are presented covering seepage detection, landfill monitoring, and soil gas investigations.

In addition to thanks to my mentors and graduate students who kept me so enthusiastic about vadose zone hydrology, I want to acknowledge several people who have contributed directly to this book by permitting the use of portions of their prior publications for use here. Jeff Havlena was the senior author for much of the discussion on the borehole permeameter. Doug Reaber and Todd Stein contributed the portion of the text on the landfill case study, and Jeff Forbes was primarily responsible for a significant part of the case study on soil gas. I am indebted to reviewers of early drafts of and portions of this manuscript including Dr. Jim Yeh, Dr. Jan Hendrickx, Dr. Michael Sully, and Jeff Havlena. I am especially grateful to Dr. Mark Ankeny for his critical comments and persistence in reviewing the entire manuscript; however, I take full responsibility for any errors or omissions. Additionally, I am happy to thank Violet Tveit, Pamela Mathis, and Deborah Salvato for their exceptional organizational skills in assisting in preparation of the manuscript, and Linda Hirtz and Jhanine Huntsman for preparing many of the tedious illustrations. And finally, I want to thank my family, Deborah, Jake, and Jordan, for their love and patience during this ordeal.

The Author

Daniel B. Stephens, Ph.D., is Principal Hydrologist and President of Daniel B. Stephens & Associates, Inc., Albuquerque, New Mexico. Formerly chairman of the Geoscience Department at New Mexico Institute of Mining and Technology (NMIMT) in Socorro, New Mexico, he began private consulting in 1976 and founded Daniel B. Stephens & Associates, Inc. (DBS&A) in 1984. Dr. Stephens is an adjunct professor of geology at the University of New Mexico in Albuquerque and an adjunct professor of hydrology at NMIMT.

Dr. Stephens received his bachelor's degree in geological science from Penn State University, his master's degree in hydrology from Stanford University, and his doctorate in hydrology from the University of Arizona. Dr. Stephens is a certified professional hydrogeologist and a registered geologist in California and Arizona. He is a member of the American Society for Testing and Materials, the American Geophysical Union, the Geological Society of America, and the American Association of Ground-Water Scientists & Engineers. He has served on ASTM committees that establish guidelines and set standards for determining hydrologic properties and monitoring the vadose zone.

Dr. Stephens is internationally recognized as an authority on vadose zone hydrology. For several years, Dr. Stephens taught the vadose zone hydrology course at New Mexico Institute of Mining and Technology. This pioneering course, at the time one of the few vadose zone hydrology courses available at any university, covered the theory and application of vadose zone characterization and monitoring. Dr. Stephens has been invited to speak in vadose zone issues to national symposia sponsored by diverse groups such as the American Geophysical Union, the Soil Science Society of America, the American Association of Ground Water Scientists and Engineers, the New Mexico Geological Society, the New Mexico Environment Department, and the U.S. Nuclear Regulatory Commission.

Dr. Stephens has been a technical director of hundreds of environmental and hydrogeological consulting projects, dozens of which include vadose zone issues such as landfill siting, seepage analysis, and monitoring systems. He has also been a pioneer in developing methods to characterize the hydrologic properties of soil. He developed the first field method that includes capillary effects to determine the saturated hydraulic conductivity of soil from a borehole permeameter. Through extensively instrumented field sites, Dr. Stephens and his colleagues have discovered new physical processes which induce significant horizontal flow components to soil water movement.

Dr. Stephens has published over 28 articles in peer-reviewed professional journals and given over 47 presentations and articles in symposia proceedings. In addition to his expertise in vadose zone hydrology, Dr. Stephens specializes in recharge in semiarid environments, application of numerical models, and aquifer monitoring and contamination problems.

This book is dedicated to my loving parents,
Dallas W. and Jean E. Stephens

Table of Contents

Chapter 3

Chapter 4

read

Chapter 5

Glossary of Common Vadose Zone Terms

Advection: fluid migration induced by hydraulic gradients.

Anistropy: a characteristic of the flow or transport properties of geologic media in which the value of the property depends upon the direction that the property is measured.

Available water: the difference between water content at field capacity and permanent wilting.

Bulk density: mass of dry soil per unit volume of bulk soil; dry bulk density.

Capillary pressure: the difference in nonwetting and wetting fluid potentials; identical to matric potential if the air, the nonwetting fluid, is at atmospheric pressure.

Capillary fringe: that part of the vadose zone immediately above a water table where the media is satiated but the water is under tension.

Contact angle: the angle created by the interface between a solid surface and fluid phases in contact with that surface.

Deep percolation: infiltration below the root zone depth, usually in the context of irrigation.

Diffuse recharge: recharge from the vadose zone that originates as infiltration of rain and snow melt over large portions of a watershed.

Diffusion: a transport process in which chemicals migrate in fluid due to concentration gradients.

Dispersivity: a characteristic of the geologic medium attributed to tortuosity and heterogeneity that affects mechanical mixing of chemicals during advection.

Distribution coefficient: the partitioning coefficient for a chemical between water and the solid phases, ratio of concentration in solid to concentration in water.

Elevation head: gravitation (elevation) potential expressed in units of potential energy per unit weight of fluid with respect to a datum.

Evapotranspiration: the process of water discharge from the vadose zone by direct evaporation of soil water and uptake by plant roots with transpiration to the atmosphere.

Field capacity: water content of a field soil 2 to 3 d following a thorough irrigation; water content at -0.1 or -0.33 bars soil-water potential.

Gravimetric water content: mass of water per unit mass of dry soil.

Gravitational potential: potential energy of soil water due to its position above a reference datum; identical to elevation potential.

Henry's law constant: the partitioning coefficient for a chemical between the gas and liquid phases, ratio of concentration in gas to concentration in liquid.

Hydraulic conductivity: volumetric rate of fluid flow per unit cross-sectional area under a unit hydraulic gradient at a prescribed temperature; ability of media to conduct a particular fluid.

Hydraulic gradient: fluid driving force per unit weight of fluid.

Hydrodynamic dispersion: a transport process causing chemical mixing in the geologic media attributable to the net effects of molecular diffusion and mechanical mixing in the advected liquid.

Hysteresis: a phenomenon describing a relationship between parameters, usually pertaining to hydraulic conductivity and water retention, in which the parameter depends on nature of change in the process, such as wetting or drying.

Imbibition: the process of wetting a geologic median with water.

Infiltration: the rate of water movement into the soil per unit area due to gravity, capillary, and pressure forces, usually associated with vertical flux density across the soil surface.

Interfacial tension: the potential energy associated with the surface area separating two immiscible fluids that is attributed to differences in cohesive forces between molecules of the respective fluids.

Local recharge: recharge from the vadose zone that originates from concentrated surface runoff in channels or seepage from impounds.

Matric potential: the component of soil-water potential relative to a reference state due to capillary and adsorptive forces that hold water in porous or fractured media.

Osmotic potential: the potential energy relative to a reference state attributed to chemical concentrations in the soil water.

Partitioning: a transport process in which chemicals are distributed between solid, liquid, and gas phases, depending upon solubility, sorption, and vapor pressure characteristics.

Perched aquifer: an aquifer within the vadose zone created by a relatively low permeable perching layer.

Permanent wilting: water content of soil at which plants become so dry that the plant cannot survive even if the soil is rewetted; water content at −15 bars soil-water potential.

Permeability: an intrinsic property of the porous or fractured media describing the fluid transmissive character.

Phreatic zone: regional zone of saturation underlying the vadose zone.

Porosity: volume of voids per unit bulk volume of soil.

Pressure head: soil-water potential expressed in units of potential energy per unit weight of fluid.

Pressure potential: the potential energy relative to a reference state due to air pressure or hydrostatic pressure.

Recharge: water that enters an aquifer.

Redistribution: the simultaneous movement of soil water upward due to evapotranspiration and downward due to infiltration.

Relative permeability: ratio of permeability at field water content to the permeability at saturation.

Relative humidity: ratio of vapor pressure at ambient conditions to saturated vapor pressure under the same conditions.

Relative hydraulic conductivity: ratio of hydraulic conductivity at field water content to the saturated hydraulic conductivity.

Retardation factor: velocity of a chemical relative to the mean porewater velocity.

Satiated: maximum saturation achievable under prevailing field conditions.

Saturation percentage: volume of fluid per unit volume of void space.

Soil-water characteristic curve: the relationship between pressure head or soil-water potential and water content of a porous or fractured media; the soil-water retention curve.

Soil-water diffusion coefficient: a property of the medium describing the chemical mass flux due to the concentration gradient; constant of proportionality in Fick's law of diffusion for soil.

Soil-water diffusivity: the ratio of hydraulic conductivity and specific water capacity; a constant of proportionality relating soil-water flux and water content gradient.

Soil-water flux: volumetric rate of fluid flow per unit cross-sectional area perpendicular to flow.

Soil-water potential: the potential energy of water in the vadose zone relative to a reference state due to the sum of matric, pressure, and osmotic potential.

Soil-water retention curve: soil-water characteristic curve.

Specific yield: volume of water released from or taken into storage per unit horizontal area of an unconfirmed porous or fractured media per unit change in water table elevation; storage coefficient of media under unconfined conditions.

Specific retention: water content at which the water phase becomes virtually discontinuous.

Specific moisture (water) capacity: the volume of water released from or taken into storage per unit horizontal area of geologic medium per unit change in pressure head; slope of the soil-water characteristic (water content versus pressure head) curve.

Thermal diffusion coefficient: constant of proportionality relating soil-water plus vapor flux and temperature gradients.

Thermal diffusivity: the ratio of thermal conductivity to heat capacity.

Total hydraulic head: total soil-water potential expressed in units of potential energy per unit weight of fluid.

Total soil-water potential: the sum of soil-water potential and gravitational potential.

Unsaturated media: porous or fractured media where the voids are occupied by both water and air phases.

Vadose zone: geologic media between the land surface and the regional water table.

Void ratio: volume of voids per unit volume of solids.

Volumetric water content: volume of water per unit volume of bulk soil; also, water content.

Water-retention curve: soil-water retention curve; soil-water characteristic curve.

Water content: volumetric water content.

Water table: surface in a geologic medium where water pressure equals atmospheric pressure.

LIST OF SYMBOLS

Symbol	Description	Example units
1	Stream tube segment closest to the source where saturation occurs	Dimensionless
ΔS	Change in soil-water storage	m^3/d
a	Hydraulic conductivity fitting parameter	Empirical; depends on ℓ
A	Cross-sectional area of the stream tube in the center of the segment	m^2
a_e	Electrode spacing	m
A_e	Water table evaporation parameter	Dimensionless
A_f	Falling head permeameter geometry factor	m
A_i	Initial radio activity	Millicuries
A_o	Observed radio activity	Millicuries
A_{oT}	Amplitude of soil surface temperature fluctuation	°C
AW	Available water	Dimensionless
b	Hydraulic conductivity fitting parameter	Empirical depends on ℓ
B	Empirical soil textural classification parameter	Dimensionless
b_d	Dimensionless constant in disc permeameter solution	Dimensionless
b_k	Klinkenberg slip flow coefficient for the gas-solid system	Pa
c	Propagation velocity of an electromagnetic wave in free space	$3 \times 10^8 m/s$
C	Specific moisture capacity	m^{-1}
C_g	Gas concentration	$\mu g/L$
C_o	Solute concentration	mg/L
Cl_p	Specific heat	$J/kg\text{-}°K$
Cl_s	Chloride concentration in precipitation and dry fallout	mg/L
C_s	Average soil chloride concentration below root zone	mg/L
d	Depth	m
D_w	Effective self-diffusion coefficient of water	m^2/s
D	Soil-water diffusivity	m^2/s
D^*	Molecular diffusion coefficient	m^2/d
D_a	Gaseous chemical diffusion coefficient in air	m^2/s
D_g	Gaseous chemical diffusion coefficient in soil	m^2/s
dH/dt	Rate of decline in head in the reservoir	m/s
D_m	Dispersion due to mechanical mixing	m^2/d
D_T	Combined gas and liquid thermal diffusion coefficients	$m^2/s\text{-}°C$
$D_{T,g}$	Gas thermal diffusion coefficient	$m^2/s\text{-}°C$
$D_{T,l}$	Liquid thermal diffusion coefficient	$m^2/s\text{-}°C$
D^v	Effective diffusivity of water vapor in air	m^2/s
D_w	Pure water diffusion coefficient	m^2/s
D'	Hydrodynamic dispersion coefficient	m^2/d
E	Evaporation	m^3/d
E_m	Evaporative flux density	$g/s\text{-}m^2$
ET	Evapotranspiration rate	cm/d
f	Conversion factor	Pa/m^3
f_{oc}	Mass fraction of organic carbon	Dimensionless
g	Gravitational constant	$32.2\ ft/s^2$ $9.8\ m/s^2$
G	Heat flux density into soil	w/m^2
G	Specific gravity	Dimensionless
h	Height of fluid column in a tensiometer	m
H	Total hydraulic head	m
h_c	Height of capillary rise	m
h_p	Water height equivalent pressure head of fluid phase P	m
H_p	Depth of the ponding	cm

	Description	Example units
h_r	Relative humidity	Dimensionless
H_s	Sensible heat flux density	w/m²
i	Infiltration rate	m/s
I	Current	Amperes
I	Infiltration	m³/d
i_f	Final infiltration rate	m/s
i_o	Initial infiltration rate	m/s
J	Hydraulic gradient	Dimensionless
k	Intrinsic permeability	m²
K	Hydraulic conductivity	m/s
$K(\psi)$	Unsaturated hydraulic conductivity of the soil	m/s
K_a	Hydraulic conductivity of soil to air	m²/Pa-s
K_{ad}	Apparent dielectric constant	Dimensionless
k_{app}	Apparent measured air permeability	m²
K_c	Crop coefficient	Dimensionless
K_{CO}	Crop coefficient with unlimited water availability	Dimensionless
K_d	Distribution coefficient	mL/g
K_e	Electrical resistivity geometric factor	m
K_F	Hydraulic conductivity of phase F	m/s
$K_H{}'$	Henry's law constant	atm-m³/mol
K_H	Henry's law constant	Dimensionless
Ki	Hydraulic conductivity at the initial water content	m/s
K_o	Hydraulic conductivity behind the wetting front	m/s
K_{oc}	Partitioning coefficient between organic liquid and organic carbon	mL/g
K_{ow}	Octanol-water partitioning coefficient	mL/g
K_o'	Conductivity extrapolated to zero pressure head	m/s
K_p	Partitioning coefficient	mL/g
k_r	Relative permeability	Dimensionless
K_r	Relative hydraulic conductivity	Dimensionless
K_s	Saturated hydraulic conductivity	m/s
k'	Empirical infiltration constant	s⁻¹
L	Length of the test interval; distance or length	m
m	Fitting parameter soil-water retention curve	Dimensionless
M	Molar mass	g/mol
n	Porosity	Dimensionless
N	Fitting parameters to water retention curve	Dimensionless
n_e	Effective porosity	Dimensionless
N_{sat}	Saturated water vapor density in air	kg/m⁻³
P	Precipitation	cm
P	Pressure; vapor pressure	Pa; mm Hg
P_a	Atmospheric pressure	Pa
P_c	Capillary pressure	Pa
PET	Potential evapotranspiration rate	cm/d
P_g	Gauge pressure	Pa
P_{nw}	Pressure of nonwetting fluid	Pa
P_o	Atmospheric pressure at sea level	Pa
P_p	Pressure in phase P	Pa
P_T	Total vapor pressure of a chemical mixture	Pa; mm Hg
P_w	Pressure of wetting fluid	Pa
q	Darcy velocity, specific discharge, fluid flux, or flux density	m/s
Q	Volumetric	m³/s
$q_{g,p}$	Advective gas flow due to total gas pressure gradients	m/s
$q_{g,T}$	Gas flow due to temperature gradients	m/s
q_m	Mass flux of a chemical	kg/s
q_v	Water vapor flux density	m/s
$q_{\ell,P}$	Liquid flow due to potential energy gradients	m/s
$q_{\ell,T}$	Liquid flow due to temperature gradients	m/s

r	Radius	m
R	Recharge	m/s
R	Molar gas constant	≈ 8.3 J/mol-°K
R_f	Retardation factor	Dimensionless
R_n	Net radiation	w/m²
R_p	Net mass transfer into the system per unit volume of porous media	kg/m³-s
R_r	Radius of reservoir	m
s	Sorptivity of soil	cm/s$^{1/2}$
S	Saturation	Dimensionless
s_c	Coil spacing in EM induction method	m
S_e	Effective saturation	Dimensionless
S_g	Gas-phase saturation	Dimensionless
S_P	Saturation percentage of phase P	Dimensionless
S_r	Rotation speed	T^{-1}
S_s	Specific storage	m^{-1}
S_y	Specific yield of an aquifer	Dimensionless
t	Time	s,d,yr.
T	Transpiration	m³/d
T	Temperature	°C; °K
t_a	Travel time or age of the ground water	Years
T_a	Average soil temperature	°C
$t_{1/2}$	Half-life	Years
u_i	Unit gravitational vector	Dimensionless
v	Travel velocity along TDR wave guide	m/s
v	Volume of the air tank	m³
V	Mean pore velocity	m/d
V_c	Velocity of the chemical	m/d
V_s	Velocity of solute front	m/d
V_{wf}	Velocity of wetting front	m/d
w	A soil parameter characterizing the moisture retention curve	Dimensionless
w'	Wind speed variation from mean	m/s
x	Horizontal space coordinate; horizontal distance	m
X_c	Mole fraction	Dimensionless
x_i	Cartesian space coordinate (i = 1,2,3)	m
y_o, y_1	Falling head permeameter manometer reading	m
z	Vertical space coordinate; depth; elevation head	m
μ	Dynamic viscosity of fluid	Poise
ℓ	Hydraulic conductivity fitting parameter	Dimensionless
e	Void ratio	Dimensionless
∇V	Voltage drop	Volts

Greek

α	Slope of luK-ψ curve	m^{-1}
α_c	Contact angle	Degrees
α_d	Dispersivity	m
α_v	Van genuchten fitting parameter to soil-water retention curve	m^{-1}
β'	Saturation index parameter	Dimensionless
β	Slope of luK-θ curve	Dimensionless
β^r	Bowen ratio	Dimensionless
γ	Psychrometric constant	Dimensionless
γ_d	Dry sample unit weight	kg/m³
γ_t	Total sample unit weight	kg/m³
δ_s	Dip of stratification	Degrees
δ	Relative difference between isotopic ratios in precipitation	Parts per thousand
δ_D	Standardized isotope ratio	Per mil
δ_D^{REC}	Standardized isotope ratio of recharge water	Per mil
ε_D	Equilibrium enrichment factor	Dimensionless
η_D	Diffusion ratio excess	Dimensionless

	Description	Example units
θ	Volumetric water content	m^3/m^3
θ_e	Effective water content	m^3/m^3
θ_{fc}	Field capacity	m^3/m^3
θ_g	Gravimetric water content	kg/kg
θ_i	Initial water content	m^3/m^3
θ_o	Water content behind the wetting front	m^3/m^3
θ_r	Residual water content	m^3/m^3
θ_s	Saturated water content; porosity	m^3/m^3
ϕ_T	Total soil-water potential	J/kg
θ_{wp}	Permanent wilting point	m^3/m^3
λ	Decay constant	yrs^{-1}
λ_p	Pore size distribution index	Dimensionless
λ_1	Correlation length scale	m
λ_v	Latent heat of vaporization	J/g
μ_o	Electric permeability of free space	henrys/m
ρ	Fluid density	kg/m^3
ρa	Air density	kg/m^3
ρ_{ar}	Apparent resistivity	ohm-m
ρ_b	Dry bulk density	kg/m^3
ρ_{rp}	Specific gravity of phase P	Dimensionless
ρ_w	Density of pure water	kg/m^3
σ	Interfacial tension	dynes/cm
τ	Tortuosity factor	Dimensionless
ϕ_g	Elevation potential	j/kg;Pa
ϕ_m	Matric flux potential	cm^2/s
ϕ_{matric}	{Matric potential of soil-water}	J/kg; Pa
$\phi_{osmotic}$	{Osmotic potential of soil-water}	J/kg; Pa
$\phi_{pressure}$	{Tensiometer pressure potential of soil-water}	J/kg; Pa
ϕ_{sw}	Soil-water potential	J/kg; Pa
Ψ	Pressure head	m
Ψ_a	Air-entry pressure head	m
Ψ_{cr}	Critical pressure or bubbling pressure head	m
Ψ_f	Mean pressure head at the wetting front	m
Ψ_i	Initial pressure head	m
Ψ_w	Water-entry pressure head	m
ω	Period of temperature cycle	d^{-1}

CHAPTER 1

Basic Concepts and Theory

Most hydrogeologists and engineers concerned about the vadose zone are typically faced with some aspect of characterizing or monitoring it. Any rationale for characterizing and monitoring in the vadose zone must be based upon fundamental concepts governing vadose zone processes and theory of flow and transport. In many instances, the motivation for vadose zone characterization is to provide quantitative information and parameters to make predictions with mathematical models. For instance, if seepage from a landfill should occur, models can be used to predict likely pathways, as well as the rate of migration and the sensitivity required for a monitoring system to detect such seepage. To gain a better understanding of what parameters require quantification during site characterization, in this chapter we present mathematical expressions governing flow and transport in the vadose zone. But first, we define the vadose zone and some basic terminology.

The vadose zone is generally defined as the geologic media between land surface and the regional water table (Figure 1). The upper part of the vadose zone commonly includes the plant root zone and weathered soil horizons. Within the vadose zone, soils and bedrock are usually unsaturated; that is, their pores are only partially filled with water.

In places, however, the vadose zone may become water saturated; consequently, the vadose zone is not synonymous with the unsaturated zone. One obvious location where this occurs is just above the regional water table, where capillary rise causes water to fill the pore spaces. Here, the pores are essentially saturated but the water is held under a tension, that is, the water pressure is less than atmospheric pressure. The thickness of this zone of tension saturation may be less than 10 cm for gravels to more than 2 m for clays. Saturated regions may also be found where the water is under positive pressures, such as above a low permeable layer where perched conditions may develop. A perched aquifer occurs within and is part of the vadose zone, and it is separated from the regional water table below by an unsaturated zone. These unsaturated zones above and below the perched aquifer are critical to the definition and field identification of a perched aquifer. In the vadose zone, saturated conditions also may develop locally beneath surface impoundments and drainages, as well as over more extensive areas near land surface during infiltration following precipitation events or flood irrigation.

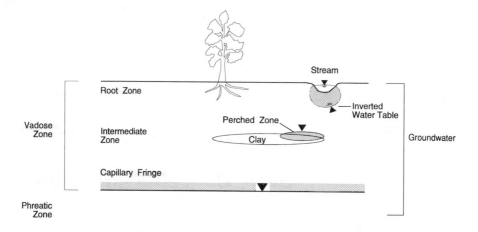

Figure 1   Conceptual model of the vadose zone.

To many hydrogeologists, groundwater occurs at a depth where water first enters a well. But Figure 1 shows that, by our preferred definition, the groundwater system includes the vadose zone as well. The groundwater system also includes the geologic medium below the regional water table, the phreatic zone, which will be either an aquifer, aquiclude, or aquitard, depending upon the permeability, well yield, and water use. The elevation of the water level in wells perforated across the uppermost of these saturated units in the phreatic zone will define the top of the water table and also the base of the vadose zone.

Note in Figure 1 that flow paths that would be drawn beneath the local impoundment or perched zone begin in saturated zones, would traverse through unsaturated media, and then enter the saturated zone again at the top of the capillary fringe. Along this flow path, the hydraulic properties vary as a function of the degree of fluid saturation. Within otherwise homogeneous media, this spatial variation in hydraulic properties, such as hydraulic conductivity, is an important difference between the vadose zone and aquifers. More detail on the nature of vadose zone hydraulic properties is presented later in this chapter.

In summary, the vadose zone has the following general characteristics:

- The vadose zone lies below the land surface and above the regional water table.
- Water pressure within most of the vadose zone is usually less than atmospheric pressure, although some areas of complete saturation and positive fluid pressure may occur.
- Flow properties are dependent upon the degree of saturation in the pore space.

In the following sections, we review the basic theory and concepts that allow us to understand the physical processes acting on the vadose zone. Section A deals with driving forces, Sections B and C address fluid storage, and Section E discusses fluid transport properties.

I. ENERGY STATUS OF POREWATER

What causes water to flow in the vadose zone? Just as in aquifers, the vadose zone water flows from areas of high potential energy to areas of low potential energy. To determine the direction of the driving force acting on water in the vadose zone, one will need to quantify the total potential energy of the soil water. The gradient of this total soil-water potential, that is, the change in potential energy with direction in space, determines the magnitude of the driving force acting on the fluid.

Mathematically, the total soil-water potential is expressed as:

$$\begin{array}{ccccc} \phi_T & = & \phi_{SW} & + & \phi_G \\ \text{total soil - water} & & \text{soil - water} & & \text{elevation} \\ \text{potential} & & \text{potential} & & \text{potential} \end{array} \qquad (1)$$

Notice that there are two primary components to the total soil-water potential, just as there are for the total hydraulic head in an aquifer.

Before defining the total soil-water potential components further, recognize that much of the discussion in this section was developed by soil physicists for problems of flow through surficial deposits having agricultural significance. Hence, the term *soil* is a common modifier of water potential energy. However, the concepts extend to lithified porous media or bedrock as well.

The first component of the total soil-water potential is called, simply, the soil-water potential, ϕ_{SW}. It accounts for soil-water movement due to gradients in capillary pressure, chemical, temperature, and electrical potential. These components of the soil-water potential have been derived from fundamental thermodynamic principals. We avoid these details here and summarize the main components of the soil-water potential:

$$\phi_{SW} = \phi_{matric} + \phi_{pressure} + \phi_{osmotic} \overset{diff. \ in \ concentration}{} \qquad (2)$$

_{↳ capillary → wet soils}
_{↳ adsorptive → dry soils}

The matric component, ϕ_{matric}, recognizes that the energy state of water, relative to the reference state of an open body of pure water at 20°C, is influenced by the mutual attraction between the liquid, gas, and solid phases of the soil. In the definition of matric potential, we assume that the other components, such as air pressure and chemistry of the pore fluids, are the same as those in the reference reservoir. Matric potential is attributed to capillary forces, which predominate in wet soils, and adsorptive forces, which are more important in dry soils. Water held by capillarity occurs in the main pore bodies, whereas adsorbed water occurs where the polar water molecules coat the charged surfaces of the solid soil particles (Figure 2). In clay soils, water can also be bound to soil particles in the electrostatic double layers. Where the soil is fully saturated and no air-water interface is present, $\phi_{matric} = 0$.

In the relatively wet range, which is generally at matric potentials greater than about −1 bar, the matric component is expressed by the capillarity equation:

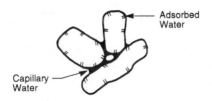

Figure 2 Matric potential is due to water held in soil by capillary and adsorptive effects.

$$\phi_{matric} = \frac{2\sigma}{r} \tag{3}$$

where σ is the interfacial tension of the liquid (MT^{-2}), and r is the radius of curvature on the air-water meniscus. If the meniscus is concave toward the air, r is negative by convention. Therefore, the matric potential is negative as referenced to a free water surface at zero potential. Equation 3 can be written in terms of the radius of a capillary tube, r_t, by recognizing that $r = -r_t/\cos \alpha_c$:

$$\phi_{matric} = -\frac{2\sigma \cos \alpha_c}{r_t} \tag{4}$$

where α_c is the contact angle made by the liquid-gas interface where it contacts the solid.

For a particular porous medium, the capillary forces are attributed to the fluid's interfacial tension and contact angle (Equation 4). Interfacial tension occurs between two immiscible fluids because of the mutual cohesive attraction of like molecules in each fluid. Recall from fundamentals of chemistry that water is a polar molecule that is attracted to other polar water molecules. Within a reservoir these water molecules are attracted equally in all directions to adjacent water molecules, so there is no net force imbalance between molecules within the reservoir. On the other hand, the molecules on the surface of the reservoir are only attracted to water molecules below and adjacent to them, because there are no liquid-phase water molecules above the reservoir. Therefore, at the air-water interface, there is a force imbalance, with the resultant force in the direction of the water. Consequently the surface contracts, thereby increasing the surface area of the air-water interface. To restore this interface to a plane would require some expenditure of energy over the surface; hence, interfacial tension has units of energy per unit area, or force per unit length acting in a direction tangent to the curved surface. Interfacial tensions can exist between various fluids, but when a liquid is in contact with air, the interfacial tension is commonly referred to as surface tension.

One method to measure surface tension is with a du Nuoy precision tensiometer or du Nuoy interfacial tensiometer following the standard test method of the American Society for Testing and Materials (ASTM), D-1950. Another method is based on measuring the height of capillary rise and contact angle in a clean glass tube. Interfacial tension values for selected fluids are shown in Table 1. As these data indicate, interfacial tension is influenced by temperature and fluid characteristics. Increasing temperature causes a decrease in interfacial tension, whereas increasing solute concentration can either increase or decrease interfacial tension. Adding to

Table 1 Interfacial Tension of Selected Fluids

Fluids	Interfacial tension (dynes/cm)	Temperature (°C)
Pure water—air	74.22[a]	10
Pure water—air	72.75[a]	20
Pure water—air	71.18[a]	30
Pure mercury—air	435.5[a]	20
10.46% (wt) NaCl solute—air	76.05[a]	20
25.92% (wt) NaCl solute—air	82.55[a]	20
10.0% (wt) acetone solute—air	48.90[a]	25
50.0% (wt) acetone solute—air	30.40[a]	25
Pure water—benzene	35.00[a]	20
Pure water—n-octane	50.8[a]	20
Pure water—carbon tetrachloride	45.00[a]	20
Pure water—tetrachloroethyene (PCE)	45.00[b]	—
Tetrachloroethene (PCE)—air	31.3[c]	—
Trichloroethene (TCE)—air	34.0[c]	—

[a] Weast, R. C., Ed., *Handbook of Chemistry and Physics*, CRC Press, Boca Raton, FL, 1974. With permission.

[b] Kueper, B. H., and Frind, E. O., Two-phase flow in heterogenous porous media, 2. Model application, *Water Resources Research,* 27(6), 1059, 1991. With permission.

[c] Cohen, R. M., and Mercer, J. W., *DNAPL Site Evaluation,* C. K. Smoley, Boca Raton, FL, 1993. With permission.

water electrolytes such as sodium chloride (NaCl) tends to increase interfacial tension because the ions increase the cohesive attraction between molecules in the solution. On the other hand, organic molecules tend to reduce surface tension of water. Natural soil solutions also contain organic acids and dissolved organic matter, which act as surfactants. Surfactants in kitchen detergents cause soap and water bubbles to spread out because they decrease the surface tension of water, thereby minimizing water spots on glassware when it dries. Oil recovery in petroleum reservoirs is sometimes enhanced by adding organic liquid surfactants to injection water which tends to dislodge trapped oil blobs. Water in the vadose zone contaminated with dissolved organic compounds from petroleum hydrocarbons and chlorinated solvents, for example, also would exhibit a decrease in interfacial tension. When free-phase organic liquids become trapped in porous media, one way to facilitate their remobilization and remediation is to flush surfactants through the porous media to decrease the interfacial tension.

Another important parameter in the capillary equation (Equation 4) is the contact angle. The contact angle is a direct result of the interfacial forces described earlier and the attraction of the liquids for the solid. The contact angle is important because it reflects the wettability of the solid to the liquid, that is, whether the liquid will tend to spread across the solid surface. If the contact angle is between 0 and approximately 70 degrees, the system is said to be water wet; if it is greater than 110 degrees, the liquid is nonwetting; and if it is between these values, it is neutrally wet (Anderson, 1986a). The contact angle can be measured by analyzing close-up photographs of the fluid in contact with a smooth glass slide, crystal, or thin section (Anderson, 1986b). Cohen and Mercer (1993) summarized four other methods to evaluate wettability. For pure water on clean glass, the contact angle is zero, (cos α = 1), so σ cos α_c in Equation 4 would be about 72 dynes/cm at 20°C (Table 1). However, for water in most earth materials, $\sigma \cos \alpha_c$ is actually about 60 dynes/cm (Corey, 1977), owing to nonzero contact angles. There are few measurements of contact angle reported in the literature, but Cohen and Mercer reported values for a variety of solvents characteristic of landfill leachate. In most flow and transport problems, the vadose zone is usually regarded as water wet.

The pressure component, $\phi_{pressure}$, accounts for either the soil air pressure or the positive hydrostatic pressure in water-saturated regions. Some scientists refer to these two components of the pressure potential as the pneumatic and submerged potential, respectively (e.g., Taylor and Ashcroft, 1972). Within unsaturated zones the pressure potential would be measured by air piezometers, and within perched saturated zones, this component would be obtained by water-level measurements in monitor wells. It should be noted that in other definitions of soil-water potential, the matric potential, ϕ_{matric}, and pressure potential, $\phi_{pressure}$, as described earlier are combined and simply called pressure potential (Mullins, 1991).

For two-phase flow problems, Mullins' definition of the pressure potential would be consistent with the petroleum engineers' capillary pressure term. Petroleum engineers have worked extensively on capillary phenomena for decades in order to enhance oil and gas production. They use the term *capillary pressure*, P_c, to represent the difference in fluid pressure between the nonwetting fluid, P_{nw}, and the wetting fluid, P_w:

$$P_c = P_{nw} - P_w \tag{5}$$

In terms of the soil pressure potential we discussed earlier in Equation 2, pressure of the nonwetting fluid would be represented by the pressure potential, $\phi_{pressure}$, and the pressure of the wetting fluid would be represented by the matric potential, ϕ_{matric}. The capillary pressure is identical to the matric potential, although opposite in sign, in a soil-water system where the nonwetting phase is air at atmosphere pressure. In soil-water systems, water is usually the wetting phase and air is the nonwetting phase. But, in some cases, such as when water infiltrates a soil containing natural organic compounds that render the soil hydrophobic, it is possible that water may be nonwetting relative to the other phase.

The interface between any two immiscible fluids is curved, with the nonwetting fluid on the concave side. The nature of the curvature always indicates which fluid has the greater pressure. In the case of water in a capillary tube, the pressure on the air side of the meniscus is atmospheric, whereas the pressure on the water side is less than atmospheric. Similarly, the pressure inside a raindrop is greater than that in the surrounding air. This important phenomenon is discussed again later in regard to capillary rise.

The osmotic component, $\phi_{osmotic}$, takes account of the chemical concentration differences that may influence the energy state of the water. Increasing the chemical concentration of the pore fluid will lower the potential energy of the liquid relative to the pure water reference state. Hence, if there is a semipermeable membrane that separates regions of different concentration, then water will tend to flow from the region of lower concentration to the area of higher concentration due to the gradient in the osmotic potential. Examples of semipermeable membranes include the air-water interfaces within porewater and the Casparian strip in plant roots. Within plant roots, the high electrolyte concentration lowers the osmotic potential and facilitates water uptake of relatively less saline soil water, which is at a greater potential. If the salinity of the soil water increases sufficiently, the plant may permanently wilt even if the soil is thoroughly wetted, owing to the lack of sufficient potential gradient across the root membrane.

Gradients in the temperature potential also cause soil water to move from areas of high temperature to areas of low temperature (Taylor and Ashcroft, 1972). Temperature components can be important, for example, in problems of soil water movement in hot, dry soils, and in performance assessments of high-level radioactive waste where heat is generated. After considering all the components of total soil-water potential in the vast majority of practical vadose zone problems, matric potential remains the most important to characterize, inasmuch as gradients of the other components of the soil-water potential are usually small relative to the soil-water potential gradient and gravitational gradient.

The components of total soil-water potential can be expressed as either a potential energy per unit weight (L), potential energy per unit mass (L^2T^{-2}), or potential energy per unit volume ($MT^{-2}L^{-1}$). The latter is simply in conventional units of pressure (e.g., bars, lbs/in^2, pascals), and hence the often interchangeable use of the term *pressure* or *tension* for potential. For computations, it is usually more convenient, however, to use potential energy per unit weight, so that the gravitational potential can be expressed in units of length and calculated simply as the elevation of the measurement point above some datum. When the soil-water potential is expressed as a potential energy per unit weight, the soil-water potential is called the pressure head, ψ. Many soil scientists and engineers use different units of pressure to express soil-water potential, so it is useful to be familiar with some unit conversions:

$$1 \text{ bar} \approx 0.99 \text{ atm} \approx 100 \text{ kPa} \approx 1017 \text{ cm H}_2\text{O} \qquad (6)$$

To convert the volume-based soil-water potential, ϕ_{sw}, to pressure head, ψ, the following equation is used:

$$\psi = \frac{\phi_{sw}}{\rho g} \qquad (7)$$

where ρ is the fluid density and g is the gravitational constant. When the potential energy of water in soil is expressed on a per unit weight basis, the total potential energy equation (Equation 1) becomes:

$$H = \psi + z \qquad (8)$$

where H is the total hydraulic head and z is the elevation head relative to an arbitrary datum. This is exactly the same equation routinely used by groundwater hydrologists to compute hydraulic head and hydraulic gradients in aquifers, except that this form is more general in that, with our definition of ψ, water movement can occur under unsaturated as well as saturated conditions.

Earlier in this section, it was pointed out that one of the key characteristics of the vadose zone is that the soil-water potential in the partially saturated soil pores is less than zero, relative to a free water surface. Keep in mind that the soil-water potential represents the work done to a unit volume of the soil water in the process of changing its conditions from the reference state to its present state; the units of pressure happen to be a consequence of this definition. The actual pressure in the water cannot be less than −1 atm, the maximum difference between a perfect vacuum (zero absolute pressure) and local atmospheric pressure; however, the value of the soil-water *potential* can indeed be much less than −1 atm due to the additive effects of all the components of the soil-water potential described by Equation 2. The soil-water potential, including pressure head, would be positive within perched zones and below the water table, whereas at the water table, the soil-water potential would be essentially zero. Elsewhere, the soil-water potential is negative. Bear in mind that there is a continuum of pressure in the vadose zone and this continuum extends into the underlying phreatic zone as well.

II. WATER CONTENT

The water content of the soil is probably the easiest property of the vadose zone to understand. However, different disciplines deal with this property in different ways. There are three different expressions for water content:

$$\text{Volumetric water content}: \quad \theta = \frac{\text{volume water}}{\text{bulk volume soil}} \quad \left(\text{cm}^3\ \text{cm}^{-3}\right) \quad\quad (9)$$

$$\text{Gravimetric water content}: \quad \theta_g = \frac{\text{mass water}}{\text{mass dry soil}} \quad \left(\text{g g}^{-1}\right) \quad\quad (10)$$

$$\text{Saturation percentage}: \quad S = \frac{\theta}{n} \times 100 \quad \left(\text{cm}^3\ \text{cm}^{-3}\right) \quad\quad (11)$$

Most hydrologists and soil scientists prefer the volumetric water content, while geotechnical engineers generally use gravimetric water content, and petroleum engineers often work with saturation percentage. The volumetric water content is typically 40 to 60% greater than the gravimetric water content because of the following relationship:

$$\theta = \frac{\rho_b}{\rho_w} \theta_g \quad\quad (12)$$

in which the dry bulk density, ρ_b, is typically 1.4 to 1.6 g/cm³, and the density of water, ρ_w, is 1.0 g/cm³. Throughout the remainder of this book, water content is calculated on a volumetric basis, unless otherwise indicated.

There are other useful relationships between water content and soil properties that are commonly employed by geotechnical engineers, such as

$$\theta_g = \frac{Se}{G} \quad\quad (13)$$

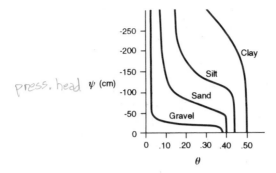

press. head ψ (cm)

Figure 3 Effect of texture on the soil-water characteristic curve.

where S is saturation percentage, e is void ratio [(volume of voids)/(volume of solids)], and G is the specific gravity of the solid (dimensionless), and

$$\theta_g = \frac{\gamma_t - \gamma_d}{\gamma_d} \tag{14}$$

specific weight

where γ_t is the total sample unit weight [(total weight)/(sample volume)] and γ_d is the dry sample unit weight [(oven-dried weight)/(sample volume)].

III. SOIL-WATER RETENTION CURVES

There is a very important relationship that exists between the pressure head (soil-water potential) and the water content, which is illustrated in the soil-water retention curve or soil-water characteristic curve (Figure 3). These curves describe the energy status relative to the volume of water stored in a porous material under variably saturated conditions.

One way to illustrate the relationship between pressure head, ψ, and water content, θ, is to represent the variable pore sizes of a field soil as a bundle of vertical capillary tubes having different diameters. If this bundle of tubes is inserted into the water table, water will be drawn up into the tubes until a static equilibrium condition is achieved (Figure 4b). The equation to compute the height of water rise in a capillary tube (Figure 4a) reveals that the smaller the tube radius, the greater the height of capillary rise. Up to a certain distance above the water table, all the tubes are filled with water. This height above the water table defines the upper limit of the capillary fringe. For clay soils, the height of the capillary fringe may exceed 1 m, and in a gravel, it may only rise 10 cm. At greater distances above the water table, notice that fewer tubes are filled with water. By looking at the proportion of the tubes that are water-filled at progressively greater elevations, it is evident that the mean water content within the bundle of tubes decreases with increasing height above the water table (Figure 4b).

Having just determined that the mean water content, θ, is constant in the capillary fringe and then decreases with increasing height above it, we now seek a relationship between height above the water table and pressure head, ψ, in order to relate water

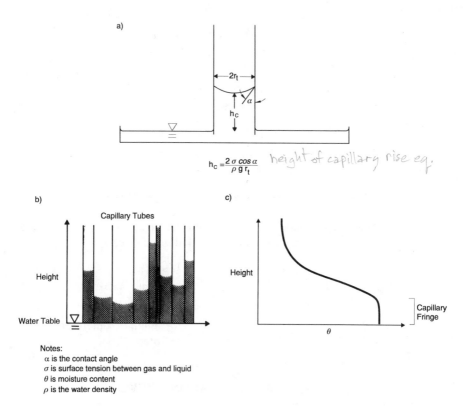

$$h_c = \frac{2\,\sigma\,\cos\alpha}{\rho\,g\,r_t} \quad \text{height of capillary rise eq.}$$

Notes:
  α is the contact angle
  σ is surface tension between gas and liquid
  θ is moisture content
  ρ is the water density

Figure 4 (a) Height of water rise in a capillary tube, (b) water rise in a bundle of capillary tubes representing a range of pore sizes, and (c) soil-water content above a static water table.

content to pressure head. There are several ways to establish this. First, in a conceptual sense, consider a point at the water table where the soil is fully saturated. At all points on the water table, by definition, the pressure head or soil-water potential is zero. And, at the water table, the maximum water content achievable equals the porosity in the absence of entrapped air. Now assume that the water table falls slowly. As it does, the point at the old water table increases in distance above the new water table and therefore is subjected to gradually increasing tension. When the water table declines by a distance just greater than the height of the capillary fringe, air enters the largest pores and soil begins to drain at the so-called air-entry pressure head. At the position of the initial water table, the pressure head gradually becomes more negative with increasing water table decline and pore dewatering; in other words, as the pressure head decreases, the water content decreases.

Alternatively, we can quantify the tension or pressure head above the water table for a hydrostatic condition by comparing the height of capillary rise equation (Figure 4a) with pressure head obtained by substituting Equation 4 into 7, assuming $\phi_{sw} = \phi_{matric}$. We see that the height of capillary rise, h_c, measured as positive above the water table is simply the negative of the pressure head in the soil, that is, $\psi = -h_c$, for the case of hydrostatic equilibrium in our bundle of capillary tubes. That the pressure head is the negative of elevation above the water table is a simple result of the assumed no-flow (zero total hydraulic head gradient) condition in which

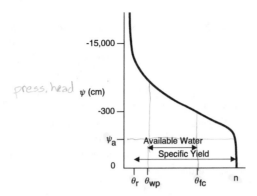

Figure 5 Indices describing the soil-water characteristic curves: θ_r is residual water content, θ_{wp} is permanent wilting point, θ_{fc} is field capacity, n is porosity and ψ_a is air-entry value.

$\Delta H/\Delta z = \Delta(\psi + z)/\Delta z = 0$, and this is valid when $\psi = -z$. Therefore, if we know the location of the water table and the profile is at a hydrostatic equilibrium (no-flow condition), the soil-water retention curve is easily obtained by measuring the water content profile and associating each measurement with the negative of the sample elevation above the water table.

The amount of water held in the soil at a prescribed pressure head or tension is dependent upon soil texture, as shown in Figure 3. Sand and gravel lose a large portion of their water just after the pressure head decreases, whereas clay tends to lose much less water with decreasing pressure head. This is because sand and gravel contain a high percentage of large pores that hold water that is only weakly bound to the soil by capillary forces. On the other hand, clays have smaller pores, and there is a larger surface area over which the water can bind to the soil and resist gravity drainage. The electrostatic bonding that can develop between polar water molecules and clay mineral surfaces is largely responsible for the high moisture retention in clay soils. Owing to differences in pore size distribution, and in mineralogy, the relationship of soil-water potential and water content is unique for each soil.

There are several features of a soil-water characteristic curve that are commonly referenced or measured (Figure 5). The pressure head at which the air first begins to enter the previously saturated soil is called the air-entry value. The absolute value of the air-entry pressure head represents the height of the capillary fringe in a soil which has reached equilibrium in the process of drainage from saturation. This also represents the air pressure required to initiate displacing water from the porous media; consequently, the air-entry pressure head is controlled by the diameter of the largest pores. The air-entry pressure is approximately equal to the bubbling pressure on capillary pressure-saturation curves described by petroleum engineers to depict relationships between water, oil, and/or gas.

Another term associated with soil-water characteristic curves is the field capacity. After 2 or 3 d of drainage following a thorough rain or irrigation, the rate of gravity drainage slows considerably, and the water content at this stage of drainage is called the field capacity. Although there is no good alternative for measuring field capacity other than in the field, field capacity often is arbitrarily quantified in laboratory tests as the water content corresponding to −100 or −300 cm pressure head (about −1/10 to −1/3 bar; Figure 5). For many agricultural soils, such pressure heads usually occur near the flattest part of the moisture retention curve, where the change in water content first begins to decrease with decreasing pressure head. With additional slow drainage or evapotranspiration to a very dry condition, the water content may become so low that many plants wilt. By convention, the water content corresponding to −15,000 cm pressure head (about −15 bars) is called the permanent wilting point, although the soil-water potential near many desert plants is even less, in places −40 to −80 bars or less. The water content difference between field capacity and permanent wilting (or −1/3 bar minus −15 bar water contents) is called the available soil water (to plants) (Figure 5). _see p.100 specific retention = same thing_

As the soil dries below the permanent wilting point, the water content approaches an asymptote, so that with decreasing pressure head no further water drains from the soil. This is called the residual water content. Some researchers argue that the residual water content represents a condition where liquid films become discontinuous or where the hydraulic conductivity of the soil water becomes zero. The residual water content is also useful in water storage problems. For example, the specific yield of an aquifer, S_y in Figure 5, can be calculated from the moisture retention curve as the porosity minus residual water content. $S_y = n - \theta r$

Compaction is an important variable affecting the pore size distribution, and thus it can affect the soil-water characteristic curve. Compaction can occur due to a wide variety of mechanisms such as heavy equipment, cattle, or even raindrop impacts on the surface veneer of soil. Compaction results in the preferential destruction of large pores, while leaving most of the smaller pores unaffected. Therefore, compacting a soil, while reducing total porosity and the number of large pores, will actually increase the number of small pores in a volume of soil. As a result, the compacted soil can hold less water at saturation and more water at lower pressure heads compared to the uncompacted soil. Likewise, when a core sample is removed from deep in the vadose zone it may expand upon release of the overburden pressure. Pore structures also can change when clay soils shrink and swell with the subtraction and addition of water. Consequently, care must be taken to recognize what condition the soil-water characteristic curve represents.

The water content of soil at a given pressure head depends upon its wetting history. This phenomenon is called hysteresis. Previously, water retention curves illustrated in Figures 3 and 5 represented water held after drainage from complete saturation. These are called the initial drainage curves. However, when the soil imbibes water (wets) from an initially dry condition, the soil behaves according to a different moisture retention curve called the main wetting curve (Figure 6). The main wetting curve is always located to the left of the initial and the main drainage curves. The moisture retention curve in Figure 6 also illustrates the behavior when the soil is only partially wetted and then drains. The intermediate paths on the moisture retention curve, bounded by the primary wetting and drying curves, are called scanning curves.

One cause of hysteresis is explained using the capillarity equation (Equation 4) applied to a simple soil comprised of two pore sizes (Figure 7). During drainage, the small-diameter water-filled pore throats do not fully dewater until the pressure head decreases (tension increases) sufficiently at the base of the throat to with-draw water against the capillary effects associated with this pore diameter. Thus, some of the

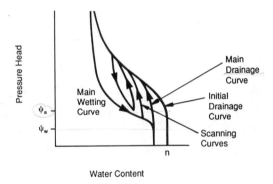

Figure 6 Effect of hysteresis on the soil-water characteristic curve: ψ_w is the water-entry value, ψ_a is the air-entry value, and n is porosity.

overlying large pores will not drain until the pressure head decreases below the threshold pressure where the narrow pores empty. This threshold pressure for a particular pore radius can be calculated from Equation 4. On the other hand, during wetting of a dry soil, the large pores remain empty at this same threshold pressure head, while water enters the small pores. The large pores do not fill until the pressure head increases to a higher threshold characteristic of the large pore radius. Therefore, if a soil has reached a given pressure head by drainage, a greater number of the large pores will be full, and if it reached this pressure head by wetting, more of the large pores will be empty (Figure 7). Consequently, at a given pressure head the water content is greater for draining than for wetting of the soil. A second reason for see Fig. 4a hysteresis is that the contact angle is greater for an advancing meniscus than for a receding meniscus; as a result, for a given water content, the pressure head in the drained soil will be less than that in the wetted soil (Equation 4).

A third factor contributing to hysteresis is entrapped air. Air becomes entrapped or encapsulated when, as the soil becomes saturated from infiltration or from a rising water table, some of the air cannot escape the advancing wetting front. As the soil wets along the main wetting curve, there is sometimes a well-defined point beyond which the water content does not increase significantly with increasing pressure. This water content typically corresponds to about 85 to 90% of full saturation, owing to entrapped air. The pressure head at this inflection, called the water-entry pressure head, is approximately one half of the air-entry pressure head (Bouwer, 1966). The water-entry pressure head also represents the pressure head at which the air phase becomes discontinuous.

For mathematical convenience, as we see in the next section, it is sometimes preferable to quantify the water retention by another parameter, the specific moisture capacity. The specific moisture capacity, $C(\psi)$, is simply the slope of the soil-water retention curve:

$$C(\psi) = \frac{d\theta}{d\psi} \tag{15}$$

The specific moisture capacity curves for sand and sandy clay loam plotted in Figure 8A are obtained by differentiating the respective soil-water release curves in

A

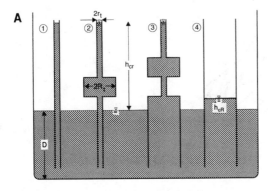

The reservoir is completely full, submerging and filling each capillary tube. The water level in the reservoir is then lowered to position D and water drains from parts of the capillary network. At equilibrium, the maximum height of the water held in the capillary network is controlled by the smallest diameter capillary.

Equilibrium after filling

B

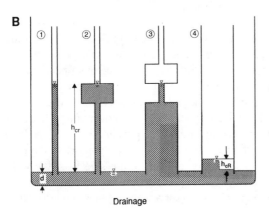

The water reservoir level is lowered to position d, and equilibrium is reestablished in all tubes. The large diameter tube in network 2 does not drain in contrast to 4, because the small diameter tube below it remains full.

Drainage

C

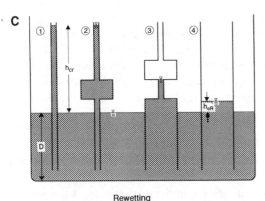

The water reservoir is raised back to the initial position. In network 3, the large diameter tube does not fill because the matric potential in the small tube is too low; to fill the large diameter portion of the tube, the reservoir level would need to rise within h_{cR} of the base of the large tube.

Rewetting

Figure 7   Hysteresis in a capillary tube network subjected to drainage and rewetting. Parts A and C illustrate that more pore space is filled with water when the soil drains than when it rewets.

Figure 8B. For noncompressible soils in the tension saturated zone, that is, where pressure head is greater than the air-entry pressure head, the specific moisture capacity is zero. At pressure heads just less than the air-entry value, the specific moisture capacity usually reaches a maximum. In Figure 8B, we see that sand will release much more water over a particular decrease of pressure head compared to

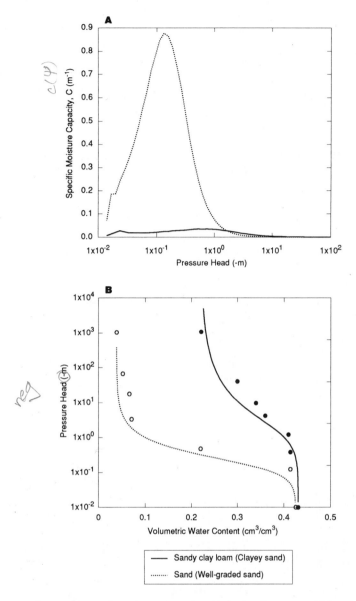

Figure 8 Graphs of specific moisture capacity. Circles in 8B are actual laboratory measurements of θ-ψ pairs.

loam — a sand, silt, clay mix. It therefore follows, as shown by Figure 8A, that the maximum specific moisture capacity of coarse, uniform soil is usually much greater than fine-textured soil. We also note that following the maximum, the specific moisture capacity decreases in the very dry range and approaches zero at the residual water content.

The specific moisture capacity is defined as the volume of water released from or taken into storage, per unit volume of vadose zone, per unit change in pressure head. This definition is essentially identical to that for specific storage coefficient in aquifers, and both terms have units of inverse length. In Figure 8, the water stored in unsaturated soil is due to changes in water content as a consequence of infiltration, drainage, capillary effects, or air-drying. In contrast, the specific storage in saturated soil accounts for water and matric compressibility (e.g., Neuman, 1973; Narasimhan, 1979). Later in this chapter, the full matrix storage properties combine compressibility effects with the specific moisture capacity for developing complete flow equations.

IV. DARCY'S EQUATION AND UNSATURATED FLOW PARAMETERS

Perhaps the most widely recognized equation among soil scientists, hydrologists, and petroleum engineers is Darcy's equation. In 1856 Henri Darcy, a French engineer, conducted laboratory experiments on porous filter materials that would be used for a sewage treatment system. These experiments were conducted under fully saturated conditions. Buckingham (1907), a soil scientist, demonstrated that Darcy's equation could be extended to unsaturated conditions as well. Darcy's equation also is used in the petroleum fields and hydrogeology for multiphase flow problems. We begin by discussing the more unfamiliar but more general form of Darcy's equation, which is relevant to a wide variety of fluid flow problems, including nonaqueous phase liquids. Subsequently, we introduce the less mathematically cumbersome equation for the flow of water in the vadose zone.

Darcy's equation for a fluid phase (i.e., liquid or gas), F, can be written as

$$q_{F_i} = -K_F(S_F)_{ij}\left(\frac{\partial h_F}{\partial x_i} + \rho_{r_F} u_i\right) \qquad (16)$$

where q_{F_i} = specific discharge of fluid F in i direction (LT^{-1}), K_F = hydraulic conductivity of phase F (LT^{-1}), S_F = saturation percentage of fluid phase F (L^{3}L^{-3}), h_F = water height equivalent pressure head of fluid phase F (L), $P_F/g\rho_w$ where P_F = pressure in phase F (ML^{-1}T^{-2}), g = gravitational constant (LT^{-2}), and ρ_w = density of pure water (ML^{-3}), x_i = Cartesian space coordinate (i,j = 1, 2, 3) (L), $\rho_{r_F} = \rho_F/\rho_W$ = specific gravity of phase F, and $u_i = \partial z/\partial x_i$ = unit gravitational vector measured positive upward in direction z.

If only water is the fluid of interest, then Darcy's equation is written as

$$q_i = -K(\theta)_{ij}\left(\frac{\partial \psi}{\partial x_i} + \frac{\partial z}{\partial x_i}\right) \qquad (17a)$$

where z is positive upward. Where the soil is homogeneous and isotropic, then in three dimensions in an x,y,z-coordinate system, Darcy's equation becomes:

$$q_x = -K(\theta)\frac{\partial \psi}{\partial x} \qquad (17b)$$

$$q_y = -K(\theta) \frac{\partial \psi}{\partial y} \qquad \text{(17c)}$$

Darcy velocity
or specific discharge
$$q_z = -K(\theta) \left(\frac{\partial \psi}{\partial z} + \frac{\partial z}{\partial z} \right) = -K(\theta) \left(\frac{\partial \psi}{\partial z} + 1 \right) \qquad \text{(17d)}$$

Darcy's equation simply states that fluid flow is a function of the driving force called hydraulic gradient (pressure and gravity terms in brackets) and a constant of proportionality called the hydraulic conductivity, K. The hydraulic conductivity accounts for the viscous flow and frictional losses that occur as a fluid moves through the porous medium.

A. HYDRAULIC GRADIENT

The hydraulic gradient in the vadose zone exhibits interesting characteristics that contrast markedly with those that hydrogeologists are accustomed to in aquifers. In aquifer systems, flow is primarily horizontal and the regional hydraulic gradient is often in the range of 10^{-4} to 10^{-3}; it is rare that the hydraulic gradient ever exceeds 0.01, although there are exceptions such as where groundwater flows across faults, across aquitards, and very close to pumped wells. But in the vadose zone, hydraulic head gradients near one are common. Unit hydraulic gradients occur in deep vadose zones with uniform texture where the soil-water content is constant with depth. The same is true if the vadose zone is stratified, when the pressure head is averaged over many layers (Yeh, 1989). Where pressure head or mean pressure head does not vary spatially, the gradient of the pressure head $(\partial \psi / \partial z)$ is zero. The only component of hydraulic head gradient that one must consider for this case is gravity, and its gradient, $(\partial z / \partial z)$, is always unity in the vertical direction when soil-water potential is expressed in units of length. Therefore, the gradient of the total hydraulic head will be one, where the pressure head is everywhere constant. A unit hydraulic gradient indicates that the soil water is flowing vertically downward. When the gradient is unity, the magnitude of the flux, q, equals the hydraulic conductivity, $K(\theta)$.

Although the hydraulic gradient is often near unity, the hydraulic gradient can be many orders of magnitude larger near sharp wetting fronts in dry soils. On the other hand, the hydraulic gradient may also be much less than unity and, in fact, is zero where no flow occurs. Hydrostatic equilibrium is one condition of no-flow flow, but this is not often encountered in the field. Another instance where zero gradient could occur is where a pulse of water percolation downward is halted by an impermeable layer or coarse-textured capillary barrier. Another example is near land surface where there is a plane above which water flows upward due to evapotranspiration and below which flow is downward due to capillary and gravity effects. This plane is usually referred to as the plane of zero flux. From these examples, it is clear that the hydraulic gradient in the vadose zone can vary substantially in response to soil-water dynamics, although in many cases the gradient can be assumed to be near unity in the vertical downward direction, especially below the root zone.

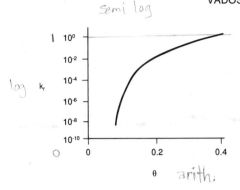

Figure 9   Relative hydraulic conductivity, K_r, vs. water content, θ. Porosity is 0.4 cm³/cm³.

B. UNSATURATED HYDRAULIC CONDUCTIVITY AND RELATIVE PERMEABILITY

The following equation further explains how the hydraulic conductivity is a function of the fluid properties, the media properties, and the water content, θ:

$$K(\theta) = \left(\frac{k\rho g}{\mu}\right) k_r(\theta) \tag{18}$$

where k = intrinsic permeability of the medium (L^2), ρ = density of fluid phase P (ML^{-3}), g = gravitational constant (LT^{-2}), μ = dynamic viscosity of fluid ($MT^{-1}L^{-1}$), and $k_r(\theta)$ = relative permeability (dimensionless, ranges from 0 to 1). In Equation 18, the quantity in brackets represents the familiar saturated hydraulic conductivity for isotropic conditions. The relative permeability, sometimes called relative hydraulic conductivity, is a dimensionless parameter that accounts for the dependence of the hydraulic conductivity on pressure head or water content, as shown in Figure 9. The maximum value of relative hydraulic conductivity is one, and at this point the pores are fully saturated with water. But in the field, the vadose zone seldom is fully saturated with water, due to entrapped air. Entrapped air is most likely to occur, for example, below a fluctuating water table or below irrigated fields and intermittently flooded arroyos. Consequently, under field conditions the maximum value of hydraulic conductivity may be only about half of the saturated hydraulic conductivity. Owing to the difficulty to achieve full saturation, the maximum field hydraulic conductivity is sometimes referred to as the satiated hydraulic conductivity.

The relative hydraulic conductivity decreases rapidly with decreasing water content. As drainage progresses, smaller and smaller pores are left holding water. As the water content decreases, the path of water flow becomes more tortuous and the cross-sectional area of water in the pores decreases. In the dry range, the relative hydraulic conductivity becomes very small, so at low water contents, the hydraulic conductivity may be perhaps more than a millionfold smaller than the saturated hydraulic conductivity. At moisture contents as small as a few percent, detailed laboratory experiments have shown liquid phase transport of water can still exist, although at this dry state vapor transport is much more important (Grismer et al., 1986).

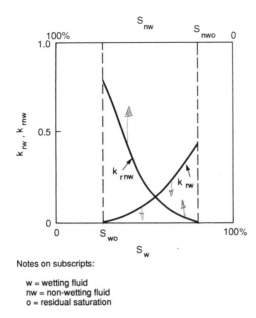

Notes on subscripts:

w = wetting fluid
nw = non-wetting fluid
o = residual saturation

Figure 10   Relative permeability, k_r, vs. saturation, S, for two fluids. Notes on subscripts: w = wetting fluid, nw = non-wetting fluid, and o = residual saturation. (From Bear, 1975.™ With permission.)

Petroleum engineers deal extensively with relative permeability data, but there are important distinctions of interest to soil scientists and hydrologists. Compare the manner in which petroleum engineers sometime represent relative permeability curves (Figure 10) with the soil physicists' perspective (Figure 9). The most significant difference between Figures 9 and 10 is that for the two-phase fluid (e.g., oil and water) system in a petroleum reservoir, each of the phases is shown to reach residual saturation where the relative permeability of a fluid is zero. In contrast, the relative permeability for water in Figure 9 does not usually become zero. In the very dry range of interest to soil scientists and hydrologists, the water may move as thin films. In this state, the relative permeability will be very small, but not actually zero. For most practical problems in reservoir engineering and petroleum production, there is no need to be concerned with film flow. Consequently, relative permeabilities less than about 0.01 or 0.001 are considered negligible in an oil reservoir. Therefore, petroleum engineers often find it more convenient to express unsaturated hydraulic conductivity and relative permeability on an arithmetic scale, whereas soil scientists and hydrologists usually use a logarithmic scale spanning many cycles. Although extensive data exist on capillary properties of oil reservoir rocks, the lower range of the relative permeability test data often does not extend to sufficiently low values to adequately characterize dry conditions. For example, one problem that can arise is in using Darcy's equation to compute recharge. If relative permeability-water saturation curves derived for a petroleum engineering application (e.g., Figure 10) are applied to obtain hydraulic conductivity where field saturation is very low, the recharge may be incorrectly predicted as zero. An understanding of the manner in which petroleum engineers deal with relative permeability can be very important to hydrologists and soil physicists, especially for problems where both soil water and vapor movement

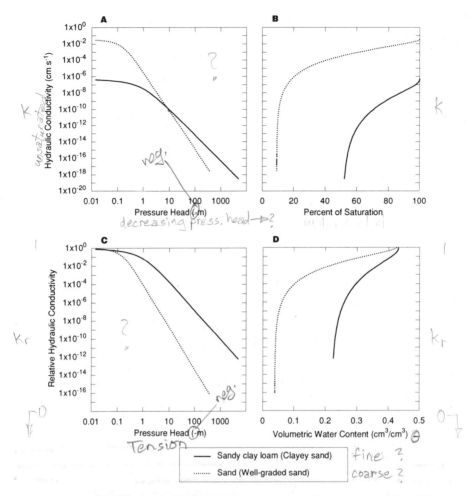

Figure 11 (A) Hydraulic conductivity, K, versus pressure head, ψ, for sand and sandy clay loam; (B) hydraulic conductivity versus water content; (C) relative hydraulic conductivity versus pressure head; and (D) relative hydraulic conductivity versus percent saturation. Water retention curves and specific moisture capacity for these soils are shown in Figure 8.

are significant or for problems of nonaqueous phase liquid migration through the vadose zone.

The hydraulic conductivity of variably saturated media is highly dependent upon soil texture (Figure 11). Hydrogeologists and engineers are well aware of the nature of spatial variability in saturated hydraulic conductivity that is attributed to variability in the intrinsic permeability (Equation 18) of the geologic material. For instance, well-sorted sand typically has a saturated hydraulic conductivity of about 10^{-2} cm/s, whereas clay may have a saturated hydraulic conductivity of about 10^{-6} cm/s. But over the range of water contents likely to be encountered in the vadose zone, the unsaturated hydraulic conductivity of a single soil sample may change by one-million- or one-billion-fold or more. There is even greater variability in the unsaturated hydraulic conductivity among samples of different soil textures.

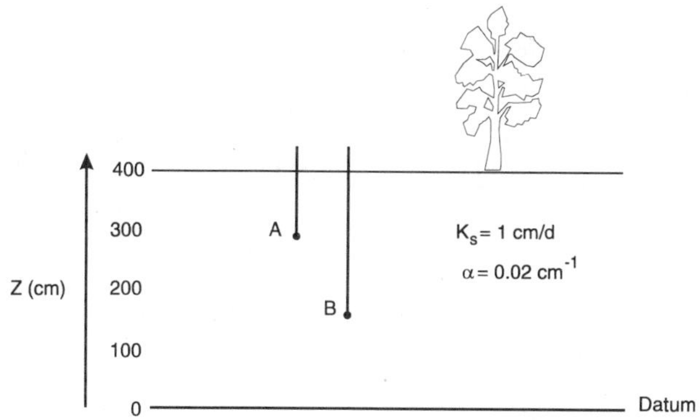

Figure 12   Example to calculate hydraulic gradient, flow direction, and flow rate.

It is especially important to recognize that at low pressure head or water content, the unsaturated hydraulic conductivity of a fine-textured soil may be greater than that of a coarse soil. Figure 11 illustrates this behavior for a sand and loam, with the loam having a greater hydraulic conductivity at pressure heads less than –10 m. This behavior arises because the unsaturated hydraulic conductivity of fine soil tends to decrease much less rapidly as pressure head decreases, in comparison to a coarse textured soil. For most hydrogeologists and engineers, this is a paradox, in that the soil with the highest intrinsic permeability (Equation 18) can have the lowest hydraulic conductivity. However, this fact can be very important in forming conceptual models about vadose zone processes of flow and transport, particularly in heterogeneous or layered media, as we demonstrate in a subsequent chapter discussing vadose zone processes.

The concepts of unsaturated flow presented thus far are summarized in the following example problem. The hypothetical problem is to determine the direction and rate of soil-water flow from *in situ* measurements of pressure head and hydraulic conductivity in a soil having a uniform texture. Figure 12 shows the location of two tensiometers for measuring pressure head. Table 2 indicates the pressure head measurements at the two depths. It has already been determined from laboratory analyses of cores that the saturated hydraulic conductivity is 1 cm/d. We assume that the unsaturated hydraulic conductivity fits the exponential model:

$$K(\psi) = K_s \, \exp(\alpha\psi) \qquad (19)$$

with $\alpha = 0.02$ cm^{-1} for this soil. (The exponential model means that on semilogarithmic paper, $\ln K$–ψ fits a straight line having a slope α and an intercept K_s.)

To solve this problem, we assume that the flow is vertical and apply Darcy's equation (Equation 17d). We also set the vertical axis as positive upward. The first step to compute the Darcy velocity (specific discharge), q_z, is to determine the hydraulic head gradient from the sum of the pressure head and total head gradients. In our problem, the pressure head decreases upward, so at first glance it may appear that flow is upward. But when the gravitational gradient is added to the pressure head gradient, the total hydraulic head decreases downward (Table 2). Recall it is the gradient of total

Table 2   Pressure Head and Total Head Measurements at Two Depths

	Measured pressure head ψ (cm)	Elevation head Z (cm)	Total head H (cm)
A	−100	300	200
B	−90	200	110

head, not pressure head, that is the water driving force. Consequently, the flow is downward and the magnitude of the total hydraulic head gradient is

$$\frac{dH}{dz} \cong \frac{H_2 - H_1}{Z_2 - Z_1} = \frac{200 - 110}{100} = 0.9 \tag{20}$$

Note that by our choice of sign convention, the higher subscript refers to the location furthest from the origin.

The second step is to compute the unsaturated hydraulic conductivity. To do this, we determine the mean pressure head in the region between the tensiometers:

$$\frac{\psi_1 + \psi_2}{2} = -95 \text{ cm} \tag{21}$$

Next, substitute this mean pressure head into Equation 19, along with our previously determined values of K_s and α. The result is $K = 0.15$ cm/d. The third step is to multiply the hydraulic head gradient by the unsaturated hydraulic conductivity to obtain the Darcy velocity:

$$q_z = -K\left(\frac{dH}{dz}\right) = -(0.15)(0.9) = -0.13 \text{ cm/d} \tag{22}$$

The negative sign indicates flow is in the direction opposite to which z increases, that is, downward.

C. HYSTERESIS IN HYDRAULIC CONDUCTIVITY

When we discussed the soil-water retention curve, we noted that the relationship was hysteretic. As one may expect, the relationship between unsaturated hydraulic conductivity and pressure head also is hysteretic (Figure 13). The simplest explanation for this hysteretic behavior is that at any given pressure head, there is a corresponding value of moisture content on the main wetting curve and a slightly greater moisture content on the main drainage curve. The wetter the soil, the greater the hydraulic conductivity. Therefore, at a particular pressure head, one may find two corresponding hydraulic conductivities, such that the hydraulic conductivity during drainage will be greater than during wetting. Near saturation, entrapped air is the primary cause of hysteresis in hydraulic conductivity. There is little evidence that the unsaturated hydraulic conductivity is hysteretic with respect to moisture content to any practical extent.

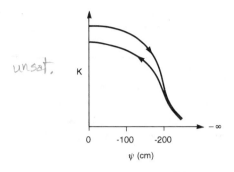

unsat.

Figure 13   Effect of hysteresis on the hydraulic conductivity, *K*, vs. pressure head, ψ, relationship.

Problems in which hysteresis may be important to consider involve periods of both wetting and drying, such as can occur during infiltration and subsequent redistribution of a pulse of infiltrated water that is drawn both downward by gravity and capillarity and also upward due to evapotranspiration. As indicated by Rubin (1967) and Hillel (1980), the downward movement of a finite pulse of water cannot accurately be modeled by assigning as input parameters either the wetting or drying unsaturated hydraulic conductivity curves. In both bounding cases, the depth of wetting will be overestimated and the amount of moisture retained near the land surface will be underestimated. However, when the process involves either only wetting or only drying, then it is appropriate to apply the corresponding wetting or drying hydraulic conductivity curve. More is presented about the importance of hysteresis in Chapter 3 on vadose zone processes.

D. ANISOTROPY

Looking back on Equations 16 and 17a presented at the beginning of this section, we subscripted the hydraulic conductivity to indicate that in its most general form the hydraulic conductivity is anisotropic. Anisotropy is a property of the medium that reflects how the hydraulic conductivity varies with direction. That is, measurements of hydraulic conductivity in the vertical direction are different from those in the horizontal direction in an anisotropic medium. By contrast, at any point within an isotropic medium, hydraulic conductivity has the same magnitude in all directions. In a three-dimensional, anisotropic system, hydraulic conductivity is a second-rank tensor or matrix having nine components:

$$K_{ij} = \begin{vmatrix} K_{xx} & K_{xy} & K_{xz} \\ K_{yx} & K_{yy} & K_{yz} \\ K_{zx} & K_{zy} & K_{zz} \end{vmatrix} \tag{23}$$

The practical significance of this representation is that it allows one to compute the component of water flow in any direction, regardless of the orientation of principal bedding directions. In contrast to an isotropic medium, in an anisotropic system the direction of flow may not be in the same direction as the hydraulic head gradient. The hydraulic conductivity tensor has nine components to account for cases in which the principal coordinate axes and bedding planes are not collinear. However, in many hydrogeologic environments the soil is horizontally stratified, so within the horizontal plane there may be no anisotropy. That is, $K_{yy} = K_{xx}$, and all off-diagonal terms in the conductivity matrix (Equation 23) would be zero, if our coordinate axes are in the horizontal and vertical direction. Consequently, anisotropy in hydraulic conductivity may be represented by the ratio of hydraulic conductivity in the horizontal to vertical direction, K_H and K_V, respectively:

$$A = \frac{K_{xx}}{K_{zz}} = \frac{K_H}{K_V} \tag{24}$$

In most cases, anisotropy is characterized as the ratio of saturated hydraulic conductivities obtained from oriented core samples. At saturation, anisotropy may commonly vary from 2 to 20, but values up to 100 or greater may occur.

In unsaturated media, hydrologists and soil scientists commonly have assumed that the anisotropy at moisture contents less than saturation is the same as at complete saturation. This assumption was questioned by Zaslavsky and Sinai (1981). Theoretical analysis based on stochastic methods (Yeh et al., 1985) suggests that in a steady flow field the anisotropy of a stratified heterogeneous soil should increase as the mean pressure head (and moisture content) of the soil decreases:

$$A(\overline{\psi}) = \exp\left(\frac{\sigma_f^2 + \sigma_\alpha^2 \, \overline{\psi}^2}{1 + \overline{\alpha} \, \lambda_1 \cos \delta_s}\right) \tag{25}$$

where σ_f^2 = variance of ln K_s (dimensionless), σ_α^2 = variance of slope of ln K-ψ (L^{-2}), $\overline{\psi}$ = mean pressure head (L), $\overline{\alpha}$ = mean slope of the ln K-ψ curve (L^{-1}), λ_1 = vertical correlation scale (L), and δ_s = dip of stratification (degrees).

Laboratory experiments have subsequently confirmed that anisotropy is moisture dependent (Stephens and Heermann, 1988; Frederick, 1988). Field and numerical model investigations by McCord et al. (1991) showed that for a uniform dune sand that was nearly isotropic at saturation, the unsaturated anisotropy was as much as 20.

The primary consequence of anisotropy is that subsurface water movement may have strong lateral flow components especially where infiltration occurs into highly stratified, dry soils. We say more about how anisotropy influences flow in the vadose zone in the next two chapters.

E. SOIL-WATER DIFFUSIVITY

The final hydraulic property we discuss here is the soil-water diffusivity, D:

$$D(\theta) = \frac{K(\theta)}{C(\theta)} \tag{26}$$

The soil-water diffusivity embodies both the unsaturated hydraulic conductivity and, through the specific moisture capacity, the soil-water characteristic curve. This parameter is analogous to the hydraulic diffusivity in aquifers and has units of length

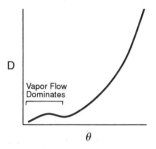

Figure 14   Soil-water diffusivity vs. water content.

squared per unit of time. Diffusivity changes much less over the range of moisture contents commonly of interest than does unsaturated hydraulic conductivity. For some problems, this smaller range makes it easier to apply numerical models based on the diffusivity. Unfortunately, very close to saturation, soil-water diffusivity approaches infinity because the specific moisture capacity becomes zero. Therefore, soil-water diffusivity is not used in equations intended to simulate both saturated and unsaturated conditions. The soil-water diffusivity is hysteretic, due to hysteresis in the soil-water characteristic curve. As shown in Figure 14, over most of the curve the soil-water diffusivity decreases monotonically with decreasing water content, due primarily to the influence of the unsaturated hydraulic conductivity that governs liquid-phase transport. But, in the dry range, the soil-water diffusivity also includes water movement by vapor diffusion. The soil-water diffusivity curve shown here is the resultant sum of the liquid (Equation 26) and vapor phase diffusivities which account for soil-water movement under isothermal conditions. Thus, the increase in diffusivity in the dry range is attributed to water conduction in the vapor phase (Philip, 1955).

The soil-water diffusivity can also be incorporated into Darcy's equation:

$$q_x = -D(\theta) \frac{\partial \theta}{\partial x} \tag{27a}$$

$$q_y = -D(\theta) \frac{\partial \theta}{\partial y} \tag{27b}$$

$$q_z = \left(-D(\theta) \frac{\partial \theta}{\partial z}\right) - K(\theta) \tag{27c}$$

Equations 27a, 27b, and 27c are obtained simply by substituting Equation 26 into Equations 17b, 17c, and 17d. From the three equations in Equation 27, it would appear that gradients of moisture content cause fluid movement. While it is generally true that water will tend to be drawn from regions of relatively high to low water content, this is not always the case.

The reason for this apparent enigma is attributed to hysteresis. To illustrate, suppose we allowed water to infiltrate into a horizontal column of uniform soil, as illustrated in Figure 15a, and then we cut off the water source when the wetting front

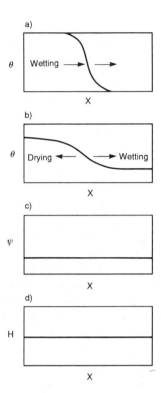

Figure 15   Horizontal infiltration and subsequent redistribution in a uniform hysteretic soil: (a) water content distribution at the end of infiltration; (b) water content distribution at the end of redistribution; (c) pressure head distribution at the end of redistribution; and (d) total hydraulic head at the end of redistribution.

reached the middle of the column. If the column were completely enclosed to prevent evaporation, and water was allowed to migrate to an equilibrium condition, the water distribution would appear similar to that shown in Figure 15b. At equilibrium there is no flow, so the pressure head, gravity head, and total hydraulic head gradients are zero across the horizontal column (Figures 15c and 15d). There is, however, a moisture content gradient (Figure 15b), but no flow. Careful examination of the process of redistribution in the soil column (Figure 15b) reveals that the wetter half of the column reached equilibrium by draining, while the drier half of the column reached equilibrium by wetting. As indicated by the moisture retention curves in Figure 6, for any given pressure head, the moisture content corresponding to a draining condition is greater than for wetting. In this example, we have a case in which the pressure head is constant along the horizontal length of the column, but there are differences in water content over the same length attributable to hysteresis. Therefore, because of hysteresis, moisture content gradients may not always be good indicators of the soil-water flow.

    The preceding illustration was for homogeneous soils, but it is even more difficult to infer flow directions from water content gradients in heterogeneous soils where each soil texture may have a different soil-water characteristic curve. Fortunately, you can be reassured that water always moves from higher to lower potential energy, provided that hydraulic conductivity is nonzero. Water content simply does not describe energy status, whereas pressure head does. For this reason, to calculate fluid velocity and flow direction at an instant in time, it is usually much better to measure *in situ* pressure head rather than water content. Unfortunately, there are many practical difficulties in measuring pressure head, especially in deep or dry vadose zones, as we see in Chapter 6. Consequently, water-content-based monitoring over time is often chosen for convenience.

V. FLOW EQUATIONS FOR VARIABLY SATURATED POROUS MEDIA

We have made numerous references thus far to the causes of fluid movement and storage in the vadose zone. At this point, we introduce the formal equations used to quantify these. The equations are presented for the purpose of allowing the field practitioner to appreciate the data needs of the modeler by examining the coefficients in these equations which require quantification. So often, site characterization activities are designed by personnel who have no idea how their data will be used in a subsequent quantitative analysis. As a result, some performance assessment or risk assessment modeling may not be based on an adequate data set. The important point is to examine these governing equations to identify the variables and parameters that require quantification.

The most complete equations used to describe vadose zone processes include an equation for each fluid phase. These equations are all derived by combining Darcy's equation with an equation of mass conservation and the assumption that the porous medium is completely incompressible. For a three-phase fluid flow problem including water, organic liquid, and air, the following equations are applied:

Water :
$$\frac{\partial}{\partial x_i}\left(K_w\left[\frac{\partial h_w}{\partial x_j}+\rho_{rw}u_j\right]\right)+\frac{R_w}{\rho_w}=n\frac{\partial S_w}{\partial t} \tag{28a}$$

Organic liquid :
$$\frac{\partial}{\partial x_i}\left(K_o\left[\frac{\partial h_o}{\partial x_j}+\rho_{ro}u_j\right]\right)+\frac{R_o}{\rho_o}=n\frac{\partial S_o}{\partial t} \tag{28b}$$

Air :
$$\frac{\partial}{\partial x_i}\left(\rho_aK_a\left[\frac{\partial h_a}{\partial x_j}+\rho_{ra}u_j\right]\right)+R_a=n\frac{\partial \rho_a S_a}{\partial t} \tag{28c}$$

and

$$S_w+S_o+S_a=1 \tag{28d}$$

where n = porosity, S_P = saturation of phase P (w = water, o = organic liquid, a = air), K_P = hydraulic conductivity of the phase, x_i and x_j are Cartesian spatial coordinates

(i,j = 1, 2, 3), and R_P = net mass transfer into the system per unit volume of porous media, a source or sink term such as water uptake by plant roots (Environmental Systems and Technology 1990).

To predict the spatial or temporal distribution of these phases would require input on the hydraulic conductivity to water, to organic liquid, and to air, such as given in Figure 10. In addition to basic fluid properties, the models also would require the user to specify the relationships between capillary pressure and fluid saturation. If the problem only has two fluids of concern, such as air and water, then only two equations are required. Whether there are two or three equations, the flow equations describing each phase are coupled in the mathematical model by recognizing that the sum of the fluid saturations of all phases must always equal 100% (Equation 28d). Properties for multiphase flow problems are usually obtained from laboratory measurements on core samples, and some of these techniques are described in Chapter 5. The flow equations just presented form the basis of a number of very sophisticated computer codes based on finite element formulations such as MOTRANS (Environmental Systems & Technology, Inc., 1990) and MAGNAS3D (Huyakorn et al., 1993). In some instances, there are analytical models that are derived from these equations (e.g., Brustkern and Morel-Seytoux, 1970).

By far, most problems of water flow in the vadose zone, such as infiltration and drainage are adequately described by assuming that pure water is the only liquid phase present, that the water is incompressible, and that the air phase is continuous and everywhere at atmospheric pressure. With these assumptions, we revise Equation 28a and eliminate Equations 28b and 28c to obtain Equation 29, the well-known Richards equation, named for L. A. Richards who first used it in 1931:

$$\nabla \cdot K(\psi)\, \nabla H = C(\psi)\frac{\partial \psi}{\partial t} \qquad (29)$$

where $K(\psi)$ = unsaturated hydraulic conductivity (LT^{-1}), $C(\psi)$ = specific moisture capacity (L^{-1}), $H = \psi + z$ = total head (L), and ψ = pressure head (L).

Here it is apparent that the unsaturated hydraulic conductivity/pressure head relationship (Figure 11) and the specific moisture capacity (Figure 8) are the only two coefficients which require quantification. The Richards equation can account for heterogeneous and anisotropic soils, when such information is available. Solutions to the Richards equation for unsaturated flow produce maps showing the pressure head distribution in time and space for specific boundary conditions, flow domain geometry, and material properties.

Where air pressure effects are significant, for example, where air is confined between a shallow water table and an advancing water front below an irrigated field, the Richards equation probably will not be suitable. For this problem, one should use mathematical models based on the two-phase (air-water) flow equations (Equations 28a and 28c).

With only slight modification to the storage term on the right-hand side, the Richards equation can be extended to apply to transient flow in both saturated and unsaturated media. As written in Equation 29, the Richards equation cannot apply to flow in aquifers where water is released from storage due to the compressibility of the aquifer or expansion of the water; it only accounts for water storage changes due to filling or draining an incompressible porous media. By adding the familiar specific storage term S_s to $C(\psi)$, the Richards equation is generally applicable to transient flow problems in the vadose zone where there may be any degree of saturation as well as in combined vadose zone-aquifer systems (Neuman, 1973):

$$\nabla \cdot K(\psi)\, \nabla H = \left[C(\psi) + \beta' S_s \right] \frac{\partial \psi}{\partial t} \qquad (30)$$

Richards eq. for transients & any degree of saturation

where $\beta' = 1$ if $\psi \geq 0$, and $\beta' = 0$ if $\psi < 0$. Narasimhan (1979) elaborated further on how to treat compressibility effects on water released from or taken into storage in a variably saturated, slightly compressible porous medium. He also discusses how to apply the effective stress concept used in soil mechanics to unsaturated media.

Upon examination of Equation 29 or Equation 30, one may begin to understand why modeling of vadose zone processes is often so difficult. The primary difficulty centers on the fact that the Richards equation is nonlinear, because the coefficients (K and C) are functions of the dependent variable (ψ). The degree of nonlinearity is typically greater for sands than for clay soils. Nonlinear equations are difficult to solve, especially analytically, at least without simplifying assumptions about boundary conditions or the hydraulic properties. To simulate infiltration into a dry sandy soil, the unsaturated hydraulic conductivity may vary by a factor of 100,0000 to more than 1,000,000 across a sharp wetting front, even though the soil is texturally uniform. To accurately account for this extreme spatial variability in computer models often requires very fine finite-difference or finite-element grids and considerable computational effort. Examples of versatile computer codes for vadose zone flow built upon adaptations of the Richards equation include UNSAT-2 (Neuman, 1973), TRUST (Narasimhan, and Witherspoon, 1977), VS2D (Lapalla et al., 1987), and others as summarized in Table 3. For a good discussion comparing the available computer models relevant to vadose zone flow, the interested reader is referred to Oster (1982).

VI. TEMPERATURE EFFECTS

Soil temperature in the vadose zone is strongly affected by the temperature on the soil surface and the diurnal and seasonal nature of the temperature fluctuations. Equation 31 (Marshall and Holmes, 1979) describes this behavior and is applicable to predict temperature in uniform soil with a constant thermal diffusivity k:

$$T(z,t) = T_a + A_o \, \exp\left\{ -\left[(\omega/2k)^{1/2} z \right] \sin\left[\omega t - (\omega/2k)^{1/2} z \right] \right\} \qquad (31)$$

where $T(z,t)$ is soil temperature at a specified depth z and time t, T_a is the average soil temperature, A_{oT} is the amplitude of the soil surface temperature fluctuation, ω is 2π divided by the period of the temperature cycle (T^{-1}), and k is the thermal diffusivity for the transport of heat ($L^2 T^{-1}$). Thermal diffusivity, the ratio of thermal conductivity to volumetric heat capacity, can be measured (Jackson and Taylor, 1986) or calcu-

Table 3 Available Codes for Single Phase (Water) Flow in the Vadose Zone

Code name	Dimensions	Method	Ref.	Code type
2DSEEP	2	FEM	OECD, 1990	Unknown
3DSEEP	3	FEM	OECD, 1990	Unknown
AMOCO	3	FDM	Odeh, 1981	Proprietary
ANGEL	3	FEM	OECD, 1990	Unknown
BETA-II	3	FDM	Odeh, 1981	Proprietary
BRUTSAERT1	2	FDM	Oster, 1982	Public domain
BRUTSAERT2	2	FDM	Brutsaert, 1971	Public domain
CMG	3	FDM	Odeh, 1981	Proprietary
DELAAT	2	FEM	Oster, 1982	Unknown
FEMWATER	2	FEM	Yeh and Ward, 1980	Public domain
FLUMP	2	FEM	Neuman and Witherspoon, 1971	Public domain
GANDALF	2	FDM	Morrison, 1977	Public domain
GPSIM	3	FDM	Odeh, 1981	Proprietary
GWHRT	3	FEM	Carlsson et al., 1983	Unknown
MOMOLS	1	FDM	Rojstocyer, 1981	Public domain
PORES	3	FDM	Oster, 1982	Unknown
REEVES-DUGUID	2	FEM	Reeves and Duguid, 1975	Public domain
SHELL	3	FDM	Odeh, 1981	Proprietary
SSC	3	FDM	Odeh, 1981	Proprietary
STGWT/MOG WT	1	FDM	De Smedt and van Beker, 1974	Unknown
SUM2	2	FEM	Oster, 1982	Unknown
SUPERMOCK	2	FDM	Reed et al., 1976	Public domain
TRIPM	2	FEM	Gureghian, 1981	Public domain
TRUST/TNN	3	IFDM	Narasimhan, 1990	Public domain
TS&E	3	Unknown	Oster, 1982	Proprietary
UNFLOW	2	FEM	Oster, 1982	Public domain
UNSAT1	1	FEM	van Genuchten, 1978a	Public domain
UNSAT1D	1	FDM	Oster, 1982	Public domain
UNSAT2	2	FEM	Neuman et al., 1974	Unknown
UNSAT-H	1	FDM	Fayer and Jones, 1990	Public domain
VERGE	3	FEM	Verge, 1976	Public domain

calculated from the weighted mean thermal properties of the air, water, and minerals comprising the soil (e.g., Hillel, 1980). Equation 31 shows that sinusoidal temperature fluctuations are increasingly damped with increasing depth z below land surface. Calculations with Equation 31 and field data show that significant diurnal temperature changes propagate to depths of less than several tenths of a meter, whereas seasonal fluctuations can propagate to depths of several meters. Below these depths, soil temperature typically increases with depth due to the geothermal gradient at a rate of about 1°C per 40 m. Temperature gradients in the soil water will affect liquid transport, because water can be transported from regions of higher to lower temperature. Consequently, during the summer within about the upper 5 to 10 m, temperatures gradients are predominantly downward. Below this depth, there is a tendency for an upward water flux driven by the geothermal gradient. The temperature-driven water flow can be in the same or different directions from the liquid gradient that is due to the matric and gravitational potential.

The magnitude of water flux under nonisothermal conditions is proportional to the thermal diffusion coefficient for the gas and liquid, D_T, and the temperature gradient (Hillel, 1980):

$$q = -D_T \, \nabla T \qquad (32)$$

Here, the gas and liquid thermal diffusion coefficients are combined into a single coefficient, D_T. The thermal diffusion coefficient for transporting water due to a temperature gradient can be calculated or measured in the laboratory on core samples (e.g., Jury and Miller, 1974; Sophocleous, 1979).

Liquid water flow in the vadose zone is also influenced by the effect of temperature on variables and parameters in the Richards equation such as matric potential (via surface tension) and hydraulic conductivity (via viscosity). For example, Taylor and Stewart (1960) found that the soil-water characteristic curve is temperature sensitive, such that the pressure head associated with a given water content increases with increasing temperature.

VII. GAS-PHASE FLOW

In very dry, thick vadose zones, conditions are favorable for potentially significant migration of gaseous contaminants, provided that air-permeable pathways exist, such as coarse porous materials and open fractures. Gas migration in the vadose zone occurs by both advective and diffusive processes. In this section, we discuss gas migration by advection, that is, process of bulk flow of fluid in the gas phase. In our discussion the term *advection* connotes transport by a moving fluid in any direction, although some definitions suggest that advection refers only to horizontal flow. Advection and convection are often used interchangeably in the literature. However the latter sometimes describes unstable fluid circulation (i.e., free convection) due to buoyancy effects that are induced by thermal gradients for example. Our reference to advection is the same as forced convection, that is, fluid migration induced by hydraulic pressure gradients.

If the vadose zone air is in equilibrium with the atmosphere, the soil gas pressure, P, can be calculated as a function of elevation with respect to sea level according to the following equation for a compressible gas:

$$P = P_0 \, \exp\left(-\frac{M}{RT} \, gz\right) \qquad (33)$$

where P_0 is the atmospheric pressure at sea level, g is the gravitational constant (9.8N/kg), M is the molar mass of air (~28.8 g/mol), R is the molar gas constant ($\approx$8.3 J/mol-K), and T is temperature (Koorevaar et al., 1983). Where the vadose-zone air pressure is in equilibrium with the atmosphere, there is no gas transport by advection. Note that at equilibrium, in contrast to water, which is relatively incompressible, the air pressure does not increase linearly with increasing depth because the soil-air density increases with increasing pressure. Nevertheless, in many practical problems, especially at shallow depths, the air pressure gradient (about 0.12 mbar/m) can be ignored, and the soil air pressure is equal to atmospheric pressure (Koorevaar et al., 1983).

Gas moves by advection when there are spatial differences in the total gas (pneumatic) pressure that cause gas to flow from regions of high to low pressure. This is in contrast to diffusive gas transport, discussed in a subsequent section, where concentration gradients cause gaseous chemicals to migrate through the vadose zone, even in the absence of air flow. Advective gas flow can be induced, for example, in a soil-vapor extraction system, or it can occur naturally, such as where there are barometric pressure fluctuations that propagate into a heterogeneous vadose zone. The cyclic nature of barometric pressure oscillations, especially from frontal storms, leads to a phenomenon called barometric pumping, in which air flows into the vadose zone during periods of high atmospheric pressure and air flows out to the atmosphere during periods of low pressure.

The rate of gas flow in the vadose zone is strongly influenced by the gas permeability (Equation 18). Usually air and water comprise a two-phase fluid system, with water the wetting phase occupying the narrowest pores (Figure 10). Clays, which often have greater *in situ* water saturations than other porous media, generally exhibit relatively low gas-phase permeability, whereas the opposite is true for uniform sand and gravel. For a given soil texture, gas-phase permeability can be highly dependent upon water content; thus, gas flow may be especially important in areas of low precipitation.

In light of the potential for bulk flow of gas in the vadose zone and our earlier discussion in this chapter on liquid water flow, we see there can be flow of both the gas and liquid phase due to pressure as well as temperature gradients. The following equation summarizes the primary fluid (liquid plus gas) flux components:

$$q = q_{\ell,P} + q_{\ell,T} + q_{g,P} + q_{g,T} \qquad (34)$$

where $q_{\ell,P} = -K \nabla H$ (liquid flow due to potential energy gradients; Darcy's equation), $q_{\ell,T} = -D_{T,\ell} \nabla T$ (liquid flow due to temperature gradients), $q_{g,P} = -K_a \nabla P$ (advective gas flow due to total gas pressure gradients), and $q_{g,T} = -D_{T,g} \nabla T$ (gas flow due to temperature gradients), where $D_{T,\ell}$ and $D_{T,g}$ are the liquid and gas thermal diffusion coefficients, respectively (m²/s-°C), K_a is the hydraulic conductivity of soil to air (m²/Pa-s), and P is gas pressure (Pa).

There are several good examples where multiple flow components have been considered in field investigations, including work at the Nevada Test Site (Montazer and Wilson, 1984), at Hanford, WA (Enfield et al., 1973), and near Ft. Hancock, TX (Scanlon, 1992). In semi-arid climates, one may likely find that the liquid and vapor fluxes are in opposite directions and that temperature-driven flow cannot be ignored where recharge is very low.

VIII. CHEMICAL TRANSPORT PROCESSES

In this section, we highlight some of the basic processes that influence the transport of chemicals within the vadose zone. The key processes include advection, diffusion, dispersion, and attenuation by, for example, ion exchange and phase partitioning. Liquid and gas advective processes, discussed in the previous section, are usually the primary means by which contaminants migrate in aquifers. The advective mass flux is obtained from the product of the liquid and gas velocity times the phase concentration, C_ℓ or C_g, respectively. The other transport processes, individually or in combination, may significantly affect the concentration of chemicals as they advect through the vadose zone. The remainder of this section discusses these other transport processes briefly, first for liquid and then for gas. For a more detailed reference, please refer to Jury and Ghodrati (1989) and Charboneau and Daniel (1993).

A. LIQUID-PHASE TRANSPORT

As water migrates through the vadose zone, dissolved chemicals will mix within the pore space in a process called hydrodynamic dispersion. The dissolved chemicals also interact with the solid phase as well as with other chemicals in the water or gas phases. In some cases, chemical transport is facilitated by movement of colloidal sized particles within the advected water. The subsequent sections address these three transport mechanisms.

1. Hydrodynamic Dispersion

Hydrodynamic dispersion, D' (L^2T^{-1}), refers to the process whereby solutes mix within the pore water due to the combined effects of molecular diffusion, $D*$, and mechanical mixing, D:

$$D' = D* + D_m \tag{35}$$

Diffusion in a bulk volume of variably saturated porous media is considerably less than that in free water, because of the interference with solid particles and soil air. The soil diffusion coefficient is related to the pure water diffusion coefficient at 20°C, D_w, by (e.g., Wilson and Gelhar, 1974):

$$D* = \frac{1}{3}\left(\frac{\theta}{n}\right)^2 D_w \tag{36}$$

As water flows through the vadose zone, it moves along tortuous paths, along and within which the velocity varies. Mechanical mixing is related to the mean pore velocity, V, by:

$$D_m = \alpha_d V \tag{37}$$

where α_d (L) is the dispersivity. The dispersivity is dependent upon the water content, inasmuch as the tortuosity of the pathways that solutes travel and the pore velocity will change as the water content changes. At an intermediate degree of saturation, some paths are rapid while others are very slow; consequently, the maximum dispersivity occurs at intermediate saturation (Krupp and Elrick, 1968). Wilson and Gelhar (1974) found that as the water content decreases from saturation, the longitudinal dispersivity (dispersivity in the direction of the mean flow) may increase more than 10-fold.

The breakthrough curves describe arrival of chemicals at a sampling point, for example, the concentration of solutes eluted from the outflow of a soil column over time. Figure 16 illustrates a breakthrough curve under ideal conditions, showing the effects of hydrodynamic dispersion to smear the front of an advancing solute initially

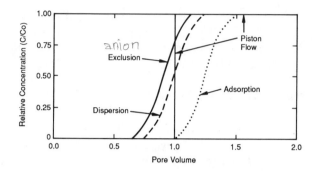

Figure 16   Solute breakthrough curves illustrating several patterns of solute elution. (From Wagenet, 1983. With permission.)

at concentration C_0 as it displaces pure water from the pores. The advection-dispersion equation can predict this type of breakthrough:

$$\nabla \cdot D' \nabla C - \nabla \cdot CV = \frac{\partial C}{\partial t} \tag{38}$$

Ideally, after one pore volume of solute has passed through the water-filled part of the pore space, the concentration at the solute front will be $C_0/2$. However, in some porous material, the initial breakthrough occurs after fewer pore volumes of water have passed when compared to the ideal conditions. Furthermore, some mixing may continue to occur over much longer times, causing a longer tail in the breakthrough curve, in comparison to the ideal behavior. This type of departure from nonideal breakthrough is attributed to diffusion from dead-end pores holding immobile water or from aggregated soils where there is relatively fast flow in the macropores and diffusion into and out of the aggregate porous matrix, (Figure 17). That is, the effective water content, θ_e, that conducts the solute is less than the volumetric water content of the soil, θ. As a result of the reduced water-filled cross-sectional area for flow, the fluid velocity at θ_e is greater than at θ, so the solutes arrive earlier (Figure 17). The long tailing at later time is attributed to some very slow path as compared to the mean pore velocity, as well as to the slow diffusion of solute into the dead-end pores that are not well connected to the flow system. We also need to recognize that the parameter α_d represents the integration of heterogeneities over the scale of the transport problem. It is usually observed that dispersivities derived from small-scale unsaturated flow experiments such as in the laboratory are much smaller than values obtained in the field where longer transport distances may have occurred. Macrodispersion refers to the dispersion of a plume over large distances in a heterogeneous media where mixing is due to velocity variations at a scale much larger than the pore scale. The scale dependence of α_d is discussed further by Gelhar et al. (1984), McElroy (1987), and Flanigan (1989).

2. Chemical Interactions

Chemicals dissolved in the porewater may be subject to concentration changes along their migration path. Processes affecting the evolution of water chemistry

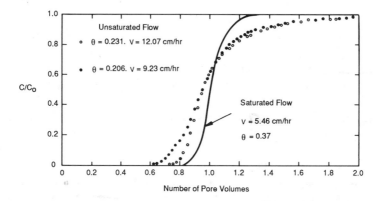

Figure 17 Displacement of water by a chloride solution in a 10.4-cm-long vertical column of glass beads (100 μm diameter) under steady-state saturated and unsaturated conditions established by a constant flow rate. Immobile water zones explain the early arrival of the chloride, the displacement of the breakthrough curves to the left of the point $C/C_o = 0.5$ at 1 pore volume, and the long tail to reach $C/C_o = 1$. (From Krupp and Elrick, 1968. With permission.)

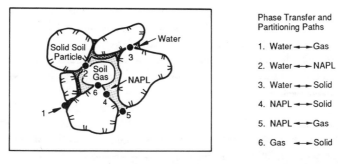

Figure 18 Schematic diagram showing phase transfer and partitioning pathways in a partly saturated soil containing water, and a non-aqueous-phase liquid (NAPL).

include oxidation-reduction reactions, dissolution or precipitation of minerals, adsorption-desorption reactions, and other phase changes. Ion exchange is the primary sorptive mechanism that causes attenuation of cations, including many radionuclides. Ion exchange occurs between charged ions or ion complexes and charged solid particles. It is particularly important in media with large specific surface areas, such as clay particles. Cations within the porewater are attracted to negatively charged solid surfaces and form a diffuse layer surrounding it. Anions, in contrast, are generally repelled from the negatively charged clay surfaces by a process called anion exclusion. Thus anions may move in at a higher velocity than the bulk porewater (Figure 16). However, some anions are attracted to sites of positive charge on the solid surfaces, which on clay particles usually occurs on the broken edges on the mineral. In general, the greater the valence of the cation, the stronger the attraction for the solid. Ions with large crystallographic (dehydrated) radii also tend to be more strongly attracted to the charged particle. These attractive forces are sufficient for the more strongly adsorbing ions to replace less strongly adsorbed ions. For example, for montmorillonite the relative adsorption potential is Th > La > Ba

$\approx$ Sr > Ca > Mg $\approx$ Cs > Rb > NH$_4$ $\approx$ K > Na $\approx$ Li (Bohn et al., 1985). However, less sorbing chemicals migrating in solution also can replace more highly adsorbed cations in the diffuse layer by the law of mass action.

Partitioning refers to the separation of chemicals among the different phases, such as the partitioning of sorbing cations between the liquid and solid phases (Figure 18). The extent of the partitioning for a specific chemical and media is described by the partitioning coefficient, K_p. For many cases, we assume that the partitioning process is instantaneous, so that there is an immediate equilibrium between, for instance, the mass sorbed to the solid as a contaminant front migrates through the vadose zone. If the partitioning of the ion concentration in the liquid and solid phases follows a linear, reversible behavior, K_p is called the distribution coefficient, K_d. Nonlinear adsorption is generally characterized by Langmuir or Freundlich isotherms. There has been little conclusive research on whether K_p is dependent upon water content, but most analyses neglect such a possibility and assume that batch mixing experiments are equally valid for saturated and unsaturated media. Where sorption does not occur instantly, then we must consider kinetic effects that control the rate of partitioning, e.g., Pignatello (1989). Schwarzenbach and Westfall (1981) conducted laboratory batch and column experiments with dissolved nonpolar organic chemicals in fine sand and found that at high percolation rates kinetic effects were significant, whereas at low percolation rates in the column experiment the partition coefficient was about the same as in the batch mixing test.

Pure organic compounds and mixtures of pure organic liquids spilled into the vadose zone typically have low solubility in water and tend to form non-aqueous-phase liquid (NAPL) blobs or pools, owing to their hydrophobic nature (Figure 19). However, these organic compounds at least partially dissolve in the pore water, to an extent depending upon their solubility. The dissolved organic molecules tend to adsorb onto natural carbonaceous material in the soil. Most dissolved organic compounds are reported to follow a linear adsorption isotherm. Therefore, we can express the distribution coefficient for organic chemicals as

$$K_d = K_{oc} f_{oc} \tag{39}$$

where K_{oc} is the partitioning coefficient between the organic liquid and the organic carbon and f_{oc} is the mass fraction of organic carbon. On the basis of numerous experiments, the distribution coefficient for hydrophobic organic compounds can also be obtained from

$$K_d = 0.63 K_{ow} f_{oc} \tag{40}$$

where K_{ow} is the octanol-water partitioning coefficient that is published in standard chemical reference books. Table 4 gives a few useful properties of organic chemicals that are common vadose zone contaminants.

The net effect of attenuation of any species, either cations or dissolved organic chemicals, is to retard the rate of migration of the species relative to the mean porewater velocity. The ratio of the velocity of the chemical, V_c, relative to the mean porewater velocity, V, is called the retardation factor, R_f.

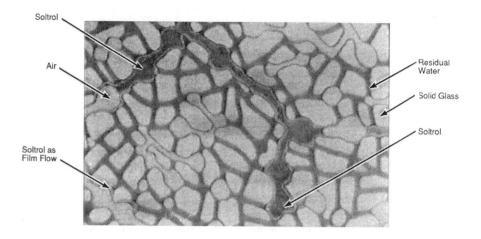

a

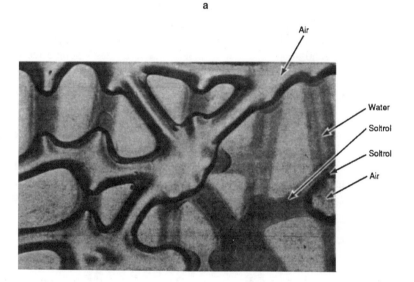

b

Figure 19 (a) Photo of an etched glass micromodel experiment showing the infiltration of an organic liquid (Soltrol-130; specific gravity 0.755; dynamic viscosity 1.45 cp). Initially, the pores of the micromodel were saturated and then allowed to drain. The organic liquid advanced by filling mostly air-filled pores and by film flow. Soltrol was dyed red (dark grey), water was dyed blue (light grey), and the air was not dyed. (b) Closeup of the micromodel experiment after Soltrol was drained under gravity, showing trapping of the organic liquid. Note that air (not dyed) is the nonwetting fluid, water (light grey) is the wetting fluid, and Soltrol (dark grey) has intermediate wetting properties. (From Wilson et al., 1990. With permission.)

$$R_f = \frac{V_c}{V} = 1 + \left(\frac{\rho_b}{\theta}\right)K_d \qquad (41)$$

where ρ is the dry bulk density (M/L^3), θ is the volumetric water content (L^3/L^3), and K_d is the partitioning coefficient of the ion between the solid and liquid phase (L^3/M)

Table 4 Physical and Chemical Data of Selected Priority Pollutants

Compound (synonym)	Formula	Molecular weight	Solubility (mg/L)	Density (g/cm³)	Boiling point (°C)	Surface tension (dyn/cm at 20°C)	Dynamic viscosity (C_p at 20°C)	log K_{ow} (ml/g)	log K_{oc} (ml/g)	Vapor pressure (mm Hg)	Henry's constant (atm-m³/mol)	Vapor density (g/L)	Relative vapor density
Benzene	C_6H_6	78.11	1.75×10^3	0.868	80.1	28.9	0.647	2.12	1.94	9.52×10^1	5.40×10^{-3}	3.19	1.172
Chloroform	$CHCl_3$	119.39	8.00×10^3	1.489	62	27.2	0.58	1.90	1.64	1.60×10^2	3.23×10^{-3}	4.88	1.664
p,p'-DDT	$C_{14}H_9CL_5$	354.49	5.00×10^{-3}	1.56	—	—	—	6.19	5.38	1.90×10^{-7}	4.89×10^{-5}	—	1.585
1,1-Dichloroethane (DCA)	$C_2H_4Cl_2$	98.96	5.50×10^3	1.18	56	38.7	0.44	1.48	1.15	1.82×10^2	4.30×10^{-3}	4.04	1.585
1,1-Dichloroethene (DCE)	$C_2H_2Cl_2$	96.94	4.00×10^2	1.22	37	24.0	0.36	2.13	1.81	4.95×10^2	2.10×10^{-2}	3.96	2.545
1,2-Dichloropropane (DCP)	$C_3H_6Cl_2$	112.99	2.70×10^3	1.159	96	28.7	0.86	2.28	1.71	4.20×10^1	2.94×10^{-3}	4.62	1.162
Ethylene dibromide (EDB)	$C_2H_4Br_2$	187.87	4.30×10^3	2.179	131	38.7	1.72	1.76	1.64	1.17×10^1	6.73×10^{-4}	7.68	1.080
Ethylbenzene	C_8H_{10}	106.17	1.52×10^2	0.867	136.2	31.5	0.64	3.13	2.20	7.08×10^0	6.60×10^{-3}	4.34	1.025
Methylene chloride	CH_2Cl_2	84.93	2.00×10^4	1.327	40	27.9	0.43	1.30	0.94	3.49×10^2	2.00×10^{-3}	3.47	1.897
Naphthalene	$C_{10}H_8$	128.18	3.00×10^1	1.145	217.9	31.8	—	3.36	3.11	5.40×10^{-2}	4.60×10^{-4}	4.42	—
PCB-1254	Various	avg. 327	5.00×10^{-2}	1.505	365	—	700	6.47	5.61	7.71×10^{-5}	2.70×10^{-7}	13.36	1.000
Phenol	C_6H_6O	94.11	8.20×10^4	1.058	182	—	12.7	1.46	1.43	2.00×10^{-1}	2.70×10^{-7}	3.24	1.001
1,1,2,2-Tetrachloroethane	$C_2H_2Cl_4$	167.85	2.90×10^3	1.593	146	36.0	1.75	2.56	2.07	5.0×10^0	3.80×10^{-4}	6.86	1.032
Tetrachloroethylene (PCE)	C_2Cl_4	165.83	1.50×10^2	1.623	121.2	31.3	0.89	2.60	2.42	1.40×10^1	1.53×10^{-2}	6.78	1.125
Trichloroethylene (TCE)	C_2HCl_3	131.39	1.10×10^3	1.464	87	29.3	0.57	2.53	2.03	5.78×10^1	9.10×10^{-3}	5.37	1.272
Toluene	C_7H_8	92.14	5.15×10^2	0.867	110.8	29.0	0.65	2.65	2.18	2.20×10^1	6.70×10^{-3}	3.77	1.029
o-Xylene	C_8H_{10}	106.17	1.52×10^2	0.880	144.4	30.1	0.81	2.95	2.11	1.00×10^1	5.27×10^{-3}	4.34	1.018

Modified from Knox et al., 1993, with permission.

(Figure 18). Degradation of contaminants also occurs in the vadose zone. For instance, radionuclides decay at a rate dependent upon the half-life of the element. Petroleum hydrocarbons are often readily degraded by indigenous microbes where oxygen and nutrients are not limiting. And chlorinated organic solvents may be transformed under aerobic and anaerobic conditions by biotic or abiotic processes (e.g., Figure 20). These degradation processes for chlorinated organics can be predicted reasonably well by a first-order decay process that is represented by a half-life, in the same manner as we treat radioactive decay.

The advection-dispersion equation for predicting the migration of a sorbing, decaying (or biodegrading or transforming) solute is:

$$\nabla \cdot D' \nabla C - \nabla \cdot CV = R_f \frac{\partial C}{\partial t} + \lambda C \qquad (42)$$

where D' is the hydrodynamic dispersion coefficient, C is the ion concentration dissolved in the pore water, V is the mean porewater velocity, λ is the decay constant, and t is time (van Genuchten and Alves, 1982). Note that $V = q/\theta_e$, where θ_e is the effective water content that participates in carrying the flow. The half-life can be obtained from standard references, and the partitioning coefficient also can be obtained from the literature as well as from laboratory batch (shaker) or column experiments. Column experiments allow one to control the water content and flow velocity to better approximate field conditions. The transport analysis with sorption may not be accurate if instantaneous equilibrium cannot be assumed in the ion partitioning or if the partitioning coefficient changes along the flow path due to precipitation, reduction, oxidation, complexing and previous adsorption reactions. In some field problems, for example, where there are dead-end pores, the partitioning does not occur instantly and equilibrium between phases cannot be assumed. Nonequilibrium mass transfer between phases where NAPL is present appears to be site specific (Miller et al., 1990; Brusseau et al., 1992; Armstrong et al., 1994).

3. Colloidal Transport

Colloidal transport has very recently been investigated by researchers as an explanation for much more rapid migration of chemicals that are expected to be highly sorbed. Colloids are small particles, usually classified as less than 1 μm in diameter, that remain suspended in solution for some period of time. Colloidal particles have charged surfaces and can attract cations or anions, depending upon the surface charge. Clays, insoluble oxides, and some organic pollutants are negatively charged in neutral pH waters (Kingston and Whitbeck, 1991). Chemicals in solution can be adsorbed onto the colloid surface by the same mechanisms we discussed earlier for attenuation by ion exchange. Contaminant transport is facilitated by mobile colloids if the chemical species is preferentially sorbed to the colloid over the bulk matrix. In the vadose zone, colloidal transport is not well documented by field data; however, it is more likely to be occurring where macropore or preferential flow occurs.

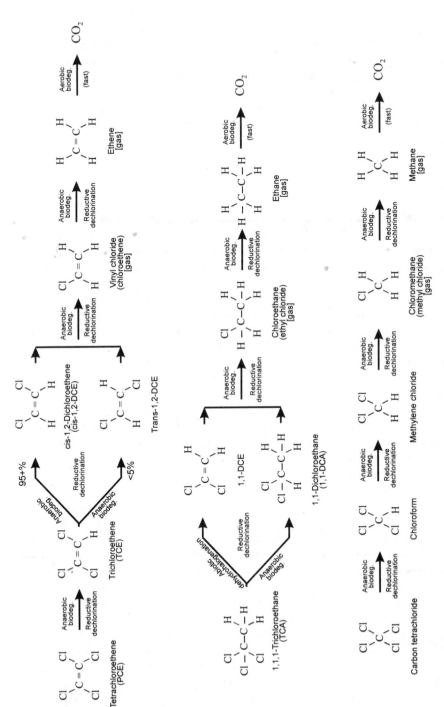

Figure 20 Chlorinated hydrocarbon degradation pathways. (From Daniel B. Stephens and Associates, Inc., 1995. With permission.)

B. GASEOUS-PHASE TRANSPORT

Gas-phase transport is becoming more important in contaminant migration investigations, especially in dry soils where liquid fluxes are very small. In addition to advective processes described in Section VII, gas migrates by diffusion. And as it migrates through the air-filled pores, the gaseous species may partition into the liquid and solid phases. In this section, we summarize the gas diffusion and partitioning processes.

1. Gas Diffusion

Gas-phase diffusion occurs where there are spatial differences in gas-phase concentration or vapor pressure. Where the water vapor in the soil pores is in equilibrium with the liquid, there is a relationship between the vapor pressure, P, soil-water potential, ϕ_{SW}, (e.g., in bars) and temperature:

$$\phi_{SW} = \rho_w \frac{RT}{M} \ln\left(P / P_o\right) \qquad (43)$$

where P_o is the saturated water pressure, ρ_w is water density, and other terms have been previously defined (see Equation 33). This equation shows that over the range of soil-water (matric) potentials encountered in the field, water vapor pressure is very nearly saturated throughout the vadose zone. For example, at 25°C the relative humidity of the soil $(P/P_o) \times 100$ is 99.99% at –0.1 bars and is still 99.26% at –10 bars soil-water potential. Consequently, water-vapor diffusion at a constant soil salinity and constant temperature is practically insignificant. However, vapor pressure is highly sensitive to temperature. A temperature change of only 1°C, from 19 to 20°C, causes vapor pressure to increase by 1.5 mbars, whereas to achieve such a vapor pressure change under isothermal conditions would require a matric potential change from about 0 to –100 bars (Hillel, 1980). Therefore, where diurnal and seasonal temperature fluctuates by even a few degrees, diffusion of vapor may need to be considered, especially if the liquid water flux is very small. To adequately account for the importance of water vapor or gaseous-phase diffusion due to temperature gradients requires an analysis of the simultaneous transport of both liquid and heat, owing to the transmission and storage of heat in soil subjected to fluctuating temperatures. That is, soil-water potential gradients can transport heat, and thermal gradients can transport water in both the liquid and vapor phases (Equation 34).

Specific gaseous components, such as radon, tritium, or a volatile organic chemical (VOC), will diffuse through the air-filled pores at a rate proportional to the gas's partial pressure (or gas-phase concentration) gradient and the gas diffusion coefficient. In a diffusive process, Fick's law of diffusion determines the mass flux of a chemical, q_m, that will move due to a gradient in gas concentration, C_g:

$$q_m = D_g \, \nabla C_g \qquad (44)$$

where D_g is the gas chemical diffusion coefficient. This equation holds for chemical diffusion in both the liquid and gas phases, but the diffusion coefficient depends upon the phase. It is possible to estimate D_g for a gas in soil from the diffusion coefficient for that gas in free air. Owing to the distribution of solids and water within the pores, the gas diffusion coefficient in many well-drained soils is approximately half of the free air diffusion coefficient published in standard references (Koorevaar et al., 1983). A common approach to better quantify the effects of the solid and liquid phases is to multiply the free air gas diffusion coefficient by a tortuosity factor, τ, where

$$\tau = n^{1/3} S_g^{7/3} \tag{45}$$

and n is porosity and S_g is the gas-phase saturation (e.g., Falta et al., 1989). The dependence of the gas diffusion coefficient on water content and void space tortuosity is often described by the Millington (Millington, 1959) model:

$$D_g = \frac{D_a (n - \theta)^{10/3}}{n^2} \tag{46}$$

where D_a is the chemical diffusion coefficient in air, n is porosity, and θ is volumetric water content (Jury, 1983). The gas diffusion coefficient in soil therefore will be relatively small in low porosity soil and where water content is high. Where variations in soil-water potential and temperature gradients affect water content for a particular soil, predicting the rate of gas transport by diffusive as well as advective mechanisms involves a complex interaction of forces that move water and water vapor. For example, we may first need to solve the three-phase flow equation (Equation 28) to obtain the gas-phase saturation, and couple this to a gas contaminant migration equation. One such code capable of simulating multiphase flow and transport under nonisothermal conditions is T2VOC (Falta and Preuss, 1991). Hence, we often seek to simplify the transport analysis by assuming gas diffusion occurs under constant soil-water potential and isothermal conditions.

2. Gas Partitioning

Gas-phase partitioning refers to the processes by which chemical components in the gas phase interact with the liquid and solid phases in the vadose zone. The tendency for a pure-phase liquid component, such as the volatile chemical benzene, to volatilize into the gas phase is strongly dependent upon the vapor pressure of the chemical, P, at a specific temperature. Chemicals with boiling points less than 150°C and vapor pressure greater than about 10 mmHg at 20°C are generally considered to be volatile and detectable in soil gas surveys (Tolman and Thompson, 1989). If a pure-phase chemical exists as an NAPL in the vadose zone, then the equilibrium gas concentration, C_g, will be

$$\left(\sqrt[g]{m^3} \right)$$

$$C_g = \frac{MP}{RT} = \frac{MW \, P^o}{RT} \tag{47}$$

where M is the molecular weight (g/mol), R is the ideal gas constant (8.2×10^{-5} atm-m³/mol-K), and T is temperature (K). Typically, the NAPL is a mixture of different chemicals such as benzene, xylene, toluene, and ethylbenzene, which comprise gasoline. In this case, Raoult's law is used to calculate the equilibrium gas concentration of a specific chemical by first computing the part of the total vapor pressure, P_T, in the soil contributed by that specific chemical:

$$P^o = X_c P_T \qquad (48)$$

where P^o is the vapor pressure of the chemical above the NAPL mixture and X_c is the mole fraction of the chemical in the NAPL (e.g., ratio of the number of moles of benzene to the total number of moles of benzene, xylene, toluene, and ethylbenzene). Now the equilibrium soil gas concentration of the specific chemical, C_g, is calculated from:

$$C_g = \frac{MX_c P_T}{RT} \qquad (49)$$

This equation would predict the equilibrium soil gas concentration adjacent to the NAPL source.

Henry's law predicts the equilibrium concentration between the gas- and aqueous-phase concentration:

$$K_H = \frac{C_g}{C_l} \qquad (50)$$

where K_H is the Henry's law constant (dimensionless). An alternative expression for Henry's law is given in terms of the gas partial pressure instead of its concentration:

$$K_H' = \frac{P^o}{C_l} \qquad (51)$$

where K_H' has units of atm-m³/mol (Table 4). The dimensionless Henry's law constant can be calculated from the chemical vapor pressure, liquid-phase concentration, and temperature according to

$$K_H = \frac{16.04\,PM}{TC_l} \qquad (52)$$

where M is the molecular weight (g/mol), T is temperature (°K), and C_l is liquid-phase concentration (Silka and Jordan, 1993). From the ideal gas law,

$$K_H = \frac{K_H'}{RT} = 41.6 K_H \text{ at } 20°C \qquad (53)$$

Henry's law implies that the partitioning relationship between gas and aqueous phase concentration is linear. However, at very low water content (less than five

molecular layers of water), the relationship may be nonlinear, with a greater partitioning onto the water phase, probably due to dissolution of some chemical into the
bound water (Silka and Jordan, 1993). Several researchers note that vapor partitioning between the solid and gas (Figure 18) can be an important process in gas transport *wrong*
in dry soils (e.g., Shoemaker et al., 1990; Valsaraj and Thibodeaux, 1988). Ong and
Lion (1991) conducted laboratory experiments of five different media that showed
that as the soil water content decreased to about five water monolayers, the partitioning coefficient for TCE decreased slightly and then with further drying the partition
coefficient increased by two to four orders of magnitude. At a particular field site,
it may also be necessary to question the validity of the assumption of instantaneous
equilibrium between the gaseous and aqueous phases that Henry's law implies (e.g.,
Smith et al., 1990).

*high partitioning coeff means contaminant
remains mostly in gas phase*

CHAPTER 2

Soil-Water Budget

In the previous chapter, we discussed the general nature of the vadose zone, as well as fluid flow and storage characteristics. For many problems, it is important to develop an accounting system to track additions, depletions, and changes of water storage in what is referred to as a water balance or water budget. A water budget is an equation of water mass conservation for a particular volume or region. In hydrology, one can derive a water budget for a surface water body, watershed, aquifer system, or vadose zone. These regions are clearly linked together, inasmuch as the output from one becomes the input to another.

This chapter presents an overview of the major components of a vadose zone water budget or what soil scientists refer to as a soil-water budget. This overview sets the stage for more detailed discussions in the subsequent two chapters of some of these soil-water budget components that are most important in hydrogeologic investigations.

The general equation for the soil-water budget is derived by considering the mechanisms by which water can enter, exit, or be stored in a predefined region of the vadose zone. For many problems, the inflow across the upper boundary of the vadose zone is infiltration, while outflow from the upper boundary is evaporation and transpiration, and outflow from the lower boundary is groundwater recharge. Net inflow (inflow minus outflow) must equal the change in soil water stored in the vadose zone. In the soil-water budget equation, for a discrete time interval, we add flows due to processes that contribute water to the vadose zone, subtract discharges and water losses, and equate this to changes in the water stored in the soil volume:

$$I - E - T - R = \Delta S \qquad (1)$$

where I is infiltration, E is evaporation, T is transpiration, R is deep percolation or recharge (all four variables in units of $L^3 T^{-1}$), and ΔS represents the change in water storage over the time interval Δt. The horizontal area in Equation 1 is problem dependent, but usually one can deal with a unit cross-sectional area (in plan view). One can divide both sides of the equation by the area, and express the water-budget components on the left-hand side as fluxes, rates, or specific discharges having units of LT^{-1}. Next we briefly describe each of the principal soil-water budget components, and present some of the methods to obtain them. For more detailed discussions, the text by Hillel (1980a,b) is excellent.

45

I. INFILTRATION

Infiltration is the volume of water that crosses the land surface to enter the vadose zone. It is usually quantified in one of three ways: by the residual from a surface water budget analysis, by field measurements, or by calculation based on soil hydraulic properties. In a surface water budget, infiltration may be computed by measurements of precipitation, applied irrigation, and surface run-on, and subtracting from this sum the surface runoff, interception of water on plant canopies, direct evaporation, and increases in surface water storage. Methods to quantify each of these components of a surface water budget can be found in standard engineering hydrology texts and reference books.

In the field, infiltration from precipitation as rainfall can be simulated by using a rainfall simulator to quantify infiltration over an area of roughly 1 to 10 m² for controlled duration and intensity storms (e.g., Zegelin and White, 1982). Under ponded conditions a ring infiltrometer, a short cylinder approximately 0.1 to 1 m² in cross section driven into the soil, can be used to determine the infiltration rate by measuring flow through the cylinder under equivalent ponded head conditions. Beneath impoundments, lakes, or reservoirs, seepage meters may provide reliable infiltration measurements (e.g., Stephens et al., 1985). Infiltration from stream channels, where evaporation can be neglected, is most readily obtained by subtracting discharges measured at stream gaging stations along the reach of the channel.

Infiltration can also be calculated from hydraulic properties of the soil using mathematical expressions which describe the infiltration process. The type of equation chosen depends upon the nature of the process, such as transient or steady infiltration and ponded or nonponded infiltration, for instance. Although more will be said about infiltration processes and equations in the next chapter, one could envision that a simple expression to calculate infiltration rate would be Darcy's equation (Equation 17a in Chapter 1). Here, the infiltration rate, designated as q, is calculated from field measurements of the hydraulic head gradient near the soil surface and the hydraulic conductivity. Methods to characterize hydraulic conductivity are detailed in Chapter 5 and methods to measure the hydraulic head are described in Chapter 6.

In the water budget analysis, it may be important to recognize that, in addition to vertical flow, lateral inflow and outflow may occur within the region of the vadose zone where the water budget is to be computed. These components are usually so small that they are neglected, as we have done in Equation 1. However, in some cases lateral water movement is significant, especially in heterogenous or anisotropic soils and where there is topographic variability, as we illustrate in the next chapter.

II. EVAPORATION AND TRANSPIRATION

Evaporation refers to the water lost from the vadose zone by vapor-phase transport from the soil directly to the atmosphere. Transpiration is the water depleted from

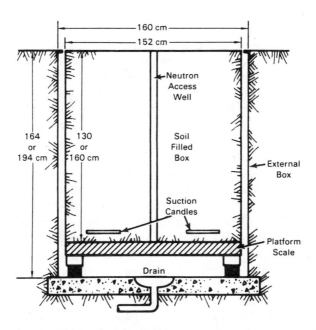

Figure 1 Cross-sectional view of weighing lysimeter. (From Kirkham et al., 1984. With permission.)

the vadose zone by plant root uptake and released to the atmosphere. For most practical problems it is both difficult and unnecessary to separate these two processes, so the two are combined and called evapotranspiration. Research on the topic of measuring evapotranspiration is extensive, beginning over 400 years ago (Sosebee, 1976), and development of new methods to accurately quantify evapotranspiration continues to be a topic of professional meetings. There are two different approaches to determine evapotranspiration, by measurement and by estimation.

Soil lysimeters provide the most accurate method to measure evapotranspiration, and in fact most estimation methods have been verified by comparing the predictions to lysimeter data. Unfortunately, soil lysimeters are usually cumbersome, expensive to construct, and require rather long periods of data collection. There are three types of lysimeters used for water balance analysis: weighing, nonweighing and floating. All soil lysimeters share the same concept. A small monolith of soil with vegetation is placed in a container and is returned to its original position in the landscape. Instrumentation is emplaced to allow measurements of precipitation, soil-water storage, and deep drainage. From these components one can compute evapotranspiration by simple arithmetic using the water balance equation (Equation 1). In a weighing lysimeter, the lysimeter is placed on a scale (Figure 1) having a capacity sufficient to determine the mass of a soil monolith with a diameter of 1 m to several meters and a depth of about 1 m. Bare soil evaporation could be determined using microlysimeters, a technique using short soil core samples which are removed from the soil for weighing over periods of 1 to 2 d (Boast and Robertson, 1982). In a nonweighing lysimeter, evapotranspiration from the soil monolith is determined by measuring the rate of water supply to the monolith container that is necessary to maintain a constant depth to water in the base of the container. In a floating lysimeter, the soil monolith is placed on a liquid-filled pillow so that water gains and losses can be obtained by measuring fluid pressure through a manometer tube.

Another approach to measure evapotranspiration is to place a canopy over the plant and measure air flow rate and water content of the inflowing and outflowing air. Sebenik and Thomas (1967) applied this technique by constructing a plastic tent over a tree. Stannard (1990) developed a portable hemispherical chamber containing fans and a psychrometer that fits over vegetation, such as grasses and small shrubs, to rapidly measure evapotranspiration over a period of less than 2 min.

Methods for making micrometeorological measurements above a vegetated surface have been developed over the past 50 years or so to determine the actual evapotranspiration (Rosenberg et al., 1983). Two such methods are the Bowen ratio method (e.g., Tanner, 1960) and the eddy-correlation method (Swinbank, 1951).

The Bowen ratio method is based upon a simplified energy-budget equation (e.g., Marshall and Holmes, 1992):

$$R_n - G - H_s - \lambda_v E_m = 0 \tag{2}$$

where R_n is the net-radiation-flux density to the surface (W/m^2), G is the heat-flux density into the soil (W/m^2), H_s is the sensible-heat-flux density into the air above the plant canopy (W/m^2), λ_v is the latent heat of vaporization (J/g), and E_m is the evapotranspiration mass flux density (g/s-m^2).

In the Bowen ratio method the energy budget equation is rearranged to

$$\lambda_v E_m = \left(R_n - G\right)/\left(1 + \beta_r\right) \tag{3}$$

where β_r is the Bowen ratio, $H_s/\lambda_v E$. R_n can be measured with a net radiometer, and G can be obtained by installing a soil heat flux plate just below the soil surface. It can be shown that the Bowen ratio may be determined by

$$\beta_r = \gamma\left(T_1 - T_2\right)/\left(P_1 - P_2\right) \tag{4}$$

where γ is the psychrometric constant, P is the water vapor pressure, and T is air temperature (Montieth and Unsworth, 1990). Therefore, measurement of temperature and vapor pressure at two elevations above the surface is required to obtain β_r. These measurements for calculating the Bowen ratio are usually obtained from averages over periods of 1/2 to 2 h (Montieth and Unsworth, 1990). The evaporation rate (LT^{-1}) is calculated as $E = E_m/\rho_w$, where ρ_w is the density of water.

The eddy correlation method is based on the principle that water vapor flux across the land surface can be measured by correlating the vertical variations of wind speed, w, with variations of vapor density, q_v (Tanner et al., 1985):

$$\overline{\left(\lambda_v E\right)} = \lambda_v\overline{\left(w'q_v'\right)} \tag{5}$$

where the overbars represent time averages and the primes represent instantaneous deviations about the time averages. The data collection requirements include an anemometer and hygrometer which are connected to a data logger.

The eddy correlation method also can provide measurements of sensible heat flux, H:

$$\overline{H} = C_P \rho_a \overline{\left(w'T'\right)} \tag{6}$$

where C_p is the specific heat and ρ_a is the density of air. Temperature fluctuations can be measured with a thermocouple connected to a data logger. In an example of the application of the eddy correlation method, Tanner et al. (1985) placed the micrometeorological instruments 1.1 m above a crop canopy to record data every 10 s on a data logger for computation of half-hour average evaporative and sensible heat fluxes over a period of about 6 d.

The so-called eddy correlation-energy budget method (e.g., Czarnecki, 1990) is used to determine the actual evaporative flux when field instrumentation accounts for net radiation and heat conduction into the ground, as in the Bowen ratio method, and when sensible heat flux, H, is determined by the eddy correlation technique.

Because of the obvious logistical difficulties associated with measuring evapotranspiration over extensive areas, estimates of evapotranspiration are usually preferred. Rosenberg et al. (1983), who present an excellent detailed discussion of evapotranspiration, delineate two broad approaches to estimate this component of the soil-water budget: climatological and micrometeorlogical methods. Among the climatological methods, some are based on air temperature (e.g., Thornthwaite, 1948; Blaney and Criddle, 1950), others are derived from solar radiation measurements (e.g., Jensen and Haise, 1963), and others incorporate both energy supply data and turbulent transfer of water vapor away from the surface (e.g., Penman, 1948).

While actual evapotranspiration (ET) is the quantity we seek, it is important to recognize that the preceding climatological methods calculate the potential evapotranspiration (PET), that is, the amount of evapotranspiration that would occur from a short green crop that fully shades the ground, exerts negligible resistance to the flow and is always well supplied with water. Potential evapotranspiration is analogous to evaporation from an areally extensive, open body of water, but PET cannot exceed lake evaporation under the same meteorological conditions. In fact, potential evapotranspiration can be estimated from lake evaporation data that may be calculated from the U.S. Weather Bureau, Class A evaporation pan measurements which are available for many regions (Linsley et al., 1975). It has been established that the potential evaportranspiration rate is approximately 0.7 times the pan evaporation rate, but this pan coefficient can vary from about 0.40 to 0.85 depending upon wind speed, fetch of green crop, and relative humidity (McWhorter and Sunada, 1977).

However, the actual evapotranspiration in a landscape is rarely equal to potential evapotranspiration, except in very humid climates, where the water table is within about 1 m of land surface or immediately after the profile is thoroughly wetted by an infiltration event. To compute the actual evapotranspiration (ET) from potential

Table 1   Estimated Plant-Water-Use Coefficients K_{co} for Native Vegetation

| | K_{co} | | | | | | | |
Vegetation	Nov. to Mar.	Apr.	May	June	July	Aug.	Sept.	Oct.
Sagebrush-grass	0.50	0.60	0.80	0.80	0.80	0.71	0.53	0.50
Pinyon-juniper	0.65	0.70	0.80	0.80	0.80	0.80	0.69	0.65
Mixed mountain shrub	0.60	0.67	0.81	0.85	0.82	0.74	0.65	0.60
Coniferous forest	0.70	0.71	0.80	0.80	0.80	0.79	0.75	0.71
Aspen forest	0.60	0.67	0.85	0.90	0.86	0.75	0.65	0.60
Rockland and miscellaneous	0.50	0.60	0.65	0.65	0.65	0.60	0.50	0.50
Phreatophytes	1.00	1.00	1.00	1.00	1.00	1.00	1.00	1.00

From McWhorter, and Sunada, 1977. With permission.

evapotranspiration (PET) when the water supply is limited requires an additional calculation based upon plant type, water availability, and vegetation coverage on the landscape:

$$ET = K_c(PET) \tag{7}$$

where K_c is a crop coefficient. The crop coefficient is usually obtained by establishing an experimental relationship between ET (measured with lysimeters) and PET (calculated by a specific method) for some brief period. Crop coefficients for agricultural crops are summarized by Doorenbos and Pruitt (1975) and values for selected native vegetation are presented in Table 1 from McWhorter and Sunada (1977). The dependence of the crop coefficient upon available water (AW) is described by a wide variety of formulations (Moridis and McFarland, 1982), one of which was developed by Jensen et al. (1970):

$$K_c = K_{co} \frac{\ln\left(\dfrac{100(AW)}{AW_{max}} + 1\right)}{\ln 101} \tag{8}$$

where K_{co} is the crop coefficient for a field where water is not limiting and

$$AW = (\theta - \theta_{wp})d \quad \text{rooting depth} \tag{9}$$

$$AW_{max} = (\theta_{fc} - \theta_{wp})d \tag{10}$$

where θ is the field water content, θ_{fc} is the water content at the so-called field capacity, θ_{wp} is the water content at the permanent wilting point, and d is the rooting depth.

Particular hydrogeological settings may make possible other methods of determining evapotranspiration. For example, Weeks and Sorey (1973) used a finite-difference array of observation wells along with aquifer properties and water level fluctuations in wells to compute evapotranspiration in the Arkansas River Valley in Colorado. Evapotranspiration near perennial streams can be determined from streamflow records (Daniel, 1976).

For very shallow water table conditions and unvegetated sites, evapotranspiration will be the evaporation from the water table. Evaporation from a bare soil can be calculated from simple equations (e.g., Gardner, 1958; Willis, 1960) using estimates of soil hydraulic properties. For the case of a homogeneous soil in which water migrates upward from the water table only in the liquid phase, the maximum evaporative flux is

$$q_{\lim} = A_c \frac{a}{d^l}$$ (11)

where a and n are obtained by fitting the equation

$$K = \frac{a}{b + |\psi|^l}$$ (12)

to K-ψ data, where d is the depth to the water table and A_c is a constant that depends on l. For most soils l ranges between 1 and 4, with the larger values for coarser soils. For example, if $l = 1.5$, $A_c = 3.77$; if $l = 4$, $A_c = 1.52$ (Gardner, 1958).

Ripple et al. (1972) developed a procedure to estimate steady evaporation rates from bare soil that takes into account the meteorological influences that affect evaporation in both liquid and vapor phases. Their results illustrate that evaporation from the water table depends not only on water table depth and soil parameters but also on vapor transfer and potential evaporation — factors accounted for by measuring air temperature, humidity, and wind velocity.

A recent investigation near Yucca Mountain, NV, provides a good example of evapotranspiration measurements as part of a hydrogeologic field problem and one in which many of the methods to obtain evapotranspiration were compared. Ground water modeling studies had showed that estimated aquifer parameters were particularly sensitive to evapotranspiration from Franklin Lake Playa (Czarnecki, 1990). To improve the reliability of the model results, Czarnecki (1990) compared several techniques to compute evapotranspiration from the 14-km^2 playa. The playa is underlain by alluvium, and the water table occurs at a depth of about 1.25 m. Upward groundwater flow in the aquifer underlying the playa was presumed to discharge as evapotranspiration and was calculated from Darcy's equation applied to the aquifer, based on the hydraulic gradient obtained in nested piezometers and saturated hydraulic conductivity from slug test measurements. Czarnecki also used temperature profiles in the saturated zone observation wells to compute the upward groundwater discharge. In the vadose zone, Czarnecki measured temporal variations in the water content profile and applied a one-dimensional finite difference model of the Richards equation to quantify evapotranspiration. Other methods included 10 empirical methods based on PET, as well as estimates based on vegetation uptake rates at similar sites, and the energy-budget eddy-correlation approach.

In the energy budget-energy correlation method, E, R_n, and H were measured at the land surface and at the top of the vegetation canopy (Figure 2); R_n was measured with a Fritschen-type radiometer, H was obtained from a sonic anemometer which obtains wind speed and air temperature data, and G was measured with a soil heat flux plate buried at various depths below land surface. The instruments and the data

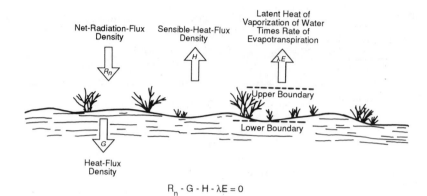

$$R_n - G - H - \lambda E = 0$$

Figure 2   The energy balance method. (From Czarnecki, 1990. With permission.)

Table 2   Summary of Evapotranspiration Estimates from All Techniques Used

Technique	Evapotranspiration Estimate (cm/d)
Energy-balance eddy correlation	0.1 to 0.3
Empirical potential evapotranspiration relations:	
Lower range (January)	0.1 to 0.5
Upper range (July)	0.5 to 0.7
Temporal changes in soil-moisture content in the unsaturated zone	Inconclusive −0.07 to 0.1
Evapotranspiration by phreatophytes (Robinson, 1958)	0.09 to 0.34
Temperature profiles	Inconclusive
Saturated-zone vertical gradients	0.06 to 0.5
One-dimensional finite-difference model	0.06

From Czarnecki, 1990. With permission.

logger operating them were established at seven locations representative of playa conditions. Data were collected for periods of 2 to 3 d at five times from June 1983 through September 1984.

A comparison of results is given in Table 2. Czarnecki (1990) concluded that all the methods produced reasonable results, except the methods based on water content changes and temperature profile data. Of all the methods applied at this site, the results from the energy-budget eddy-correlation technique were considered the most reliable for the conditions tested because the results were based on the most direct measurements of evapotranspiration.

III. WATER STORAGE AND DEEP PERCOLATION

The remaining two major components of the water budget equation are the water storage change and the rate of deep percolation from the base of the profile during the time interval selected for the water budget. To quantify water storage changes requires repeated measurements of water content within the soil water budget volume.

Over a year or several years, the water content change is usually small, and for some sites and climatic conditions, the long-term change is negligible.

Deep percolation is the water that moves downward below the root zone. Most references to this term are in the context of irrigation, but this term should not be so restrictive. If the depth of the water budget element is greater than the rooting depth, then the deep percolation will likely become groundwater recharge when it reaches the water table. Because deep percolation and recharge are so often at the heart of vadose zone problems, we present here a rather complete survey of methods to compute recharge, and later a separate chapter is devoted to a review of processes that affect recharge.

A. STORAGE

Water storage in the vadose zone is simply the volume of water stored in the soil or rock to a particular depth. Inasmuch as it is the change in water content that is required in the water budget, the preferred method is to measure water content *in situ* repeatedly at exactly the same depths, usually using a geophysical method such as neutron probe logging. Neutron probe logging, and other methods described later in Chapter 6, afford a means to nondestructively measure water content changes at the same depths without introducing the sampling bias accompanying destructive soil sampling techniques. From the discrete water content measurements, θ, the volume of water in storage is calculated as

$$\Delta S = \frac{1}{\Delta t} \left[\int_0^D \theta \, dz \right] \cdot (\text{Area}) \tag{13}$$

The rate of change in water storage is calculated by subtracting the water storage to depth D at two different time periods and dividing this by the time between monitoring events. In weighing soil lysimeters, the change in water storage within the monolith can be simply obtained from the change in mass divided by the water density.

B. DEEP PERCOLATION AND RECHARGE

Groundwater recharge is water that enters either a perched aquifer or the phreatic zone. One of several ways water can enter aquifers is by migration of deep percolation through the vadose zone. In the following discussion, we summarize methods to calculate deep percolation and recharge to an aquifer from the vadose zone by physical and chemical methods applied to both the vadose zone and groundwater.

1. Physical Methods

a. Soil Lysimeters

Soil lysimeters for the purpose of collecting deep drainage and estimating recharge are constructed by excavating soil to the desired depth, installing a casing with a sealed base, and repacking the casing with soil to the *in situ* bulk density. Water percolating under a tension can be collected at the bottom of the lysimeter by extracting it from porous ceramic cups or tubes which are subject to a vacuum that exceeds the soil-water potential. Alternatively, the water accumulated at the base of the lysimeter can be obtained by measurements of water content change with a neutron probe or by piezometers to measure the depth of saturation. Most soil lysimeters are less than about 2 m deep, but one at the Hanford, WA site is 18.5 m deep and is possibly the deepest one anywhere in the world (Gee et al., 1994). A simple variation of the soil lysimeter, called the pan lysimeter, is a trough or impervious pan installed beneath or buried within a soil plot. The water collected in the pan drains laterally through an inclined pipe to a collection vessel located in a trench adjacent to the lysimeter.   *measures percolation*

The principal advantage of soil lysimeters is that they are direct and precise measures of soil-water flux. Gee and Hillel (1988) indicate that the deep drainage collected in a soil lysimeter can be determined with a precision as low as about 1 mm.

For deep water table conditions, we can capture the infiltrated water in a lysimeter constructed below the effective depth of evapotranspiration and infer that this water would ultimately reach the water table. The rate of soil water collected per unit area of the lysimeter would be only an estimate of the recharge.

Although soil lysimeters are the only direct methods available, they still produce recharge estimates that are subject to some uncertainty. For instance, virtually all soil lysimeters contain disturbed soil. It is nearly impossible to preserve completely the *in situ* water content, pore geometry, stratification, and macropore structures that can strongly influence soil-water movement. Months or years may be required to completely reestablish vegetation with the same canopy and rooting characteristics as the surrounding native soils. Furthermore, as we discussed in the previous section, under deep water table conditions, unless the lysimeters are completed below the rooting depth, the water collected will only comprise an upper bound on the actual recharge to the water table below. Additionally, the lysimeter data will represent recharge estimates only over the period of measurement, and therefore, depending on the prevailing precipitation and climatic conditions during this time, the estimates may not represent long-term behavior. Disadvantages to consider in lysimetery include the considerable expense for construction, the disturbance of site facilities, operations and soils, and the long-term commitment to monitoring required to obtain representative data.

need climatological data

b. Water Balance

One of the most widely adopted approaches to determine deep percolation or groundwater recharge is the water balance method. In this approach, recharge is calculated as the residual in the soil-water budget equation (Equation 1), with precipitation, evapotranspiration, and change in water storage computed by the methods described earlier. Gee and Hillel (1988) pointed out that precision in precipitation is seldom less than ±5%, and the precision for evapotranspiration is typically at least ±10%. Because of the propagation of errors in the water balance analysis, there is large uncertainty in the calculated recharge when recharge is a small fraction of precipitation. They conclude that in arid climates, the error in recharge calculated by water balance methods could exceed the calculated value of recharge itself. Hence, water budget methods to quantify deep percolation and recharge are more appropriate in humid climates.

c. Plane of Zero Flux

A plane of zero flux develops during redistribution of a pulse of infiltrated water. Above this plane or surface, soil-water movement is upward due to evapotranspiration, and below this plane water moves downward in response to capillary forces ahead of the wetting front and to gravity. Water below the plane would eventually become recharge, although bear in mind that the depth of the plane is not constant in time. To determine the water flux below the plane of zero flux, the following equation from Dreiss and Anderson (1985) is applied to volumetric water content profiles measured over different time periods $[\theta(z,t)]$:

$$q = \frac{1}{\Delta t} \int_{D_0(t)}^{D} \theta \, dz \qquad (14)$$

Recharge is calculated by summing the water content changes over the interval from the zero flux plane depth, D_0, to the depth of the monitored profile, D. Implementing the method requires instrumentation of soil-water potential sensors to locate where the hydraulic gradient is zero. Alternatively, the zero flux plane can be inferred from water content data collected by neutron probe, frequency domain reflectometry probe, time domain reflectometry probes, or other in situ water content measuring devices described in Chapter 6. These same devices monitored over time provide the requisite data to compute the recharge rate. The zero flux plane method was applied by Dreiss and Anderson (1985) to quantify deep water percolation beneath a land treatment facility.

Allison et al., (1994) pointed out that the zero flux plane method breaks down when the hydraulic gradient is downward throughout the profile, and this occurs at a time when there is significant recharge. During these periods, the Darcy flux method could be applied.

d. Darcy Flux in the Vadose Zone

Darcy flux calculations also comprise a physical means to calculate recharge from the vadose zone. The components in Darcy's equation for unsaturated vertical flow include the hydraulic conductivity at the field water content (or potential) and the hydraulic gradient. In Chapter 5, we discuss field and laboratory techniques to determine the unsaturated hydraulic conductivity, and in Chapter 6, we summarize the various methods to quantify soil-water potential. In soils below the root zone where the matric potential is nearly constant with depth, it is a good assumption that the hydraulic gradient is approximately unity, and flow is downward. Consequently, where this assumption is reasonable and where vapor-phase transport downward is negligible, the recharge rate is approximately equal to the in situ vertical, unsaturated

$$q = k(\theta)\left(\frac{d\phi}{d\ell}\right) \quad ?$$

hydraulic conductivity. The precision in the recharge rate computed by Darcy velocity calculations is therefore not less than that for the unsaturated hydraulic conductivity. Unless considerable care is taken in the conductivity analysis, it is possible that errors in the recharge rate by this method could range from a factor of less than two or three in wet soils to more than an order of magnitude in dry soils. In Chapter 1 we presented a sample calculation of soil-water flux by Darcy's equation for a case where the hydraulic gradient was calculated from *in situ* pressure measurements.

Good examples of the Darcy flux method are the field studies by Nnyamah and Black (1977) applied at a Douglas-fir stand in British Columbia, by Sophocleous and Perry (1985) in Kansas, and by Stephens and Knowlton (1985) at a sparsely vegetated sandy site in New Mexico.

e. Soil Temperature

Soil temperature gradients develop due to both heat flow from the natural geothermal gradient and atmospheric temperature fluctuations. The actual soil temperature profile depends on the geothermal gradient, the period and amplitude of the atmospheric temperature changes, the thermal properties of the vadose zone, and the water flux through the vadose zone. Several different techniques have been developed to provide estimates of the soil-water flux or recharge.

Bredehoeft and Papadopulos (1965) developed a type-curve method to compute the flux by fitting steady-state temperature profile data to the theoretical type curves. This approach has been applied by Cartwright (1970, 1979), Boyle and Saleem (1979), and Sammis et al. (1982), for example. Taniguchi and Sharma (1993) indicate that applications of the method have been limited to areas where the temperature gradient is upward, that is, more than several meters below the surface, where significant seasonal soil temperature changes do not occur.

Stallman developed an analytical solution for computing the steady downward flux from sinusoidally varying surface temperatures. Taniguchi and Sharma (1993) built upon Stallman's analysis and computed recharge from *in situ* measured changes in temperature. After applying the method to forested sites in Western Australia, they concluded that their temperature difference method produced reasonable results when the annual recharge rate is low and the temperature is measured at depths less than several meters.

A third temperature method for estimating soil-water flux was developed by Wierenga (1970) based on a steady-state soil heat balance. When water infiltrates to a particular depth, the mean temperature of the soil within this depth will change. By knowing the heat capacity of the soil and by measuring *in situ* the initial and final soil temperature as well as the temperature of the infiltrating water, the amount of infiltrated water is obtained. Inasmuch as infiltration-induced temperature changes are not readily detected below about 2 m, the method is best suited to determining soil-water flux rather than recharge (Taniguchi and Sharma, 1993).

Although the instrumentation requirements for implementing temperature methods are quite simple, the methods have not been widely used to date. One reason for this may be the difficulty and uncertainty associated with determining thermal properties of the media.

f. Electromagnetic Methods

There have been few attempts to determine recharge rates from electrical conductivity measurements. In principle, for a given soil texture, the maximum recharge should occur where the water content is greatest, that is, where the electrical conductivity is greatest. Changes in soil salinity also affect the electrical conductivity measurement, and such changes can have practical significance to recharge. For example, at some sites changing land use patterns, such as clearing forests for farmland, lead to increased recharge, owing to reduced evapotranspiration and increased permeability of the surface soils. Where high recharge occurs, the soil salts *prob.* are leached, thereby decreasing the electrical conductivity of the soil. Cook et al. (1992) evaluated frequency-domain and time-domain electromagnetic measurements, as well as the direct current resistivity method, at a sites in southeastern Australia. By comparing the geophysical analyses with independent estimates of recharge, they concluded that soil texture is the principal reason for the correlation between electrical conductivity and recharge. Airborne electromagnetic surveys were also conducted in this same area to identify recharge zones (Cook and Kilty, 1992). From limited testing it appears that the electromagnetic methods are suited for reconnaissance-level investigations to screen sites where more quantitative methods for recharge should be applied.

g. Darcy Flux in Aquifers

Darcy's equation also can be applied to aquifers to compute recharge in some hydrogeologic settings. As an illustration, assume that vadose-zone percolation is the only source of recharge that is uniformly distributed over a basin having well-defined, impermeable lateral and lower boundaries. Also assume that at some downgradient location, the groundwater flow rate, Q, leaving this part of the basin, that is, the underflow out of the basin, is obtained from a form of Darcy's equation:

$$Q = -T \left(\frac{\Delta h}{\Delta \ell} \right) w \tag{15}$$

where T is the aquifer transmissivity, $\Delta h / \Delta \ell$ is the hydraulic gradient of the aquifer (a negative quantity when the head loss is taken in the direction of flow), and w is the width of the aquifer where underflow is calculated (Figure 3). Then the average recharge rate for the basin would be

$$R = \frac{Q}{A} + S_y \frac{\Delta h}{\Delta t} \tag{16}$$

where Q is obtained from Equation 15, A is the upstream surface area of the watershed where recharge could occur, S_y is the specific yield, and $\Delta h / \Delta t$ is the average head change during the time interval. The method has the potential to work well if the basin boundaries and areas of recharge are well defined and the transmissivity can be determined with reasonable accuracy. Theis (1937) employed a steady-state

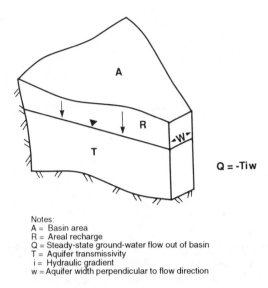

$$Q = -Tiw$$

Notes:
A = Basin area
R = Areal recharge
Q = Steady-state ground-water flow out of basin
T = Aquifer transmissivity
i = Hydraulic gradient
w = Aquifer width perpendicular to flow direction

Figure 3   Diagram of the relationship between recharge and groundwater basin underflow.

version of Equation 16 to calculate recharge to the Ogallala aquifer in eastern New
Mexico and Texas. Maxey and Eiken (1951) applied this technique to 22 groundwater basins in Nevada that they assumed were in hydrodynamic equilibrium, such that
the recharge simply equaled the outflow or discharge from the basin. They then
correlated the recharge with the elevation and mean annual precipitation of the basin
and developed relationships that they suggested could be useful to estimate recharge
for other basins simply on the basis of basin elevation. However, Watson et al. (1976)
indicated that geology, hydrologic characteristics of the consolidated and unconsolidated rocks, antecedent soil moisture, and vegetation strongly affect recharge, and
that unless these are incorporated, considerable uncertainty in results from the
Maxey-Eiken method would remain. More recently, Avon and Durbin (1994)
concluded that the Maxey-Eiken method produced recharge estimates comparable
to estimates obtained independently by water budget and groundwater model
methods.

h. Water-Level Fluctuations

In undeveloped aquifers water levels are often observed to fluctuate seasonally,
primarily in response to recharge from precipitation and discharge by basin outflow
and evapotranspiration. Usually, the slope of the water table and the transmissivity
remain nearly constant during normal seasonal water level fluctuations in most
aquifers. Throughout the year, water levels decline during periods of little to no
recharge, and they rise when recharge exceeds basin outflow, but year after year, the
mean water level in undeveloped aquifers remains virtually unchanged. For this case,
Theis (1937) suggested that the annual recharge could be determined during periods
of no recharge by multiplying the annual rate of water level decline by the specific

yield, inasmuch as over the long term the annual basin outflow equals the annual inflow. Similarly, during periods of actual recharge, the rate of water level rise times the specific yield gives the recharge rate, provided that one takes into account the rate of water level decline over the period that would have occurred in the absence of recharge.

Sophocleous (1991) suggested a simple modeling approach to obtain recharge from precipitation records, vadose zone water balance analysis, and water-level fluctuations in wells. In this analysis, called the hybrid water-level fluctuation method, one establishes field instrumentation for determining recharge from a soil-water balance at one or more locations. At these same locations within the basin, the water-level response to precipitation is determined from a monitor well hydrograph. Knowing recharge from the soil-water balance and water table response, one can obtain the aquifer specific yield, if basin outflow during the period is neglected. Multiple instrumented sites within a basin provide some basis for determining a mean specific yield or its spatial variability. Recharge throughout the basin at sites without soil-water balance instrumentation, but ones with monitor wells, can be calculated simply from the water-level fluctuation and the estimate of specific yield. Sophocleous (1992) applied the hybrid water-level fluctuation method to the Great Bend Prairie of central Kansas to identify zones of similar recharge within the region.

i. Stream Gaging

Stream gaging data can be very valuable in quantifying recharge. Stream hydrographs are generally characterized by a series of peaks followed by recessions. The peaks generally represent surface runoff, interflow, bank storage, and groundwater discharge, whereas the recession curve represents primarily groundwater discharge to the river. In groundwater basins that discharge to perennial streams, the groundwater recharge is approximately equal to the surface-water discharge in the absence of direct surface water runoff. The analysis assumes that this aquifer discharge is due to diffuse recharge from rainfall over the surface water drainage basin.

There have been several approaches to quantify recharge from streamflow measurements (e.g., Meyboom, 1961; Rorabaugh, 1964; Daniel, 1976). Figure 4 illustrates a method to compute recharge from a runoff event, based on the upward displacement of the recession curve at some critical time, T_{cr}, following a flood peak. According to Rutledge and Daniel (1994), the critical time is calculated from

$$T_{cr} = 0.2144 K_i \tag{17}$$

where K_i is the recession index, the time in days required for groundwater discharge to the stream to decline by one log cycle of flow. By computing the flows at time T_{cr} from the recession curve preceding the flood peak, Q_1, and the recession curve following the flood peak, Q_2, recharge from the rainfall event is

$$R = \frac{2(Q_2 - Q_1)}{2.3026} K_i \tag{18}$$

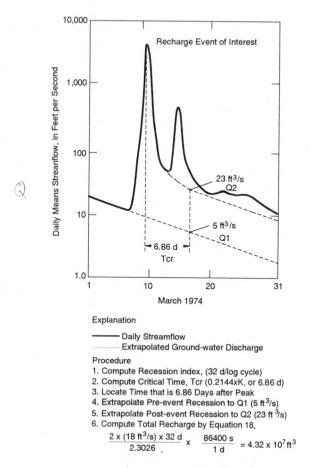

Figure 4   Procedure for using the recession-curve-displacement method to estimate recharge in response to a recharge event. (From Rutledge & Daniel, 1994. With permission.)

This method seems to offer the greatest potential for success where streams are not managed to control water storage and diversions, where the streams are well connected to the aquifer, and where there is little recharge due to snowmelt. Rutledge and Daniel (1994) indicated that an automated version of this method greatly reduces the labor to evaluate records and has successfully been applied to 15 streamflow gaging stations in the eastern United States.

Chiew et al. (1992) developed an integrated surface water and groundwater model to compute recharge for a nonirrigated area in southeastern Australia that is tributary to the Murry River. The surface water model was HYDROLOG and the aquifer model was AQUIFEM-N. The model was calibrated against heads and streamflow, with recharge estimated as an output from the calibrated model.

In areas of low precipitation many stream reaches are losing flow to the groundwater. Recharge from stream flow can be quantified by subtracting flow at the downstream gage from flow at the upstream gage. Where permanent stream gages are not available, a flow-meter survey can be conducted to measure instantaneous flow rates and channel cross section at discrete locations along the stream. Alternatively, stream-bed infiltration can be measured by isolating a small reach of the channel and measuring the discharge needed to maintain a particular stage.

2. Numerical Models of Soil-Water Flow

There are two types of numerical models we consider relevant to calculating deep percolation and recharge: water-balance models, and the models based on the Richards equation.

The water-balance models include codes such as HELP (Schroeder et al., 1984), GLEAMS (Leonard et al., 1989), and PRZM-2 (Mullins et al., 1993). Additionally, Bauer and Vaccaro (1987) developed a soil-water balance model that has been applied to determine recharge for regional groundwater models that cover extensive areas of the Pacific Northwest (Bauer and Vaccaro, 1990) and Kansas (Hansen, 1991).

All these vadose-zone water-balance models partition precipitation into runoff and infiltration. Infiltration is further separated into components such as evapotranspiration, lateral drainage or interflow, soil water storage, and deep percolation by applying deterministic and empirical equations which describe each of the processes. Actual evapotranspiration is computed from climatic data (e.g., precipitation, temperature, solar radiation) input from on-site measurements or from default daily historic data for the nearest location stored in the program library. Other factors such as the vegetation cover and rooting characteristics also enter into the evapotranspiration analysis. Water that cannot be held in storage or extracted by the plants becomes available for deep percolation. Some models such as PRZM-2 take the deep percolation output from the water balance in the root zone and also route this through the deeper vadose zone using the Richards equation for one-dimensional, unsaturated flow.

One of the algorithms often used to compute recharge in the soil-water budget models is based on the concept of field capacity. Field capacity for modeling purposes is typically defined as the water content the soil can hold against gravity (e.g., Phillips et al., 1993). Percolation below any soil layer is allowed in the models only if the water content exceeds the field capacity. That is, if the water content is less than field capacity, then it is assumed either that the plants will consume all the water or that the water would remain in the layer and no deep percolation could occur. There is significant research to demonstrate that water percolation can occur under gravity at water contents much less than field capacity (Stephens, 1985; 1994). Thus, at poorly vegetated sites where water content is almost always less than field capacity, recharge could not occur; but Stephens and Knowlton (1985) found that this is certainly not the case at a semi-arid site in New Mexico. At a humid site in England, Wellings (1984) concluded from field measurements that the soil-water profile continuously changed over time, and no evidence supported the validity of the field capacity and available water concepts as they relate to recharge. If recharge rates are low and the period of water balance accounting is too long, water balance models are likely to underestimate recharge because they only roughly approximate the physics of unsaturated flow.

The water-balance model developed by Eagleson (1979) improves significantly in the application of unsaturated flow physics and provides a probabalistic approach to recharge. The model considers uncertainty in storm characteristics, soil physical properties and one-dimensional vertical flow, evapotranspiration via Penman's equation, and groundwater flow out of the basin. Cumulative probability density functions for recharge by this method have been developed for a subhumid site, Clinton, MA, and an arid site, Santa Paula, CA (Eagleson, 1978). These functions provide the return period or probability of occurrence of recharge events of a particular magnitude.

There are a large number of numerical models for simulating soil-water processes, including finite difference and finite element forms, based on one-, two-, or three-dimensional forms of the Richards equation (Chapter 1, Table 3). There are many other codes that include both flow and transport, such as VAM2D (Huyakorn et al., 1989) and TRACR3D (Travis, 1984). To account for infiltration and evapotranspiration in these codes, in lieu of detailed meteorological information, the upper boundary of the model and/or the root zone is usually specified as either a constant or time-varying flux or pressure head. In contrast to some of the water-balance models, the numerical models allow the user to more realistically represent the physical properties of the porous medium, including complex geology with spatially varying hydraulic conductivity and water retention characteristics. When the lower boundary of the model is specified as the water table, the water flux out the base of the model represents the groundwater recharge.

3. Groundwater Flow Models

Groundwater flow models also can predict recharge when other hydrologic information is known or assumed. In finite-difference numerical models, for example, a rectangular grid of cells is laid over the domain of interest, and representative values of aquifer transmissivity and storage are assigned to the aquifer materials. The concept here is to calibrate the model to observed water levels (hydraulic heads) by adjusting the model parameters until model predicted and observed heads reach suitable agreement. The flux input to the water table necessary to produce this agreement is then assumed to represent recharge.

Inasmuch as there is always some uncertainty associated with aquifer properties and hydraulic heads, there is also uncertainty in the recharge estimate. Calibration of numerical models of large aquifer systems by trial and error is quite tedious and does not readily permit quantification of the uncertainty in the recharge. However, automatic procedures, such as MODFLOWP (Hill, 1992) are now available to facilitate calibration and parameter estimation. Regardless of the calibration procedure, errors in model input parameters will be accumulated in the back-calculated recharge estimate.

Simpson and Duckstein (1976) developed a discrete-state compartment, or mixing cell, model that partitions the aquifer into cells, within which conservation of mass is applied. Tracers are used to calibrate the model, including carbon-14 (Campana and Simpson, 1984), tritium (Campana and Mahin, 1985), and deuterium (Kirk and Campana, 1990).

There are numerous groundwater model studies that have focused on recharge. For example, Theis (1937) applied a steady-state analytical solution for a sloping water table aquifer to predict recharge to the Ogallala aquifer by adjusting the recharge in the vertical slice model until the predicted water table reasonably well matched the observed water-table shape. Su (1994) developed analytical solutions for certain boundary conditions to calculate transient recharge from precipitation, water-level hydrographs, and soil and aquifer properties; the analysis takes into account the sloping water table and groundwater drainage. Glover (1960) summarizes analytical solutions relevant to estimating recharge from the rate of rise and shape of groundwater mounds beneath spreading basins or channels. Lohman (1972) presented a graphical method to calculate diffuse areal recharge where monitor wells were located in a five-spot pattern, much like a finite-difference cell. Based on the one-dimensional groundwater flow equation applied to water table fluctuations for a project on Long Island, Jacob (1943) developed a predictive relationship between a weighted average annual precipitation and recharge.

Besbes et al. (1978) combined numerical modeling with a convolution method to compute recharge for regional aquifers. Their approach uses existing well hydrographs for obtaining the unit response functions in groundwater from, for instance, a known flood event, and the numerical model is used to predict the water-level response in the absence of recharge, so that the recharge is calculated from the changes in the volume of groundwater in storage during a long period of record. Moench and Kiesel (1970) showed that where well hydrographs are available along a losing stream and aquifer hydraulic properties are known, local recharge from the stream can be calculated by the convolution approach and simple analytical solutions for aquifer flow.

Duffy et al. (1978) evaluated the recharge from an ephemeral stream using stochastic methods that analyzed the frequency of the water-level response in hydrographs of wells adjacent to the stream. Where some information exists on recharge and its variability at some locations within a groundwater basin, a stochastic inverse method may be useful to provide optimal estimates of the recharge throughout the entire basin (Graham and Tankersley, 1994). Graham and Neff (1994), who applied this analytical approach in conjunction with groundwater flow modeling of the Upper Floridian Aquifer in northeast Florida, indicated that some of the assumptions in the analytical method may need modification before the method is widely applicable.

These are only a sampling representing a small fraction of the hydrogeologic investigations where groundwater flow models in the phreatic zone have assisted in quantifying recharge.

4. Chemical Methods in the Vadose Zone

Among the chemical methods for calculating recharge, there are stable and radioactive isotopes that serve as a means to compute recharge, including tritium, chlorine-36, chloride, and oxygen-18 and deuterium. One advantage of some of these methods is that the analysis may represent an integration of hydrologic events over decades or even tens of thousands of years. Another advantage is that the data are derived from *in situ* sampling, without need for field instrumentation for monitoring. These can be important considerations in selecting an appropriate method, especially if quantifying long-term natural recharge is the objective.

Any tracer-based analysis approximates recharge only if the tracer has migrated below the root zone; otherwise this analysis represents a soil-water flux that may exceed the actual recharge. In fact, Tyler and Walker (1994) demonstrated that tracer methods applied within the root zone significantly overestimate the deep soil-water flux, because the roots induce nonconstant soil-water velocities within the root zone. They suggest that tracers, such as tritium, can be effective when the recharge exceeds 10% of the annual precipitation.

The following discussion focuses on the chemical methods that are based on soil-water data.

a. Tritium

Tritium, a radioactive isotope of hydrogen with a half-life of about 12.4 years, is well suited as a hydrologic tracer because it is part of the water molecule. During the atmospheric nuclear testing beginning in the 1950s, the tritium in the atmosphere increased substantially over a relatively short time, culminating in the period 1963 to 1964 (Phillips et al., 1988). This tritium pulse rapidly circulated world-wide, with primary deposition in the northern hemisphere where most of the testing occurred.

After the vapor condensed and fell as precipitation, the record of tritium has been preserved in atmospheric water that infiltrated the soil profile. Recharge, or, more precisely, net infiltration, is obtained from the depth to the center of mass of the tritium pulse, L, with the following equation:

$$R = \theta \frac{L}{\Delta t} \tag{19}$$

where θ is the mean water content through depth L. Because tritium can move in the vapor as well as in the liquid phase, strong temperature gradients may influence the tritium peak due to thermally driven vapor-phase migration (Knowlton et al., 1992).

b. Chlorine-36

Another tracer of soil-water flux or recharge is chlorine-36. This is a radioactive isotope with a half-life of about 300,000 years, produced as a by-product of thermo-nuclear testing near the oceans in the 1950s (Bentley et al., 1982). Chloride is very stable in the environment and enters the hydrologic cycle as the chloride ion dissolved in water and as a component of dust fallout. Because it is soluble and nonvolatile, chloride is an excellent tracer for liquid-phase transport. Recharge can be determined from the depth of the chlorine-36 peak, in the same manner as described earlier for tritium. One slight disadvantage of the method is that it requires analysis by a tandem accelerator/mass spectrometer, which is highly specialized equipment available to commercial users at only a few research institutions. Phillips et al. (1988) first applied the chlorine-36 method, along with the tritium technique, to study both liquid and vapor transport and to evaluate the suitability of these tracers for quantifying recharge in very dry desert soils.

c. Chloride Mass Balance

The chloride mass balance method relies upon the slow accumulation in the soil profile of natural chloride that dissolves in precipitation and infiltrates. The concentration of chloride typically decreases with increasing inland distance from the coasts. The expected chloride pattern in the soil profile, at least in areas of modest precipitation, is that chloride is concentrated with increasing soil depth, as water is extracted by the plant roots; below the root zone the chloride concentration is expected to be constant where the deep percolation migrates toward the water table.

To interpret chloride distribution in the vadose zone, three fundamental assumptions are made: all chloride in the vadose zone originates from atmospheric deposition; the only long-term sink for chloride is downward advection or dispersion; and chloride behaves conservatively during soil-water transport. If dispersion and macropore flow are neglected, a simple piston-displacement model is derived in which the chloride concentration increases in proportion to the ratio of precipitation to recharge (Allison and Hughes, 1978):

$$R = P \frac{Cl_p}{Cl_s} \tag{20}$$

where P is the average precipitation rate, Cl_p is chloride concentration in precipitation and dry fallout, and Cl_s is the average soil chloride concentration below the root zone.

Chloride patterns that depart from this model may produce a bulge in the chloride concentration at some depth in the profile, and below this the concentration decreases to approach a near constant value. Phillips (1994) discussed possible explanations for this behavior in desert soils such as preferential or bypass flow in macropores, but suggested that the low concentration of chloride at depth most likely reflects greater recharge during a wetter paleoclimatic period when the indigenous plants were less efficient at capturing the soil moisture. An alternative hypothesis only recently advanced suggests that the chloride bulge is best explained by ultrafiltration, a process that causes chloride anions to migrate more slowly than the water (Ankeny et al., 1995). If ultrafiltration does indeed affect migration of chloride anions in the vadose zone, many previous infiltration studies based on the chloride mass balance method may underestimate the recharge component. Additional research is required to investigate this phenomenon, especially in arid environments.

d. Stable Isotopes

An isotope is a variation of an element produced by differences in the number of neutrons in the nucleus of the element, and hence by differences in mass. The two stable, or nonradioactive, isotopes of hydrogen (^1H and ^2H or deuterium, D) and the three stable isotopes of oxygen (^{16}O, ^{17}O, and ^{18}O) form part of the water molecule, and analyses of their concentrations in natural waters have long been used to trace movement of water in the subsurface. It is well established that the isotopic compo-

movement of water in the subsurface. It is well established that the isotopic composition of precipitation at a particular location will vary seasonally and with individual storms. The isotopic composition of precipitation will also vary between locations, depending upon climate and elevation. Nevertheless, the composition of all precipitation generally falls on a straight line of a plot of δD versus $\delta^{18}O$, where δ is the relative difference of the isotopic ratios in precipitation versus standard mean ocean water (SMOW) expressed in parts per thousand. This line is called the meteoric line.

The stable isotope concentration of the precipitation can be modified subsequent to infiltration, and this signature of the soil water reveals important information about recharge. Evaporation of soil water leads to a fractionation of the stable isotopes deuterium and oxygen-18. When water evaporates, the heavier atoms tend to remain behind in the liquid phase, thus leading to an enrichment in the concentration of the heavier isotopes. When the water vapor condenses, the condensed liquid is more concentrated in the heavier isotopes than the residual vapor. The isotopic enrichment and shape of the isotope profile in the soil depends upon the net infiltration and evaporation rates, soil-water status, and diffusive properties of the liquid and vapor. Based on these and other considerations, Barnes and Allison (1983) developed a theoretical model to predict how the stable isotopes should be distributed in the soil. Knowlton (1990) built on this theoretical analysis and developed an equation for recharge based on the measured deuterium isotope concentrations:

$$R\left(\delta_D - \delta_D^{Rec}\right) = \left(\frac{h_r N_{sat} D^v}{\rho}\right) \left(\varepsilon_D + \eta_D\right) \left(\frac{d\left[\ln\left(h_r N_{sat}\left(\varepsilon_D + \eta_D\right)\right)\right]}{dz}\right)$$

$$- \left(D_w + \frac{h_r N_{sat} D^v}{\rho}\right) \frac{d\delta_D}{dz} \qquad (21)$$

where R = recharge rate (m s^{-1}), δ_D = standardized isotope ratio at any depth (per mil), δ_D^{Rec} = standardized isotope ratio of the recharge water at depth (per mil), h_r = relative humidity, N_{sat} = saturated water vapor density in air (kg m^{-3}), D^v = effective diffusivity of water vapor in air (m^2 s^{-1}), D_w = effective self-diffusion coefficient of water (m^2 s^{-1}), ρ = density of liquid water (kg m^{-3}), ε_D = equilibrium enrichment factor, η_D = diffusion ratio excess, and z = depth below land surface (m). The processes involved in isotope distribution in the vadose zone are illustrated in Figure 5. This method holds promise, but has only recently been developed and requires additional applications in the field.

Allison et al. (1984) used another technique to assess recharge with stable isotope measurements of pore liquids collected from the vadose zone. They assumed that if the rainfall events are uniform throughout the year and the evaporation rate is constant, then isotopic enrichment increases linearly with the square root of the time since the last rainfall. From this, they developed a relationship between recharge and the magnitude of enrichment or shift between meteoric line and composition of deuterium in soil water. The method apparently has not been widely used thus far and requires further validation for a variety of soils. Nevertheless, it may be a useful and simple tool to estimate recharge in areas of low precipitation.

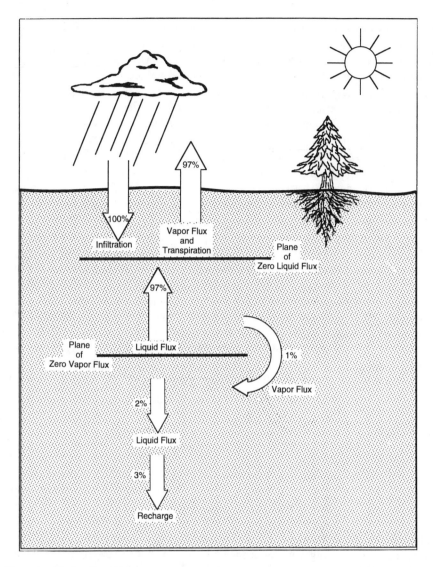

Figure 5 Mechanisms of water transfer a few meters below the surface of a sandy site on the Sevilleta National Wildlife Refuge north of Socorro, New Mexico. (From Knowlton, 1990. With permission.)

5. Chemical Tracers in Aquifers

Chemical tracers commonly detected in aquifers that permit quantification of recharge include tritium, carbon-14, chlorine-36, and chlorofluorocarbons. In principle, these tracers determine the age of the groundwater, which in turn permits calculation of groundwater travel time. The recharge rate then can be calculated by

$$R = Vn_e = \frac{Ln_e}{t_a} \tag{22}$$

where V is the mean porewater velocity, n_e is the effective porosity, L is the distance along the flow path, and t_a is the travel time or age of the groundwater at the distance L. The velocity of groundwater can be determined based on an age date at one location in the aquifer if the distance from the outcrop or point of recharge entry into the aquifer is well established. In this analysis, the travel time through the vadose zone is considered negligible compared to that in the aquifer; however, where it is known, the travel time through the vadose can be added to the travel time through the aquifer. Two or more wells located along the flow path can be sampled to obtain groundwater ages and calculate the mean porewater velocity. The effective porosity is usually estimated based on lithology of the formation.

The use of chemical tracers for determining groundwater ages and recharge rates has several common approaches and difficulties. The simple models for evaluating recharge from groundwater age assume that the tracers move in a piston displacement process that neglects dispersion. In fact, however, there is mechanical mixing and diffusion within the porous media, which decrease the input concentrations. Unless hydrodynamic dispersion is taken into account by mathematical modeling, the calculated age will exceed the true age; however, in permeable aquifers where advective transport dominates, dispersive effects are not significant, especially if the tracer input is relatively constant in time (Solomon and Sudicky, 1991). The true age may also be over- or underestimated due to commingled water samples obtained, for example, by pumping from different parts of the aquifer where the groundwater has different ages.

The following is a brief summary of some of the currently used methods for groundwater age dating.

a. Tritium

Tritium in aquifers is derived from both natural and anthropogenic sources. Tritium is produced naturally when cosmic rays interact in the upper atmosphere with nitrogen. Precipitation naturally contains approximately 5 TU (tritium units) (Mazor, 1991); however, precipitation that entered the soil prior to 1952 would have decayed by now to concentrations near or below the analytical detection limit. The thermonuclear testing that began in the 1950s generated peak concentrations in precipitation ranging from about a few hundred to about 10,000 TU. An example of measured tritium in precipitation for the Delmarva Peninsula on the eastern coast of the United States is shown in Figure 6A. The apparent age, t_a, of the water sample is

$$t_a = -t_{1/2}\ \ln(A_i / A_o) \tag{23}$$

where $t_{1/2}$ is the half-life in years, A_i is the activity of the sample at the time the precipitation or surface runoff entered the subsurface, and A_o is the observed activity in the sample.

Owing to the rather short half-life of tritium, its usefulness as a dating method will soon expire. Even where detectable, there is considerable uncertainty in the calculated age. The tritium distribution at the source of recharge is rarely known, but

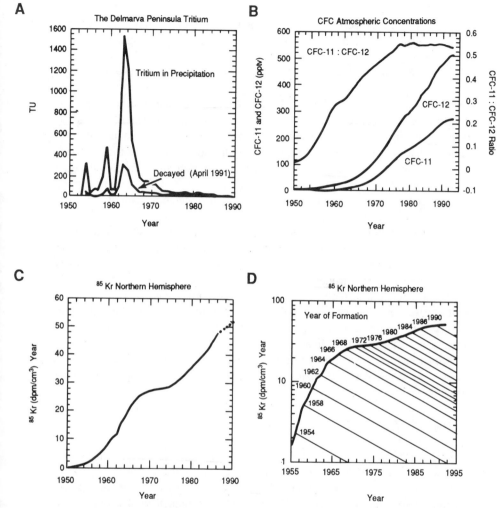

Figure 6 (A) Average tritium concentration in precipitation on the Delmarva Peninsula. The tritium curve was based on the assumption that recharge from precipitation occurred only from November through April. As a result, the concentrations obtained are lower than if a weighted yearly average was used. Residual tritium concentration due to radioactive decay is calculated for April 1991. (B) Atmospheric concentrations of CFC-11 and CFC-12 in parts per trillion volume per volume air and the ratio CFC-11:CFC-12. (C) Krypton-85 specific activity (i.e., the ratio of ^{85}Kr to stable krypton in disintegrations per minute per cubic centimeter krypton) in the troposphere of the northern hemisphere between 40° and 55°N as a function of time. The ^{85}Kr specific activity is extended to pass through the atmospheric specific activity measured at Locust Grove in November 1991. (D) Same curve as in (c), but plotted on a logarithmic scale. Diagonal lines represent radioactive decay after groundwater is isolated from the atmosphere. (From Ekwurzel et al., 1994. With permission.)

where it has been measured, tritium varies considerably with location, year, and season. Additionally, the groundwater flow path taken by the tritium is almost always complex, and some samples may represent composite paths. For these reasons, tritium, where detectable, usually has only semiquantitative significance. That is, detectable tritium suggests that the groundwater sample contains at least some portion of water derived from precipitation that fell after 1952 (Mazor, 1991).

b. Tritium/Helium-3

Helium-3 is an inert gas that is naturally produced primarily in the atmosphere and in the deep subsurface. It is also generated by the decay of tritium derived from nuclear weapons testing. Thus, tritium-bearing recharge will produce dissolved helium-3 that increases in concentration along the flow path as the tritium concentration diminishes. The ratio of tritium to helium-3 concentrations forms the basis for dating the age of the groundwater, provided that other sources of helium are negligible or can be measured in order to quantify the tritiogenic portion of the helium-3 (e.g., Ekwurzel et al., 1994).

The tritium/helium-3 method is a significant advance over the tritium method because the initial concentration of tritium in the atmosphere does not need to be estimated. Furthermore, the tritium/helium-3 method enables the effects of dispersion to be distinguished from radiodecay (Phillips, 1995). As we discussed earlier, the tritium concentration record in the atmosphere contains many spikes, at least prior to the conclusion of atmospheric testing. But most importantly, the record is rarely known at locations where recharge is to be calculated. Solomon and Sudicky (1991), who evaluated the effects of hydrodynamic dispersion and the nature of the tritium input function on the reliability of ages, concluded that the tritium/helium-3 method can be used to accurately date shallow groundwater with ages ranging from 0 to about 50 years.

c. Krypton-85

Krypton-85, a radioactive noble gas with a half-life of 10.76 years, is produced in the atmosphere by the interaction of cosmic rays with krypton-84. By far the greater source, however, is from nuclear weapons testing and reprocessing nuclear fuel rods (Ekwurzel et al., 1994). The atmospheric production of krypton-85 has increased steadily since about 1950. After precipitation infiltrates and no longer contacts the atmosphere, the krypton undergoes decay, but, owing to its inert characteristics, it does not interact chemically with the aquifer materials. Figure 6C illustrates the krypton activity in the northern atmosphere, and Figure 6D shows how to graphically determine the age of the water sample based on the sample collection date and the measured activity in groundwater.

d. Carbon-14

Carbon-14 is a radioactive isotope of carbon produced in the upper atmosphere by cosmic ray interactions with nitrogen. The carbon becomes part of the carbon dioxide molecule that dissolves in the water, enters the soil gas, and becomes part of the animal or plant tissue. When the water or soil gas no longer is free to exchange with the atmosphere, and when the animal or plant dies, the carbon-14 activity decreases at a rate controlled by its half-life, 5730 years. The concentration or activity ratio of carbon-14 to carbon-12, expressed as a percent, is called the percent modern carbon, pmc. Except for input from thermonuclear testing in the 1950s and 1960s, this ratio has remained relatively constant, varying only by a factor of about two, in the atmosphere over the last 100,000 years (Phillips, 1995). Inasmuch as carbon-14 is detectable to about 1 pmc, the potential usefulness of the age dating method is about 40,000 years (Mazor, 1991), based on Equation 23, but in most groundwater investigations the practical limit is about half this (Phillips, 1995).

Unfortunately, there are a number of sources of uncertainty in carbon-14 age dating. The most significant of these is due to the interactions of carbon-14 with elements in naturally occurring minerals. Most of the carbon-14 in groundwater occurs in the bicarbonate ion. However, some of the carbon in the bicarbonate ion is derived from the dissolution of geologically old carbonate minerals, such as calcite and dolomite. Reactions such as the oxidation of organic matter or the reduction of sulfate in the presence of methane (CH_4) also release dead carbon to groundwater, thereby lowering the $^{14}C/^{12}C$ ratio. The sources of dead carbon make the calculated age older than the true age. Thus there have been considerable efforts to correct the carbon-14 age for the geochemical effects, but thus far, the different correction methods have produced wide variations in predicted carbon-14 age dates (Domenico and Schwartz, 1990). The principal difficulty seems to lie in our inability to reconstruct with confidence the geochemical and hydrological processes that have influenced the carbon-14 concentration in the sample. Another potential difficulty may arise if the sample age is post 1952, because the same nuclear weapons testing that produced tritium also generated carbon-14 concentrations of a few tens to about 200 pmc (Mazor, 1991). Therefore, care must be taken to use other methods, such as tritium, in combination with carbon-14, to assess whether the sample may be mixed with very young water. As with tritium, owing to the uncertainty in assessing the true age, carbon-14 is generally regarded as only a semiquantitative technique for recharge analysis.

e. Chlorine-36

Based on its long half-life, chlorine-36 is potentially useful to date groundwater as old as about 2 million years. In many respects, chlorine-36 is an ideal tracer for dating old groundwater. Unlike carbon-14, chlorine-36 does not interact appreciably with most aquifers, although dead chloride can be dissolved from some salt-bearing natural formations to increase the apparent groundwater age. In clay-rich deposits the anion exclusion process may cause chloride ions to migrate slightly faster than the water. Chlorine-36 is readily detected in very small concentrations by a tandem accelerator-mass spectrometer, but this equipment is available for commercial use at only a few research institutions. The ratio of chlorine-36 to stable chloride is used to determine the groundwater age from Equation 23. This method has been applied to date groundwater in Canada (Phillips et al., 1986) and Australia (Bentley et al., 1986).

f. Chlorofluorocarbons

Chlorofluorocarbons (CFCs), including CFC-11 (CCl_3F) and CFC-12 (CCl_2F_2), are chemically stable manmade volatile compounds that have been manufactured since the 1940s and 1930s, respectively, for use as aerosol can propellants, foaming agents in plastics, refrigerants, and solvents (Dunkle et al., 1993). Their release into the atmosphere, documented worldwide, produced a steady increase in concentration in the atmosphere (Figure 6b). The atmospheric concentrations of CFCs partition into water according to Henry's law. Because the CFCs are relatively stable in the atmosphere and subsurface, the CFC concentration in groundwater recharge should increase over time as a result of the increasing atmospheric production. The age of a groundwater sample analyzed for CFCs is determined simply by comparing the measured concentration in groundwater with a graph of the atmospheric water concentrations over approximately the past 50 years or so (e.g., Busenberg and Plummer, 1991).

A number of factors should be kept in mind when interpreting groundwater ages from CFCs. First, the temperature of the atmosphere must be determined because temperature affects the CFC partitioning between the atmospheric gas and water phases. As the dissolved CFCs migrate through the vadose zone, there is opportunity for additional phase partitioning, depending upon the CFC partial pressure and gas-phase advection of CFCs. Also in the vadose zone, the CFCs may be sorbed by organic carbon in the soil. The same sorption processes may occur in the aquifer as well. Biodegradation may also reduce the CFC concentration in the aquifer, especially under anaerobic conditions (Busenberg and Plummer, 1992). Furthermore, contamination of the sample must always be guarded against. This is particularly important for older waters that have exceedingly low concentrations. Sample contamination for determining age may occur due to improper cleaning of sampling and analytical equipment, sample exposure to the atmosphere, or contamination of the aquifer by migration of CFCs released from industrial facilities. In spite of these potential concerns, in a field study on the mid-Atlantic coast, Ekwurzel et al., (1994) found that ages determined by the CFC method, the tritium/helium-3 method, and the krypton-85 method all agreed within about 2 years.

CHAPTER 3

Physical Processes Relevant to Deep Soil-Water Movement

In this chapter, we present a summary of the physical processes commonly encountered by groundwater hydrologists: infiltration, drainage, and redistribution. These topics are emphasized because they are most relevant to assessing recharge, often the most difficult component to quantify in the water balance, and usually the most important to groundwater hydrologists, especially in areas of low precipitation. The subsequent chapter (4) deals exclusively with studies on the occurrence of recharge. The discussion of physical processes in this chapter is also intended to establish an appreciation for the applications of methods of vadose-zone characterization (Chapter 5) and monitoring (Chapter 6). For a more in-depth analysis of physical processes and for a broader spectrum of topics, the texts on soil physics by Hillel (1980a,b) are excellent.

I. INFILTRATION

Perhaps the most widely studied subsurface hydrological process has been infiltration. Infiltration is the process whereby water enters the soil from surficial sources, such as rainfall, snowmelt, flooding, irrigation, liquid waste spills, etc. Thus, infiltration rate is the volume rate of water flowing into a unit area of soil surface. Infiltration is usually regarded as a process that occurs predominantly in the vertical direction when water is applied over extensive areas of the land surface. But even where the infiltrated water has been applied over extensive areas, the local flow paths usually depart from the vertical in most soils, owing to heterogeneity in permeability, as we discuss later in this chapter. Nevertheless, for many practical problems of large-scale significance, the mean flow path from infiltration over a wide area is approximately one-dimensional in the vertical direction, especially in relatively homogeneous media. One-dimensional horizontal infiltration, called absorption, rarely occurs in the field. Nevertheless, absorption has been extensively analyzed mathematically and studied in the laboratory by soil physicists.

There are other situations in which infiltration from a localized source area is best described in a two- or three-dimensional sense. Infiltration is often analyzed as in a

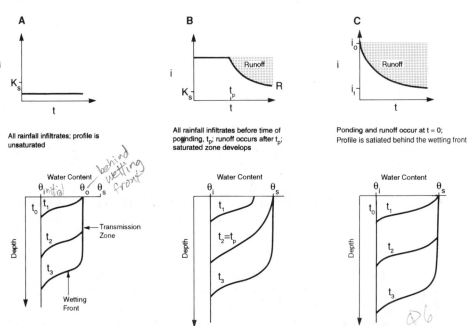

Figure 1   Infiltration rate dependence on rain intensity, *i*, and water content profiles during (A)
low, (B) moderate, and (C) intense precipitation.

two-dimensional sense along the perimeter of a linear channel of a canal or ephemeral stream, especially one with a small width-depth ratio. Three-dimensional infiltration behavior occurs where water enters the soil through a water-filled auger hole created to test soil permeability, where runoff fills a root channel and then soaks into the surrounding soil, and where seepage occurs through a puncture in a synthetic liner of liquid waste impoundment.

A. FACTORS AFFECTING INFILTRATION

The infiltration process in any geometry is controlled by many factors, such as the rate of water application, antecedent moisture in the soil, soil hydraulic properties, topography, and others. First consider a thick, uniform-textured, permeable soil that is subject to a gentle but constant rainfall rate over a flat-lying terrain (Figure 1A). In this case, we are describing a one-dimensional infiltration process in which the rainfall intensity is less than the magnitude of the saturated hydraulic conductivity of the soil. All the water supplied is free to infiltrate, so none is available for runoff. Because the capacity of the soil to conduct water is greater than the rate of supply, the soil will be unsaturated. And, if the rain continues steadily for a prolonged period to allow the deep penetration of the wetting front, the unsaturated hydraulic conductivity of the wetted soil will approach the rain intensity.

Next consider the case in which a constant rainfall rate in excess of the saturated hydraulic conductivity is applied to initially dry soil. The soil initially takes up the rainwater as rapidly as it falls (Figure 1B), owing to the strong capillary forces in the dry soil. As the soil pores fill with water, the influence of the capillary forces on

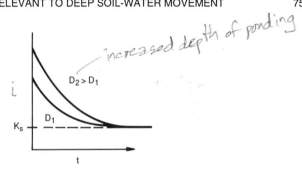

Figure 2   Effect of ponding depth, D, on infiltration rate.

infiltration diminishes gradually. Therefore, as the surface soil becomes wetter and as the depth to the wetting front increases, the hydraulic head gradient decreases. Consequently, the infiltration rate decreases over time. If the rain persists, ponding on the soil surface will occur at time t_p, when the soil cannot transmit water at a rate greater than the rainfall rate. When ponding occurs, the soil surface layer is saturated, except perhaps for entrapped air. After ponding, the infiltration rate continues to diminish and approaches the field-saturated hydraulic conductivity of the soil. Finally, for the case where storm intensities are high relative to the saturated hydraulic conductivity, it is probable that ponding will begin almost instantly (Figure 1C). Figure 1 also illustrates the water-content profiles associated with the low and moderate intensity rain infiltration as well as ponded infiltration into a uniform soil.

There have been a number of researchers who have attempted to empirically fit power function curves or exponential equations to infiltration data. For example, for a constant depth of ponding, Horton's (1940) equation describes the decrease of the infiltration rate, i, with time:

$$i = \left(i_o - i_f\right)e^{-k't} + K_o \quad \text{where } K_o = K_{sat}. \tag{1}$$

where i_o is the initial infiltration rate, i_f is the final infiltration rate (or field saturated hydraulic conductivity), t is time, and k' is an empirical constant for the particular soil. Although these empirical infiltration equations do a reasonably good job of fitting observed data, the parameters of fit lack physical meaning and are not readily measured independently.

When ponding occurs during runoff or in a surface impoundment, hydrologists often need to predict how the depth of ponding affects the infiltration rate. For a constant depth of ponding, the infiltration rate is sensitive to ponding depth primarily at early time. To illustrate, suppose there is an intense storm during which ponding occurs instantly and remains at some constant depth. The infiltration rate curve would follow a pattern similar to Figure 1C. Now, assume the soil subsequently dries to the prestorm moisture and then another, more intense storm produces a greater ponding depth; the early time infiltration rate would be greater for the second storm than for the first one (Figure 2). The increased infiltration is attributed to an increase in the hydraulic head gradient caused by the increased depth of ponding. As the wetting front penetrates deeper into the soil, however, the hydraulic gradient gradually decreases and approaches unity at late time. Thus, if either the shallow or deep ponding conditions persisted, they would both approach the same late time, steady

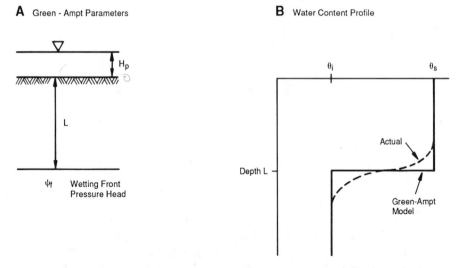

A Green - Ampt Parameters

H_p

L

ψ_f Wetting Front
Pressure Head

B Water Content Profile

θ_i θ_s

Actual

Depth L

Green-Ampt
Model

Figure 3   Transient, one-dimensional infiltration following the Green-Ampt approach.

infiltration because in both cases the hydraulic head gradient approaches one as the wetting front depth gets deeper.

To describe this one-dimensional vertical infiltration process, Green and Ampt (1911) developed the following simple but very useful equation:

$$i = K_0 \left[\frac{(H_P + L) - \psi_f}{L} \right] \tag{2}$$

where K_0 is the hydraulic conductivity behind the wetting front, H_P is depth of ponding, L is depth to the wetting front, and ψ_f is the mean pressure head at the wetting front (Figure 3). Note that $\psi_f < 0$. This equation is virtually identical to Darcy's equation, with the term in brackets representing the hydraulic gradient. Unlike the usual application of Darcy's equation to flow in soil columns or aquifers, L is time dependent in the Green and Ampt equation. Inspection of Equation 2 shows that for constant depth of ponding, as the depth to the wetting front increases, the gradient eventually approaches one regardless of the magnitude of the ponding depth. Consequently, the late time, steady infiltration rate is independent of the ponding depth. As an example of practical application, McWhorter and Nelson (1979) showed how this simple infiltration model can be applied to problems of seepage from surface impoundments to predict when the wetting front would reach the water table.

Undoubtedly, the most widely referenced mathematical equation on infiltration is that by Philip (1957). Where the depth of ponding is very thin, the simplified form of Philip's transient infiltration equation is

$$i = \frac{1}{2} St^{-1/2} + K_0 \tag{3}$$

where s is called the sorptivity of the soil, t is time since infiltration began, and K_0 is the hydraulic conductivity of the soil's upper layer. Sorptivity is a soil hydraulic

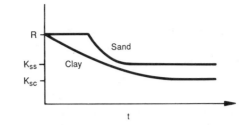

Figure 4 Effect of soil texture on infiltration with a constant rainfall rate, R, where the saturated hydraulic conductivity of sand is K_{ss} clay is K_{sc}.

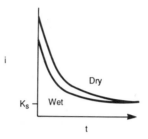

Figure 5 Effect of antecedent water content on ponded infiltration rate.

property that embodies several factors describing the capacity of the soil to imbibe water at early time, such as the initial dryness of the soil (e.g., Brutsaert, 1977). Note that as $t \to \infty$, the late time infiltration rate approaches K_0, just as we discussed previously for the Green-Ampt equation. Solutions to Philip's equation can be used to predict not only the transient infiltration rate, but also water-content profiles. This is in contrast to the Green-Ampt equation, which only predicts depth to the wetting front, in addition to the transient infiltration behavior.

As one might expect, the soil properties also have a strong influence on the infiltration rate. The most important property is the saturated hydraulic conductivity (Figure 4). The greater is the saturated hydraulic conductivity, the greater is the final infiltration rate under ponded conditions. In the example in this figure, we consider rain falling at a constant rate on both a thick clay and sandy soil. The rain intensity is so large that ponding occurs immediately above the clay. The sand, in contrast, takes up water as rapidly as it is applied, but eventually ponding occurs. The final infiltration rates are governed principally by the respective field-saturated hydraulic conductivities of each soil, which may differ by several orders of magnitude.

The moisture content of the soil prior to the rainfall also is an important consideration in predicting the infiltration rate or cumulative infiltration over time. Surface-water hydrologists are well aware that the frequency between storms is an integral part of flood forecasting. When the soil pores are full of water, there is little storage available, and therefore more of the rainfall is likely to run off. The high initial water content is represented by a reduced wetting front pressure head in Equation 2 and a reduced sorptivity in Equation 3. The effect of antecedent soil moisture is illustrated in Figure 5.

There are a number of other factors that affect infiltration that are more difficult to quantify for predictive models. For instance, during a storm the impact of rain-drops on the soil surface can cause compaction and reduced permeability at the soil surface, thereby causing infiltration to decrease more rapidly than one may expect. Where soils contain clays with expanding lattice structures, the pores may swell shut as the soil wets so that the saturated hydraulic conductivity and infiltration rate decrease over time. Although desiccation or other mechanisms may create macropores that lead to rapid initial infiltration at early time, the silt entrained in the infiltrating water may gradually fill the macropore, thus leading to a sharp reduction in the infiltration rate. This filtration process also can form surface crusts in unfissured soils, for instance, as suspended sediments or chemical precipitates accumulate in the upper soil layers beneath surface impoundments. Soil surface crusts can also form by chemical cements and mineralization from evaporation; however, when water subsequently percolates through these crusts, the infiltration rate may increase over time as the mineralization is dissolved or eroded.

B. INFLUENCE OF SOIL AIR

Air entrapment and air compression also can influence infiltration. During ponded infiltration, the downward-percolating water may bypass air in some of the soil pores, but over time this entrapped air can dissolve in the infiltrating water. Dissolution of entrapped air is enhanced if the infiltrating water is cooler than the soil, inasmuch as air solubility in water increases with decreasing water temperature. Additionally, the entrapped air bubbles eventually may be translocated or dislodged by the infiltrating water. The net effect is that entrapped air initially reduces infiltration because air bubbles in the pores decrease the hydraulic conductivity to water (Figure 6). Subsequently, the infiltration rate may increase as the air bubbles are removed and the soil becomes more water saturated. Owing to the expansion and contraction of the entrapped air or other gas bubbles with temperature fluctuations, the infiltrating water may cause the infiltration rate to fluctuate inversely with temperature (e.g., Robinson and Rohwer, 1959; Stephens et al., 1983). On the other hand, the effect of increasing temperature or decreasing water viscosity may lead to the fluctuations in infiltration rate that oppose the effects of temperature on infiltration when there is air entrapment or gas generation (e.g., Jaynes, 1990; Constantz, 1982).

When entrapped air cannot dissolve or translocate during ponded infiltration, it may become compressed by the hydrostatic forces in the water outside of the air bubble. Much of this trapped air may occur ahead of the wetting front in pores above an impervious layer (e.g., Vachaud et al., 1973) or in a macropore, for example. Linden and Dixon (1973), who measured water levels beneath and adjacent to a flooded border irrigation field, noted that during infiltration the water table, which initially was about 2 m below land surface, was depressed beneath the center of the plot and raised on the borders. They concluded that recharge was greater at the edges of the flooded plot because air pressure near the center of the field increased ahead of the wetting front as it approached the water table, thereby reducing infiltration. Until the entrapped air is dislodged or migrates upward due to a buoyant effect, infiltration slows (Figure 6) or can even cease (Adrian and Franzini, 1966). The displacement of entrapped air explains why bubbles often appear above the soil during infiltration following rapid ponding.

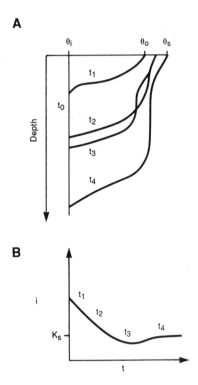

Figure 6   Effect of entrapped air on infiltration under ponded conditions, showing (A) transient water content profiles and (B) infiltration rate. Counter flow of air upward occurs between time t_2 and t_3, reducing both the water content and infiltration rate.

C. EFFECT OF INFILTRATION ON SOIL-WATER STATUS

The preceding discussion centered mostly on aspects of transient infiltration behavior at the soil surface. Most of the discussion in this section addresses the response of the soil to constant infiltration, as manifested by the steady-state profiles of pressure head and water content. In some field situations, infiltration is sufficient in duration that the wetting front reaches the water table and the entire profile is at steady state. But even before the wetting front reaches the water table, the water content may stabilize above the wetting front; that is, the soil above the wetting front does not store additional water, but rather it merely transmits water to the wetting front. In uniform soils subject to continuous ponding over extensive areas, the entire profile would be saturated when the wetting front reached the water table. If the same soil is subjected to a constant but low-intensity infiltration, the steady-state moisture profile would be unsaturated if the water application rate is less than the saturated hydraulic conductivity of the soil (Figure 7A). In this case, the water table will rise in response to the recharge; the extent of the rise of the groundwater mound would depend upon the application rate and aquifer hydraulic properties (e.g., Glover, 1960).

As illustrated in Figure 7A, for a particular soil, the degree of water saturation and pressure distribution vary as a function of the water application rate, q. To predict the

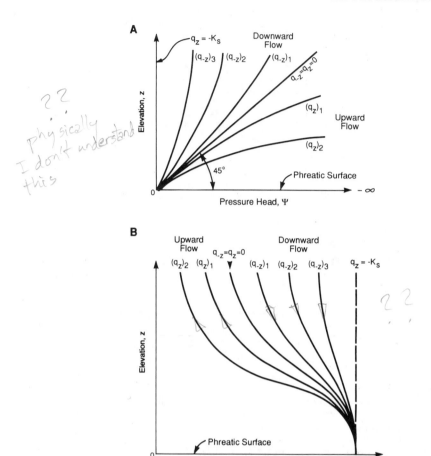

?? physically I don't understand this

Figure 7 (A) Pressure-head distributions and (B) water-content distributions in steady vertically downward (q_{-z}) and upward (q_z) unsaturated flows in uniform soil. (Modified from Bear, 1975. With permission.)

steady-state distribution of pressure head above a water table, the following equation (e.g., Zaslavsky, 1964; Warrick and Yeh, 1990) is appropriate:

$$z = -\int_0^\psi \frac{d\psi}{1 + \dfrac{q}{K(\psi)}} \tag{4}$$

where z is the elevation above a constant water table (datum) and $K(\psi)$ is the unsaturated hydraulic conductivity of the soil. If $q = 0$, we have a hydrostatic condition and Equation 4 shows that the pressure head is equal to the negative of the elevation above the water table; hence, the slope of this line is -1. If $q < 0$, then flow is downward, and if $q > 0$, flow is upward from the water table. Note in Figure 7B that as the steady infiltration rate increases, the soil becomes wetter relative to the

↳ q is water applic. rate, these #'s > or < 0 seem to be reversed

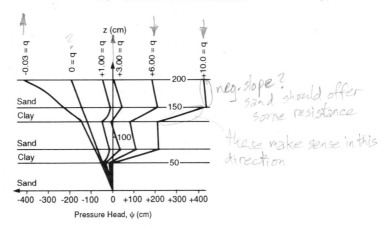

Figure 8 Steady-state pressure head profiles in a layered soil for different values of the specific discharge, q (cm/d). (From Bear, 1972. With permission.)

static equilibrium moisture profile where $q = 0$. Also recognize that, for mathematical convenience only, these examples unrealistically assume that the water-table position does not vary with infiltration rate. Figures 7A and 7B also show the pressure-head and water-content profiles, respectively, for the related case of steady upward liquid transport from the water table. Here, the soil profile is drier than the static equilibrium case. For additional evaluation of evaporation from a water table, refer to Chapter 2, Section II, as well as Willis (1960) and Ripple et al. (1972).

Because many soils are stratified, constant-flux infiltration into layered soils can lead to significantly different pressure-head profiles at steady state, as illustrated in Figure 8 for a range of infiltration rates through a horizontally continuous layer of sand overlying a clay. As the flux increases from zero, the hydrostatic case, the steady-state pressure head increases throughout the profile. In this example perched conditions develop at $q = 3.0$, and before $q = 6.0$, pressure head exceeds zero over the entire profile. These pressure-head profiles illustrate that the perched zone develops at the interface of the sand and clay and extends into both layers; the top of the perched aquifer is in the sand and the base of the perched zone is in the clay. The presence of a perching layer (clay in this example) and an unsaturated zone beneath it are both necessary to characterize the saturated zones as perched. Warrick and Yeh (1990) described analytical and numerical steady-state results based on Equation 4 for a wide range of boundary conditions in homogeneous and layered soils. For a similar two-layer case shown in Figure 9, Aylor and Parlange (1973) used numerical models, and Srivastava and Yeh (1991) developed analytical solutions to predict soil-water profiles under one-dimensional transient vertical flow conditions.

D. THE DISPLACEMENT PROCESS

The downward movement of soil water below the root zone often occurs as a piston displacement process. This is perhaps best illustrated by Horton and Hawkins (1965) who showed that infiltrating water displaces antecedent moisture. Warrick et al. (1971) also pointed out that the wetting front is not an indicator of the position of

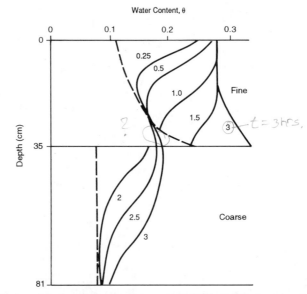

Figure 9   Numerical model simulation of water content distribution during one-dimensional infiltration into a soil consisting of an upper fine and lower coarse layer. Dashed lines are initial water content, and solid lines show water content profiles in hours after infiltration began. (From Aylor and Parlange, 1973. With permission.)

the infiltrated water, because of the displacement of antecedent water (Figure 10). That is, the water that infiltrates lies above the wetting front because the infiltrated water, in effect, dislodges the initial porewater and pushes it ahead. For one-dimensional, gravity-dominated flow systems when a constant water content is maintained near the land surface, the rate of advance of a wetting front can be approximated as

$$V_{wf} \approx \frac{K(\theta_0)}{\theta_0 - \theta_i} \tag{5}$$

and the rate of advance of the infiltrating water or nonreactive solute front can be approximated as

$$V_S \approx \frac{K(\theta_0)}{\theta_0} \tag{6}$$

where θ_0 is the water content behind the wetting front and θ_i is the initial water content. For very dry soils, $V_S \approx V_{wf}$, but as the antecedent water content, θ_i, increases, the velocity of the wetting front increases relative to that of the infiltrated water or solute front.

The piston displacement process is important when interpreting pore-liquid and water-content monitoring data. At any depth, water-content increases are likely to be detected before contaminants dissolved in the percolating water are detected, except where θ_i is very small. If a buried water-content monitoring device is triggered and a core sample is subsequently collected at this depth to assess fluid chemistry, the soil

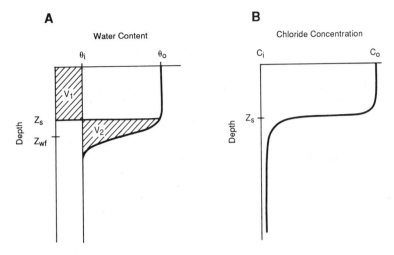

Figure 10   Water content and chloride profiles after chloride-laden water infiltrates a uniformly moist soil. The depth to the wetting front, Z_{wf}, is deeper than the depth of the infiltrated water, Z_s, represented by the chloride tracer added to the infiltrated water. The volume of water ahead of the solute front, V_1, is shown equal to the antecedent water displaced by the infiltration, V_2. (Modified from Warrick et al., 1971. With permission.)

sample is likely to be clean. Where piston displacement occurs, look for chemicals in soil above the depth where water-content increases may have been detected.

Another mechanism involved in predicting the downward movement of infiltration is the influence of immobile water (e.g., Krupp and Elrick, 1968). Immobile water occurs in pore spaces that are poorly connected to the main pore network, which conducts most of the flow in the soil (van Genuchten and Wierenga, 1976). As a consequence, the soil may contain in isolated pores antecedent water that is not displaced by the new pulse of infiltration. Because some of the total pore space does not participate in the displacement process, immobile water reduces the transit time through the vadose zone. Owing to immobile water, we expect that potential recharge water and contaminants would be transported more rapidly. In addition to the usual processes of molecular diffusion and hydrodynamic dispersion, chemicals in the infiltrating water will slowly diffuse into stagnant water zones within the immobile pores, thereby diluting the concentration of the advancing solute. During subsequent infiltration of fresh water, the contaminant is flushed from the main pore network; however, the chemicals in the immobile pores will diffuse slowly back into the percolating fresh water (Figure 11).

E. HETEROGENEITY AND MULTIDIMENSIONAL FLOW

Multidimensional flow mechanisms in the vadose zone must also be considered in deep seepage assessments. The downward vertical path of seepage can easily be diverted by heterogeneities such as lenses or sloping interfaces between layers having contrasting hydraulic properties. The study by McCord and Stephens (1987) of soil-water movement near Socorro, NM, also shows that, even on a soil as uniform

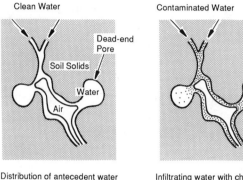 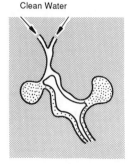

Distribution of antecedent water

Infiltrating water with chemical
diffusing into immobile pore liquid

After drainage of chemical-laden
water from main pore network,
natural flow is re-established
with diffusion of chemicals from
the immobile pores.

Figure 11   Conceptual model of variable saturated flow and transport in a pore network.

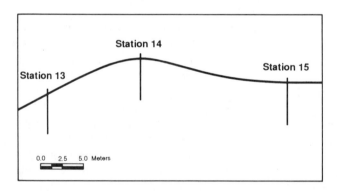

Figure 12   (A) Cross section of sand dune in vicinity of soil-monitoring stations on the sand
dune at the Sevilleta site in New Mexico. (From McCord and Stephens, 1987. With
permission.)

as dune sand, there can be significant lateral components to soil-water movement
caused by macroscale anisotropy. They found sharply contrasting moisture content
profiles at different topographic positions on the dune, which suggested that water
within the dune propagated downward, but with a distinct horizontal component
(Figure 12). Chemical tracer experiments confirmed this (Figure 13) and numerical
simulations (McCord et al., 1991) revealed that state-dependent anisotropy caused
soil water to accumulate in the topographic low areas. Winter (1986) also noted
significant recharge in dune lowlands that affected the shape of the water table in the
Sand Hills of Nebraska, as discussed by McCord and Stephens (1987). Theoretical
work by Zaslavsky and Sinai (1981) and laboratory experiments by Frederick (1988)
support the field observations that topography can contribute to local variability in
the direction of soil-water movement.

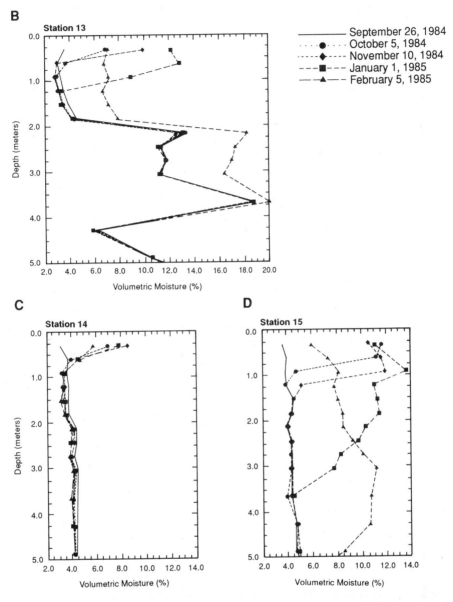

Figure 12   Selected moisture content profiles are shown at (B) Station 13, (C) Station 14, and (D) Station 15 on the sand dune at the Sevilleta site in New Mexico. Note: Rainfall for this period consisted of 4.8 cm for October 3, 1984, to October 24, 1984, and 5.4 cm for October 24, 1984, to January 24, 1985. (From McCord and Stephens, 1987. With permission.)

The importance of local-scale stratification on lateral flow components was demonstrated by laboratory studies by Heerman (1986). He conducted a sand-tank experiment with point-source infiltration into 2-cm-thick alternating layers of coarse and fine dry sand. The entire flow field was unsaturated, in as much as the pressure head was less than about –10 cm behind the wetting front. He observed that horizontal

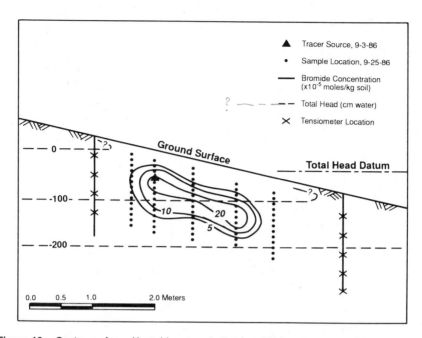

Figure 13   Contours of equal bromide concentration from hillslope tracer experiment on a sand
dune at the Sevilleta site, New Mexico. Note: Bromide concentration units: $\times 10^{-5}$
mol/kg soil; total head contours presented represent the average of measurements
taken on September 5, 12, and 18, 1986; approximately 2.6 cm of rain fell during
the time between tracer emplacement and sampling. (From McCord et al., 1991.
With permission.)

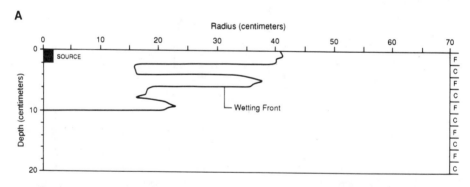

Figure 14   (A) Wetting front position observed after 5 min of water infiltration at 20 cm³/min
from a point source in alternating layers of air-dry, fine dune sand (F) and medium
blasting sand (C). (B) Unsaturated hydraulic conductivity of dune sand (K_F) and
medium blasting sand (K_C). (From Heerman, 1986. With permission.)

flow occurred to a greater extent in the fine rather than coarse sand layers (Figure
14A). In this test, the buried coarse layers were impediments to downward percolation and enhanced lateral water retention and migration in fine layers. Thus, the
alternating layers of fine and coarse sand effectively formed a soil system that was
highly anisotropic at a macroscopic scale. This behavior was attributed to the
unsaturated hydraulic conductivity contrast between the two porous materials.

B

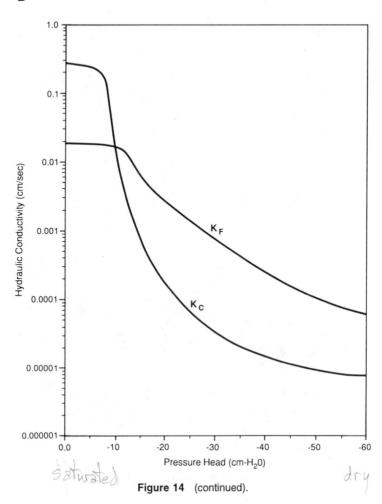

Figure 14 (continued).

Although the coarse sand is more permeable than the fine sand at full saturation, just the reverse was true for the laboratory experimental conditions (Figure 14B). Another application of this concept to deep vadose-zone flow and recharge is that faults filled with fine-textured gouge may be more conductive to water than faults with open apertures, provided that the faults are not in contact with perched or ponded water.

The importance of heterogeneity on water and tracer migration is dramatically illustrated in another sandbox experiment shown in Figure 15. In this experiment, we packed a box having dimensions of 1 m × 1 m × 0.3 m with dry, fine-textured sand and other materials having distinct shapes. The front panel of the box consists of Plexiglas. In the upper part of the box, there were three boomerang-shaped lenses, which, from left to right, were composed of an equal mix of fine and coarse sand, coarse sand, and coarse sand with the apex upward. At the middle depth, on the left we placed an inclined layer of silica flour, and on the right we placed a lens of coarse

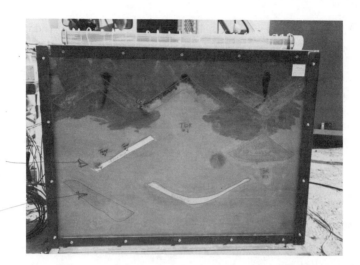

silica
flour

coarse
sand

A

B

Figure 15   Sandbox experiment showing the importance of heterogeneity on water and tracer
migration. Constant infiltration along the top of a box packed with fine sand and with
local zones of coarse sand, silica flour, and clay. A dye slowly injected at discrete
locations maps the flow pattern: (A) wetting front after approximately 20 min, and
(B) wetting front after approximately 60 min.

sand. On the lower left was a dipping layer of coarse sand, and on the right half was
a concave upward layer of bentonite. Water was applied uniformly across the upper
surface of the sand box at a rate of approximately one-half of the saturated conductivity of the fine sand. A red dye was injected at discrete locations into the sand from
the rear of the box.

The wetting front is clearly shown after about 20 min in Figure 15A and after 60
min in Figure 15B. Note that at the upper level, the middle coarse sand lens wets
slower than the other lenses and water tends to be diverted to its sides. The unit on

the left does not divert water or tracer because the amount of fine sand in this mixed coarse-fine layer renders its hydraulic properties effectively the same as the uniform fine sand surrounding it. On the upper right side, water is focused toward the base of the downward pointing coarse sand lens, but water and tracer are diverted around the underlying upward-pointing coarse sand wedge. The inclined coarse sand layer in the lower left has the effect of shedding water downward and along the interface slope. We also see in Figure 15B that although the silica flour has wetted, tracer injected above the silica flour clearly avoids the silica flour zone. In fact, the path of the tracer in the sand above the upper end of the silica flour zone bends at an angle greater than 90 degrees. Very pronounced changes in flow path are exhibited near some of the coarse sand units as well. The low permeable bentonite at the base of the box behaved as a basin, allowing water to accumulate within the basin before spilling over; this is the so-called "bathtub" effect associated with clay liner designs. The dye paths reveal that because of the heterogeneity in the soil textures, the flow field in this experiment is considerably two-dimensional, in spite of the uniform rate of infiltration along the upper surface and the constraints on lateral movement by the sides of the box.

The primary control on the flow-field behavior is hydraulic conductivity (Figure 16). Obviously, the uniform coarse sand has the greatest hydraulic conductivity at saturation, but owing to the slope of the interface, the large pore size, and the extremely low conductivity at the initial water content, water is relatively slow to enter it. Over time, the uniform coarse unit in the upper center of the box does wet and water and tracer do migrate through it, as evidenced by breakthrough just to the right of the apex of this unit in Figure 15B; nevertheless, this coarse unit appeared to have a significant influence on the flow field even after the profile was thoroughly wetted. Inasmuch as the silica flour is often used for backfill around buried vadose zone monitoring sensors, the behavior of the flow field near this unit is a bit surprising. However, from Figure 16 we find that the unsaturated hydraulic conductivity of the silica flour at the *in situ* pressure head (about −20 to −30 cm of water) behind the wetting front is less than in the surrounding fine sand. With respect to designing a monitoring system to intercept seepage, this experiment illustrates the importance of site characterization, especially mapping the geometry of geologic units and quantifying their unsaturated hydraulic properties.

F. PREFERENTIAL FLOW

Infiltration along preferential flow paths has received considerable attention over the past several decades and has been highlighted recently by Gish and Shirmohammadi (1991). Preferential pathways are open conduits or macropores that potentially "short-circuit" all or part of the path to the water table. Preferential paths can develop as a consequence of desiccation cracks, root channels, and animal burrows in shallow soils. On a finer scale, pore spaces between soil structure units, called peds, are also preferential flow paths. Fractures, joints, faults, bedding planes, and karst features can comprise preferential pathways in both soil and rock. These preferential paths contribute to deep seepage or recharge where they are vertically continuous over significant depths and where they are connected to a free surface, such as ponded water during a precipitation event (Figure 17).

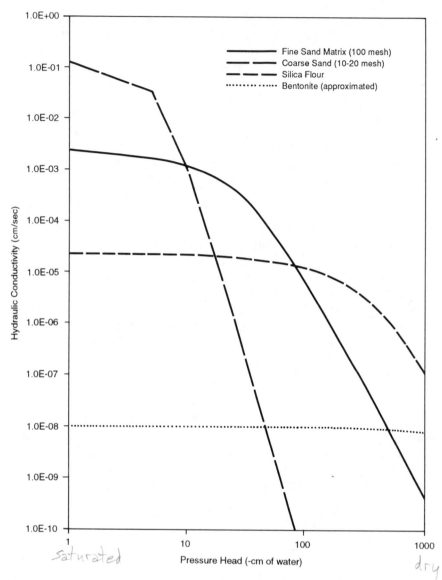

Figure 16 Hydraulic conductivity of materials in the sand box in Figure 15.

Macropore flow is usually associated with ponded conditions or positive water pressure because substantial water flow into the macropore will only occur when the fluid pressure exceeds the water-entry pressure (Figure 6) that is characteristic of the macropore. Water that infiltrates and fills a macropore propagates downward rapidly at a rate that is controlled mostly by the macropore diameter or fracture aperture. The downward movement of water in the macropore may be impeded, and could in fact cease, if air pressure ahead of the wetting front increases sufficiently. Water in the macropore will also diffuse into the porous matrix due to capillary and adsorptive forces. In many instances, water percolating through a macropore may be completely

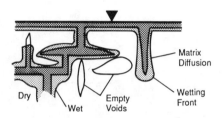

Figure 17   Where continuous macropores are connected to a ponded source of water on land, the macropores rapidly conduct water through the vadose zone. Soil matrix imbibes water from the macropores. Large voids unconnected to the source of ponding remain empty.

absorbed by the surrounding matrix before reaching the water table. Macropores that terminate before reaching the water table may contain water at positive pressure for prolonged periods, if the porous matrix is relatively low in permeability (Figure 17). In this case, results of vadose zone monitoring networks with sensors in both the water-filled macropores and the unsaturated matrix would be interpreted as local perched systems. Additionally, macropore geometry and hydraulic characteristics can change over time, due to freeze-thaw action, root growth, seismic activity, etc. One of the greatest challenges for researchers in vadose-zone hydrology is to develop reliable methods to quantify the continuity and hydraulic conductivity of macropores *in situ* for a range of field moisture conditions, at a scale and depth sufficiently large to be useful for applying predictive models of water flow or solute transport. The problems relating macropore characterization and flow modeling are further described by Beven (1991). Sharma et al. (1987) concluded, from studies below native vegetation in coastal western Australia where there is a mediterranean climate, that macropores could account for about 50% of the total recharge in a sandy soil profile and more than 99% of recharge in a highly structured lateritic soil.

Macropores are important recharge pathways, but can eventually become plugged due to the accumulation of calcium carbonate or clay. This is clearly evident in trenches in the southern New Mexico desert (Gile et al., 1981) which reveal complex patterns of caliche development that appear to reflect variability in soil-water movement due to macropores, local geology, and drainage. Consequently, the spatial distribution of deep soil-water movement may gradually change with time as soil profiles are modified.

As part of a site investigation for a low-level radioactive waste repository, preferential flow (Figure 18) through a single earth fissure was evaluated at a site in Hudspeth County, east of El Paso, TX (Scanlon, 1990). The fissure is located on basin fill alluvium where the depth to groundwater is about 150 m. The landscape is very gently sloping and is covered mostly by creosote bush. Gullies have eroded the fissure to a width of about 30 cm and to a depth of more than 1 m in places. This feature was readily identifiable as a lineation in low sun angle air photography (George Beckwith, 1989, personal communication). A 6-m-deep trench through the fissure showed vertical continuity throughout, with fine sand and silt filling the fissure below a depth of about 1 m. Chloride analyses of soil cores and a ponded infiltration test showed that the fissure was a preferential pathway for downward liquid transport through at least the excavated depth, which included rather dense calcareous paleosols. On the other hand, soil-water potential measurements in the

Figure 18   Earth fissure in a thick vadose zone east of El Paso, TX.

fissure zone showed evidence for upward liquid transport. Evidently, both upward and downward soil-water flow can occur at different times in a single macropore. Whether recharge occurs via these fissures has not been determined; however, the groundwater age dates suggest that significant rapid recharge probably did not occur by this process directly. On the other hand, infiltration penetrated much deeper than would have occurred by porous media flow alone.

Another interesting aspect of preferential soil-water movement that is relevant to deep percolation is the development of unstable flow (e.g., Parlange and Hill, 1976; Hillel, 1987). Unstable flow conditions lead to the downward propagation of the wetting front as discrete fingers, instead of as a uniform plane (Glass, 1991). Yao and Hendrickx (1995) noted that instability of the wetting front is predicted to occur under the following conditions: (1) during infiltration of ponded water with compression of air ahead of the wetting front, (2) during redistribution of water in the soil profile, (3) in water repellent soils, (4) where water content increases with depth, and (5) where there is continuous nonponded infiltration. They observed fingers in a laboratory experiment of slow infiltration into a 1-m-diameter column perched with a very uniform quartz sand (0.841 to 1.41 mm diameter). The water source comprised of 1100 needles was continuously moving to provide a random raindrop pattern. The fingers are clearly evident in a section cut 15 cm below the top of the column (Figure 19). Unstable flow also is likely to develop beneath the interface of a fine porous

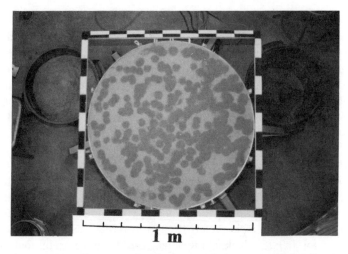

Figure 19   Dye patterns showing evidence of unstable flow where slow infiltration (0.5 mm/ min) was applied to a 14–20 mesh sand. Pattern is 15 cm below top of the lysimeter. (From Yao and Hendrickx, 1994. With permission.)

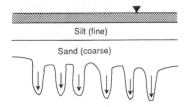

Figure 20   During ponded infiltration, wetting front instability can occur in coarse layers that underlie fine textured layers.

material overlying a coarse material when ponding occurs above the fine layer (Figure 20).

The work by Glass (1991) showed that after infiltration ceased and the profile drained, the subsequent infiltration followed the same finger-like pathways which were developed in the initial infiltration. As shown in Figure 20, fingers may occupy only a portion of the horizontal cross sectional area of the porous media. This led Thomas and Phillips (1979) to conclude that recharge of groundwater can occur long before the soil is thoroughly wetted. Baker and Hillel (1991) demonstrated that finger width and distinctness decreased with increasing antecedent moisture content. Yao and Hendrickx (1994) indicated that finger width also decreased in coarser soils and decreased as infiltration rate decreased. Under deep water-table conditions, it is also possible that the fingers gradually blend together with increasing depth due to moisture diffusion. This would be more likely in stratified soils, which would cause the fingers to spread laterally. At sites in arid climates with a deep water table where surface ponding is relatively brief, the mechanisms causing unstable flow may dissipate before the fingers reach the aquifer. It is important to point out that there has been little fieldwork to verify the importance of unstable flow where there are highly stratified deposits of texturally different strata, deep water-table conditions, and arid climate.

Theoretical (Raats, 1973) and experimental work (Hendrickx and Dekker, 1991; Hendrickx et al., 1993) leaves no doubt that infiltration in dry water-repellent soils always results in unstable wetting fronts. Soils can become water repellent to varying degrees, primarily due to the effect of naturally occurring hydrophobic organic compounds, which may coat portions of the soil particles or occupy part of the soil matrix (Bond, 1964). It is not yet clear what factors affect infiltration in non-wetting soils. Although laboratory experiments (Glass, 1991; Yao and Hendrickx, 1994) clearly demonstrate that unstable wetting may occur in wettable soil, field evidence so far has been found only in beach-related dunes in the Netherlands, Germany, and Canada (Hendrickx and Dekker, 1991). However, no unstable wetting has been observed in the Sevilleta sand dunes near Socorro, NM (e.g., Figure 12). It is likely that in many sparsely vegetated desert environments, the organic matter in the soil is insufficient to create water-repellent conditions to an extent that would induce unstable flow and create rapid preferred pathways. Methods to identify the water repellency of soil (e.g., Mallick and Rahman, 1985) may be useful tools to screen sites for the potential for preferential flow by this mechanism.

II. DRAINAGE AND REDISTRIBUTION

Infiltration ceases when water is no longer moving into the soil at the soil surface. The cessation of infiltration, however, does not mean that water movement in the soil profile ceases. The general processes describing water movement after infiltration are called drainage and redistribution.

A. GENERAL PROCESSES

Drainage in the soil profile occurs where the soil water content is decreasing over time (Figure 21A). Where the profile was initially completely satiated throughout, the drainage process is referred to as internal drainage, as illustrated in this figure. Most frequently, infiltration from precipitation only partially wets the vadose zone, as shown in Figure 21B at time t_1. In this case, after infiltration ceases, the water behind the wetting front can move upward in response to evaporation or it can move downward due to gravity and matric potential gradients. Note in Figure 21B that at some time, t_2, after infiltration stopped, the upper part of the profile has drained while the lower part of the profile has wetted. This process is called redistribution. It is highly probable that most groundwater recharge occurs during redistribution, especially where the water table is deep and precipitation events are brief.

There are also some field data that suggest that there can be net upward movement within some deep vadose zones. At a site in Hudspeth County, near El Paso, Texas, in ephemeral stream and interstream settings, Scanlon (1992) concluded from *in situ* pressure head measurements with psychrometers installed to a depth of 15 m that there was a net upward liquid flux. Likewise, at a site near Needles, CA, Lapalla and Ohland (1991) used psychrometers and heat dissipation sensors buried to depths of 32 m and found that there was a net upward movement of water from the water table more than 200 m below land surface. Similar results were obtained by Enfield

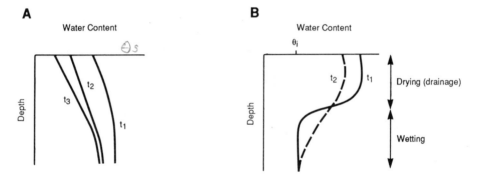

Figure 21   (A) Drainage following a thorough, deep wetting of the soil and (B) redistribution following a brief period of infiltration.

et al. (1973) with psychrometers at the Hanford site in Washington, but below about 25 m and to at least 90 m depth, the hydraulic head gradient appeared to be downward. If a condition of net upward liquid transport is representative of long term soil-water movement, then diffuse recharge (Chapter 4) would be nil.

Geologic heterogeneity must be kept in mind when interpreting infiltration and redistribution from water-content profiles. In a soil comprised of fine and coarse textured layers, the water content usually will be greatest in the fine layer. Thus, even if there is no flow, bulges in water content usually appear in the soil profile that are attributed to heterogeneity in the soil-water characteristic curve, rather than to redistribution (Figure 22).

Until recently, quantitative analyses of drainage and redistribution based on the Richards equation were done mostly using numerical simulations. However, to predict drainage and redistribution Warrick et al. (1990) have developed water content-based analytical solutions which require input data on soil water diffusivity and hydraulic conductivity. Earlier numerical simulation work by Rubin (1967) showed the importance of hysteresis in the soil hydraulic properties to accurately predict redistribution. Rubin (1967) assumed that each of three different cases began with the same initial depth of wetting, but the input hydraulic properties were slightly different. The soil was the Rehovot sand, but in one case the hydraulic properties were represented as the main drainage curve, in another by the main drying curve, and in the third by the full hysteretic properties, which included both the main wetting and drying curves. There is a tendency for many modelers to want to avoid the complexity of hysteresis by using one of the main curves to bound the analysis for the problem at hand. However, as Figure 23 illustrates, simulations using either the main wetting or drying curve overestimate the depth of wetting and poorly reproduce the water content profile shown by the hysteretic analysis. In general, the effect of hysteresis is to retard redistribution, so that more water is retained near the surface.

B. PROPAGATION OF PULSES OF INFILTRATION

There are a number of other aspects of infiltration and redistribution that may be important to consider that are relevant to vadose-zone monitoring or modeling. For

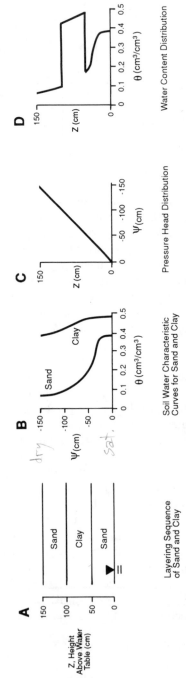

Figure 22   Effect of layered heterogeneity on water-content and pressure-head profiles under hydrostatic equilibrium: (A) layering sequence of sand and clay, (B) soil-water characteristic curves for sand and clay, (C) pressure-head distribution, and (D) water-content distribution.

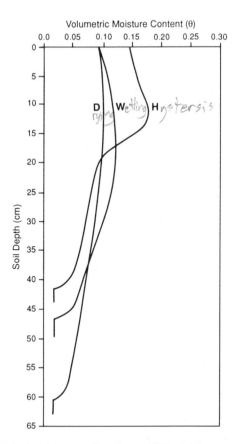

Figure 23   Calculated soil moisture profiles after redistribution in a sandy soil, illustrating the effect of hysteresis on retention of moisture in the upper layers. *Note:* Rehovot sand, $n = 0.387$, $\theta_{fc} = 0.066$; H = hysteretic redistribution, W = nonhysteretic, wetting soil characteristic curve, and D = nonhysteretic, drying soil characteristic curve. (From Hillel, 1980. With permission.)

example, within the root zone during the growing season, the redistributed water may be depleted almost entirely by vegetation, especially in semi-arid climates. In the absence of evapotranspiration, if the soil is initially nearly air-dry, it may be possible that the pulse of water will be completely retained in the soil by adsorptive and capillary forces before it reaches the aquifer, as described by Smith (1967). During redistribution, pulses of infiltrated water are also naturally damped in a diffusion-like process (Kirkham and Powers, 1972). In fact, in some soils, a series of inputs is so well damped that the water content of the soil is essentially constant, so that below some depth the soil-water content appears to be at steady state (e.g., Wierenga, 1977). On the other hand, transient effects due to natural infiltration have been observed even deep in the vadose zone. For example, Nixon and Lawless (1960) mapped pulses of moisture in soil profiles in a coastal California sand where translocation occurred to depths of at least 6 m long after rainfall ceased. Mean annual rainfall at that site is about 34 cm, but the particular rainy period they studied produced nearly 61 cm of rain. On a swale in a sand dune north of Socorro, NM, McCord and Stephens

(1987) showed a pulse of moisture, generated by about 10 cm of rainfall over a 4-month period, had propagated to more than 5 m by the end of the rainy period (Figure 12). Whether pulses of soil moisture can be measured at a particular site depends upon a variety of factors including soil hydraulic properties and initial moisture content, as well as precipitation intensity, duration, and frequency, and the depth of observation in the soil profile.

The preceding examples, as well as model studies (e.g., Evans and Warrick, 1970), show that it is quite possible for significant pulses of moisture to propagate completely through the vadose zone and reach the water table in a transient process. However, as the depth to the water table increases, as the rain intensity decreases, and as the infiltration events become more frequent, the water content deep in the profile would tend to become more uniform over time. Unsaturated hydraulic conductivity and specific moisture capacity of the soil also have important controls over whether infiltration events are readily observed as increased moisture content. The implication of this is that for some soils, an order of magnitude change in soil-water flux may produce only a very small change in moisture content. If you consider that the popular method to monitor water content by neutron logging only has a water-content resolution of about 1/2 to 1%. Water-content monitoring to detect translocating pulses of soil water may not always be an effective means to quantify temporal variability in seepage. The absence of detectable changes in water content should not be construed as indicating that there is no seepage. To the contrary, as pointed out previously (Figure 7), at a steady downward flux no temporal change in water content should be expected.

Owing to heterogeneity, the downward percolation of water during infiltration or redistribution may virtually cease where the infiltrated water migrating through a fine soil encounters a dry and relatively uniform, coarse-textured layer (Figure 24). This occurs when the pressure head in the water pulse is not sufficiently great to force water to enter the large pores of the coarse soil. Figure 24 illustrates a pulse of water above the interface between the fine and coarse layers. For water to enter the large dry pores, the pressure head in the coarse layer must exceed the critical threshold or water-entry value, ψ_{wc}. This will occur when

$$\psi_{wf} + L \geq \psi_{wc} \tag{7}$$

where ψ_{wf} is the critical water-entry pressure head for water to enter the fine pores and L is the depth of water accumulation above the interface (Iwata et al., 1988).

C. FIELD CAPACITY AND WATER STORAGE

Redistribution experiments by Gardner et al. (1970) using laboratory soil columns indicate that the rate of drainage is about the same at any depth (Figure 25). They also found that the rate of redistribution increases as the depth of the initial wetting decreases, owing to the increased effect of matric potential gradients to disperse the pulse of water. Perhaps one of the most significant results of these and many other similar experiments is that above the initial depth of wetting, the soil profile continues to drain, even after more than 30 d. This result is relevant to the

redistribution rate ↑ as depth to initial wetting ↓

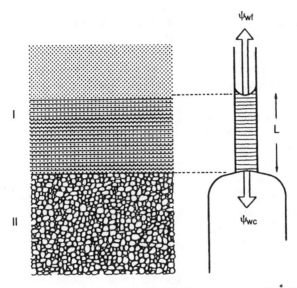

Figure 24   Suspended water in a layered soil. (Modified from Iwata et al., 1988. With permission.)

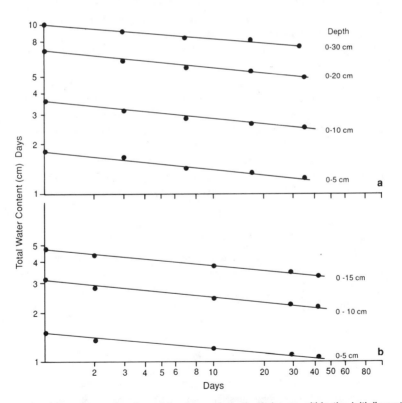

Figure 25   Total amount of water retained in various depth layers within the initially wetted zone of a fine, sandy loam during redistribution following irrigations of (a) 10 cm and (b) 15 cm of water. (From Hillel, 1980. With permission.)

concept of field capacity, which unfortunately is often misused to represent the water content below which water movement ceases (e.g., U.S. Environmental Protection Agency, 1986). Agricultural scientists coined the term field capacity to represent the water content after 2 to 3 d of drainage following a thorough irrigation. They realized that the water content in the root zone decreased rapidly prior to this time by gravity drainage, and this rapidly draining water in excess of the field capacity was unavailable to plants (Veihmeyer and Hendrickson, 1931). In order to conserve water and diminish pumping costs, the irrigators attempted to add only enough water to bring the water content in the dry soil up to field capacity. For many uniform soils, the field capacity roughly corresponded to the water content at about −0.3 to −0.1 bars of matric potential, and the water held between the field capacity and the permanent wilting percentage (−15 bars) is called the water available to the plants, or simply the available water (Chapter 1, Figure 5; Chapter 2 Equation 4a,b). Clearly, even from the practical point of view of the irrigator, most of the soil water movement to plant root systems occurs at water contents less than the field capacity.

Perhaps one reason for the misuse of the term *field capacity* by hydrologists is due to incorrectly equating it to specific retention (e.g., Everett et al., 1984). Specific retention refers to the asymptotic water content approached as the soil drains toward the point where liquid-phase transport is nil. Field capacity, on the other hand, is well within the range of water contents where significant liquid-phase transport can and does readily occur. In fact, liquid-phase water flow has been demonstrated at potentials of several tens of bars. For example, a laboratory study by Grismer et al. (1986) found measurable liquid transport to water contents as low as 0.04 m^3/m^3, corresponding to −525 bars soil water potential in a silt loam. The same study found liquid transport within a loamy sand, even at water contents as low as about 0.01 m^3/m^3. Consequently, specific retention from the asymptote of the θ-ψ curve, not field capacity, more closely represents the threshold below which liquid transport ceases.

Application of the correct definitions of specific retention and field capacity are also important in evaluating the amount of water that the soil can hold when there is static equilibrium, as discussed by McWhorter (1985). If one incorrectly assumes that water movement cannot occur at water contents below field capacity, when defined as the water content held at −0.1 to −0.3 bar, then the amount of water potentially stored in the vadose zone will be overestimated, because over long periods of time water will drain from the profile until the specific retention is reached. To calculate the water storage capacity of the vadose zone at static equilibrium, that is, in the absence of water movement, evaporation, or recharge, the following equation is suggested:

$$V_w = \int_{\psi=0}^{-z} \theta \, d\psi \tag{8}$$

where z is the depth to the water table and the θ-ψ relationship is derived from analyses of the soil-water characteristic curve for drainage. Stephens (1985) describes the precautions that should be taken in using the water storage capacity in the vadose to calculate permanent waste disposal of seepage.

Although field capacity may be a reasonably practical parameter for irrigation management of a given field, the field capacity is not a good choice for hydrologists and engineers seeking to quantify field soil-water holding capacity. In a field soil, the water content remaining in the soil a few days after infiltration of irrigation or rainfall will depend on many factors described previously that affect the hysteretic behavior of the soil, such as soil texture, antecedent water content, and depth of wetting. The nature of stratification of the soil profile can also drastically influence the water held in storage at a particular depth. For example, if the profile is uniform, deep, and sandy, with the exception of a thin clay layer at about the 1 m depth, redistribution of a shallow pulse of water will be markedly slowed by the low permeable clay layer, in comparison to redistribution in the completely uniform sandy profile. Consequently, the water content remaining at the 50 cm depth would be greater in the layered soil than in a uniform soil, and furthermore, the water content would undoubtedly exceed that corresponding to $-1/3$ or $-1/10$ bars, the so-called field capacity of the sand. After infiltration, both spatial and temporal variability in evaporation, transpiration, and root distributions can also influence the amount of retained water, especially at shallow depths. Because a good, physically based definition of field capacity remains to be identified, the term should be avoided by hydrogeologists.

D. EFFECTS OF VEGETATION ON SOIL-WATER MOVEMENT

During drainage and redistribution, multidimensional flow mechanisms may be induced within the root zone by vegetation, especially where the root distributions are nonuniform or sparse. At most sites, this is a local-scale phenomenon. Roots move in the soil as the plant grows (e.g., Brown and Scott, 1984), so it is expected that water-content distributions in the soil change in time and space as soil water is drawn toward the roots. In one example of native desert plants, *in situ* pressure-head measurements were made with 29 tensiometers surrounding an isolated lavender bush (*Dalea scoparia*) on a sand dune north of Socorro, NM. The hydraulic head fields revealed complex three-dimensional flow patterns dominated by the roots (Kickham, 1987) during the spring and summer (Figure 26). In contrast, during late fall and winter the soil-water movement was nearly one-dimensional downward. At another site a couple hundred meters away on an alluvial sand where the water table depth is shallower, about 5 m, 160 tensiometers and 12 neutron probe access tubes were placed in a poorly vegetated, 24-m^2-area situated between dense four-wing salt bush (*Atriplex canescens*) stands. This instrumentation revealed that the flow of soil moisture was nearly one-dimensional downward throughout the year (Stephens et al., 1991) (Figure 27). The latter study suggests that, at least for relatively large, flat areas vegetated by four-wing salt bush and native grasses, it is safe to neglect multidimensional flow mechanisms in mathematical models of deep soil water movement.

A similar field investigation was conducted north of Tucson, AZ, by Sammis and Weeks (1977). Neutron probe access tubes were installed at selected locations around the desert plants, creosote (*Larria divaricata*) and bursage (*Ambrosia deltoidea*), as well as in open space. They found significant spatial variability in moisture content from neutron probe logging at different times, only around the bursage. Differences

A

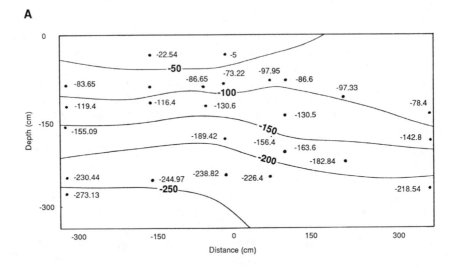

B

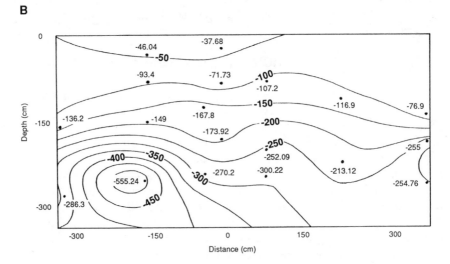

Figure 26   Vertical cross section of total head through a lavender bush (plant located in center of plot). Soil is dune sand located at the Sevilleta National Wildlife Refuge north of Socorro, NM: (A) February 5, 1986, and (B) June 10, 1986. (From Kickham, 1987. With permission.)

in moisture content profiles were attributed to root distributions; the creosote plant has an extensive root system, whereas the roots of the bursage plant did not extend far beyond its canopy.

There is really insufficient information to draw general conclusions about the role of vegetation on flow patterns; nevertheless, transpiration is often the largest term in a water balance. It is clear that vegetation can play a dominating role in the uptake of infiltration and redistributed water that otherwise would become recharge. For instance, at Hanford, WA, water-balance studies in coarse base soils or coarse soils

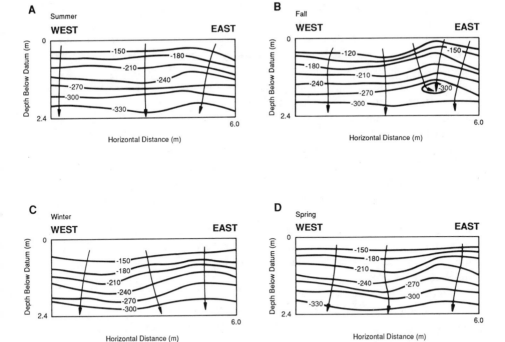

Figure 27   Total hydraulic head (cm) and flow lines in a poorly vegetated plot of sandy alluvium that is surrounded mostly by four-wing salt bush: (A) July 20, 1988; (B) October 15, 1988; (C) December 9, 1988; (D) April 21, 1989. (From Stephens et al., 1991. With permission.)

vegetated with sparse grass cover [such as cheatgrass (*Bromus tectorum*) or native bluegrass (*Poa sandbergii*)], a significant portion of precipitation becomes deep drainage (Gee et al., 1989). In contrast, Gee et al. (1989) found that deep-rooted plants (i.e., shrubs and weedy species with roots deeper than 1 m) appear to be highly effective in removing soil water.

E. PNEUMATIC AND THERMAL EFFECTS

Water movement in the vadose zone can also be impacted by soil air pressure and thermal effects (Chapter 1, Sections VI and VII). Pneumatic pressure gradients are particularly important to consider in thick vadose zones comprised of fractured bedrock. Temperature gradients, to the extent they relate to deep vadose-zone processes, may also require evaluation in thick vadose zones. For example, Ross (1984) concluded for a deep water table condition that upward vapor flux driven by a geothermal gradient would be relatively insignificant where the net recharge was at least 0.03 mm/year, and Rose (1976) reported that field data suggest when temperature gradients are greater than $0.5°C$ cm^{-1}, water transport in the vapor phase can be comparable to the transport in the liquid phase, even at pressure heads as great as

–200 to –300 cm. In general, however, both pneumatic and thermal effects are only likely to have a significant impact on deep liquid flow and recharge in areas of low precipitation.

Barometric pressure fluctuations at the land surface may penetrate to several meters in thick, unsaturated soils, but normally the depth of atmospheric pressure influence is much less than a meter (Wood and Petraitis, 1984). Although barometric pumping rarely plays a significant role in the natural ground-water recharge process, barometric pumping is potentially important for transport of radioactive gases and volatile contaminants. During periods of relatively high atmospheric pressure, fresh air will be drawn into permeable air pathways such as fractures and poorly sealed wells. And, because of the concentration gradient, gaseous contaminants and water vapor will diffuse from the porous matrix to mix with the fresh air. During low pressure, the contaminated soil gas will be expelled from the fracture to the atmosphere. The process is repeated at a regular frequency, with the net result being a transfer of gaseous contaminants from the soil to the atmosphere. Another result may be a gradual drying of the vadose zone near the gas pathways that transport vapor having a very low relative humidity. The time scales for advection of gases through hundreds of meters of rock by barometric pumping are on the order of months, in comparison to an equivalent transport by molecular diffusion alone, which may take decades (Nilson et al., 1992).

Migration of gas in the vadose zone can also be attributed to winds, temperature fluctuations, and gas density differences (buoyancy). At sites of low relief, wind blowing across the soil surface can cause gas movement to a depth of several centimeters (Wood and Petraitis, 1984). Where unvegetated mountain ridge lines generally trend perpendicular to the predominant wind direction, the atmospheric pressure on the windward face may exceed that near the crest and along the leeward face, thereby inducing advection of air within the vadose zone (Weeks, 1991). Convection cells of vapor can be generated in thick, gas-permeable vadose zones by temperature differences produced by the geothermal gradient. At depth, where temperatures are greater, the saturated water vapor is less dense and tends to migrate upward, especially along fractures. As this vapor rises, it cools, increases in density, and then tends to move downward. Gases more or less dense than water vapor at the ambient vadose zone temperature will also move by advection along permeable pathways (Montazer and Wilson, 1984).

If the gas migration rate through preferential paths is too rapid, the gas or water vapor may not be in equilibrium with water vapor or gas in the porous matrix (Montazer and Wilson, 1984). In this case, there will be mass transfer between the fracture and matrix, the direction of transfer being dependent upon the concentration gradient.

The advective gas flux rate can be determined from Darcy's equation applied to the gas (Chapter 1, Equation 34). Key parameters are the pressure gradient and the permeability of the medium to gas. Gas permeability is enhanced in coarse porous deposits and in bedrock with open fractures. On the other hand, zones of high water content or perched conditions will inhibit or prevent gas transport by advection as well as by diffusion.

We can gain some understanding perhaps of the nature of both water and water vapor transport from the conceptual model developed by Montazer and Wilson

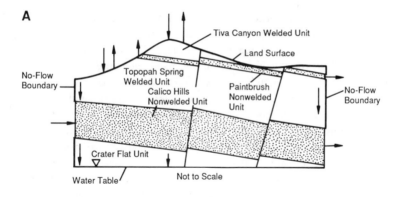

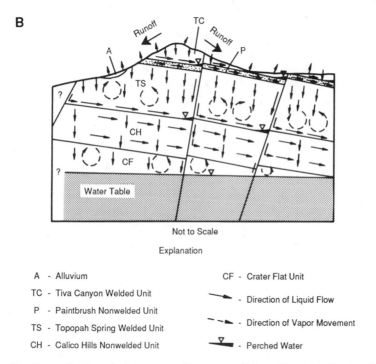

Figure 28   Generalized geologic cross section across Yucca Mountain showing (A) flow
boundaries and inferred water flux directions (arrows) at the boundaries and (B)
flow regime under baseline conditions; lengths of arrows show relative magnitude
of fluxes. (From Montazer and Wilson, 1984. With permission.)

(1984) for Yucca Mountain, NV (Figure 28). This area of southern Nevada receives
only about 20 cm of precipitation, and summer temperatures frequently exceed 40°C.
Yucca Mountain is at an elevation of about 1500 m and the land surface is very
sparsely vegetated. The lithology consists predominantly of unwelded tuff derived
from volcanic ash flows. In their conceptual model, the vast majority of the infiltration is returned to the atmosphere, especially that which falls on alluvium or unfractured
and porous tuff. Water can move downward rapidly along fractures and other

structural pathways, but this downward movement is slowed by several mechanisms. Where fractures terminate against unwelded, porous, and unfractured layers, a capillary barrier may form that prevents significant liquid migration into the lower porous layer. However, the dip of the interface and the nature of stratification within the porous layer cause lateral flow migration down-dip above the interface. If this laterally migrating water reaches a much less permeable structural feature such as a fault zone, perched conditions will likely be created. On the other hand, if the vertical structural feature is permeable, then water could be routed downward from the base of the fractured and welded unit and through the underlying porous layer (e.g., see Topopah Springs and Calico Hills units in Figure 28). Water percolating along the faults or the sparse fractures that penetrate the porous units will diffuse into the porous matrix; after fracture flow ceases, water may diffuse from the matrix back to the fracture, where the water most likely will be transported in the vapor phase. Once in the vapor phase within the fractures, some water will be carried to the land surface, where it will discharge the vadose zone. Temperature gradients and hydrogeologic conditions within the thick welded fractured tuffs appear to be favorable for establishing convective vapor circulation cells, and in places some water may be transported to the atmosphere by vapor diffusion. In spite of these impediments to downward percolation, there is a net recharge to the aquifer underlying Yucca Mountain amounting to about 0.5 to 4.5 mm/year (Montazer and Wilson, 1984).

CHAPTER 4

Recharge

Groundwater recharge from the vadose zone occurs when soil water crosses the water table of a perched aquifer or the phreatic zone. Recharge is important where hydrologists may be required to quantify the perennial yield of a groundwater basin in order to assess, for example, the potential viability of an irrigation project, the adequacy of a municipal water supply, or the feasibility of dewatering a mine pit. Undesirable consequences of pumping in excess of recharge to a groundwater basin may include loss of water supply, land subsidence, intrusion of saline water, increased pumping costs, or impairing existing surface and groundwater rights. Quantifying recharge is also highly relevant to contaminant transport and remedial action programs. To assess the performance of a proposed remedial action or to assess the risk of a no-action alternative, recharge is probably the single most important parameter in the analysis.

In humid climates, recharge can be 20 to 50% or more of the mean annual precipitation. But in areas of low precipitation, recharge is usually a small part of the hydrologic budget for an aquifer. Some scientists and engineers believe that recharge, at least from areal precipitation, is virtually nil in desert environments. They give explanations such as (1) the evapotranspiration potential far exceeds the mean annual precipitation, or (2) the field moisture content is less than the so-called field capacity. For example, Mann (1976) stated, "It is generally recognized that there is no direct recharge by rainfall through the vadose zone of arid regions." Falconer et al. (1982) stated that in regions where evaporation plus runoff exceeds precipitation, solutes are retained in the vadose zone. The fallacy of the first explanation is clearly articulated by Stephens and Knowlton (1986) and Gee and Hillel (1988). The inaccuracy of the second explanation is addressed by Stephens (1985a) and by McWhorter (1985). This chapter will highlight some of the numerous studies that report the widespread occurrence of recharge, in even the most arid deserts.

Recharge to groundwater occurs as diffuse recharge, localized recharge, or recharge from mountain fronts, leakage, and underflow (Figure 1). These different types of groundwater recharge are distinguished by source of water and by the path the water takes to enter the saturated zones within a groundwater basin. Diffuse recharge is natural recharge derived from precipitation that falls on large portions of the landscape and percolates through the vadose zone to the aquifer. Diffuse recharge

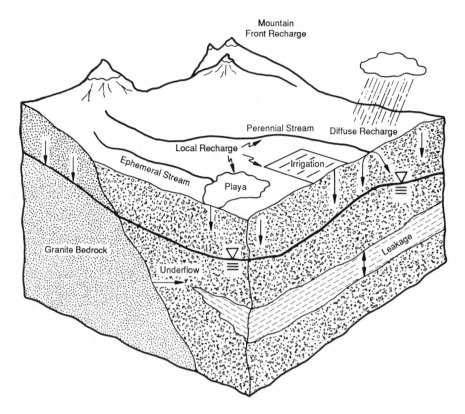

Figure 1 Conceptual model of diffuse and localized recharge.

probably dominates in humid climates. Localized natural recharge occurs mainly where there is ponding within a basin, along losing stream channels and playas, for example. In comparison to diffuse recharge, local recharge is probably the most important source of natural recharge in arid and semi-arid lands. Mountain front recharge typically involves complex processes of unsaturated and saturated flow in fractured rocks, as well as infiltration along channels flowing across alluvial fans. On a large scale, mountain front recharge through fractured bedrock is primarily a diffuse recharge process, whereas infiltration from mountain streams is considered a local recharge process. Vertical leakage across aquitards and underflow from adjacent aquifers can be important sources of recharge, but typically they do not involve the vadose zone; consequently, they are not discussed further.

Diffuse and, to a lesser extent, local recharge are relatively large-scale processes within a groundwater basin. That is, the aquifer surface area over which the recharge occurs is usually a significant part of the basin area, perhaps on the order of one to more than hundreds of square kilometers. Diffuse and local recharge are affected by processes that occur at much smaller scales, such as preferential flow in macropores, unstable flow, and effects of spatial variability in media properties. For this reason, in Chapter 3 we presented a discussion of these smaller scale processes that are necessary to critically evaluate the larger scale recharge mechanisms and the water resources in a groundwater basin. Each process, whether at the large or small scale,

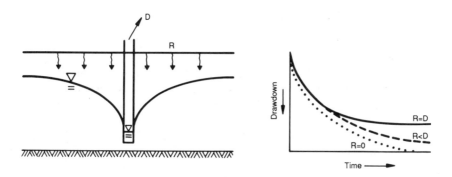

Figure 2 Water-level decline in a well pumping from an aquifer that receives within the cone of depression less recharge, *R*, than discharge, *D*.

ultimately involves water movement through the vadose zone. Flow in variably saturated, heterogeneous porous and fractured media involves interacting complex processes such as flow from fractures into the matrix and vapor phase transport, for example. Therefore to fully understand recharge processes, an appreciation for both geology and soil physics is necessary.

Additionally, it is critical to couple one's knowledge of aquifer response with vadose zone information when interpreting both diffuse and local recharge. For instance, one should bear in mind that the water table may not necessarily rise in response to recharge (Freeze, 1969). If the recharge is less than groundwater discharge by evapotranspiration or pumping from the aquifer, water levels will decline, even though recharge is occurring (Figure 2). If the recharge is balanced by the discharge from the system, the water table will remain steady.

In the following discussion, the cited examples are derived primarily from natural recharge in arid and semi-arid lands for two main reasons. First, most cases where vadose-zone recharge is a key issue for hydrogeologists are in areas of low precipitation. Second, natural recharge is often so small and difficult to accurately quantify in these areas that many doubt its existence. The processes leading to recharge nevertheless are relevant to any climate and to both natural and culturally modified recharge.

I. DIFFUSE NATURAL RECHARGE

Diffuse recharge is distributed over extensive areas and is derived from the direct infiltration of precipitation over the landscape. Diffuse recharge, in contrast to local recharge, has not been channelized into depressions, impoundments, fields, or drainages. To soil scientists, diffuse recharge is soil water lost as it exits the lower part of the soil profile below the root zone, but to a hydrogeologist, diffuse recharge represents water gained by the aquifer across a free surface on the top of the water table. The two views are identical.

In humid climates, most storms have potential to induce recharge, owing to a number of factors, such as long storm duration, relatively high storm frequency, low precipitation intensity, low potential evapotranspiration, high antecedent soil moisture, and shallow water table depth, among others. In most humid climates, recharge could occur throughout most of the year, although in very cold regions, frozen ground could prevent significant infiltration for several months.

In arid and semi-arid climates, precipitation generally occurs in two forms, intense summer thunderstorms and frontal showers in the winter and/or spring. However, the relative proportion of summer and winter precipitation will vary from one desert region to another (MacMahon, 1985). At least within the hydrogeological and civil engineering disciplines, many believe that virtually all of this precipitation is taken up by plants with extensive root systems, because over the year, potential evapotranspiration exceeds precipitation. Regardless of the precipitation and evapo-transpiration rates, there is the potential for recharge where the infiltration or redis-tributed water penetrates below the plant root zone. Winter precipitation, especially in a series of storms, is most likely to contribute to deep soil-water movement and recharge. During the winter storm periods when evapotranspiration is at a minimum, a soil-water budget over the wet period is likely to indicate that infiltrated water would penetrate below the root zone and potentially could become recharge.

A few studies illustrate that annual mean precipitation (P) and potential evapo-transpiration or gross lake evaporation (E) are not adequate indicators of whether recharge occurs in areas of low annual precipitation. At Los Alamos, NM, where P is about 45 cm and $P/E = 0.4$, Hakonson et al. (1992) found that deep percolation can occur during snowmelt, primarily because the antecedent soil moisture content is high when evapotranspiration is low. In a field study of soil-water movement in a poorly vegetated sandy site near Socorro, NM, where mean annual precipitation is about 20 cm and $P/E = 0.11$, Stephens and Knowlton (1986) used tensiometers and neutron probe measurements of moisture content to show that soil-water movement occurred to depths of more than about 2.5 m in response to both winter and summer precipitation (Figure 3). Nichols (1987) found pulses of soil moisture propagating to about 2 m on an alluvial fan near Beatty, NV ($P/E = 0.04$). He concluded that there was potential for deep percolation because the late March precipitation exceeded the seasonal evaporation demand. These and other examples in Chapter 3 are only some of the quantitative studies of diffuse recharge in areas of low precipitation that suggest that the recharge potential is best evaluated from a soil-water balance analysis conducted over only a period of several weeks to perhaps a few months, rather than on the basis of an annual average water balance.

Most of the detailed recharge investigations have not addressed the variability of recharge over periods of decades, which is the time scale of interest for water-supply evaluations. Research is needed to determine whether significant recharge events could be correlated to the recurrence interval of extreme precipitation events. There is also little research on variability in recharge over periods of thousands of years that are relevant to predicting performance of radioactive waste repositories in desert regions. Some existing studies do indicate that long-term recharge and climate variability are well correlated. For example, Phillips (1993) used the chloride mass balance method to show that the soil profile has preserved information that reflects significant changes in diffuse recharge in the Southwest since pluvial times. Calf (1978) investigated groundwater in Alice Springs, Australia (P = 25 cm), using carbon-14 and oxygen and deuterium isotopes to conclude that recharge to the sandstone aquifer system there occurred at intervals of a few millennia. Calf further

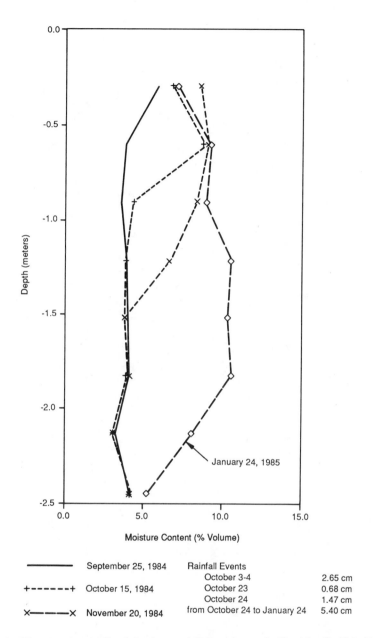

Figure 3   Water-content profiles following precipitation at a sandy site at the Sevilleta National Wildlife Refuge north of Socorro, NM. Depth to groundwater is about 5 m. (From Stephens et al., 1986. With permission.)

speculated that the increased recharge from these geologic times of more humid climate may be worldwide.

Table 1 summarizes results of some diffuse recharge studies in arid and semi-arid areas. What is most surprising from data in this table is the remarkably small variability in recharge, especially among sites in the United States, considering the

Table 1   Compilation of Diffuse Recharge Analyses in Areas of Low Precipitation

Site	Mean annual precipitation (mm/year)	Potential evapotranspiration (mm/year)	Soil	Recharge (mm/year)	Ref.
1. Socorro, NM	190	1780	Sandy alluvium	4.0–9.0	Stephens et al. (1991)
2. Socorro, NM	190	1780	Sandy alluvium	7.0–37	Stephens and Knowlton (1986)
3. Socorro, NM	190	1780	Sandy loam	2–2.6	Phillips et al. (1988)
4. Sunland Park, NM	200	1780	Medium sandy alluvium	0.05–1.9	DBS&A (1994)
5. Las Cruces, NM	230	1780	Sandy loam	1.5–9.5	Phillips et al. (1988)
6. Hudspeth County, TX	280	1960	Gravel/clay loam	0.01–1	Scanlon et al. (1991)
7. Curry County, NM	444	1156	Clayey and very fine sand (playa)	2.8	Stone (1986)
8. Curry County, NM	444	1156	Fine to very fine sand (pasture)	0.2	Stone (1986)
9. Curry County, NM	444	1156	Fine sand and caliche (dunes)	1.2	Stone (1986)
10. Hanford, WA	160	1400	Loamy sand, sand	0–100	Gee et al. (1989)
11. Beatty, NV	74	1900	Sand and gravel	0.036	Nichols (1987)
12. South Australia	300	No report	Dune sand	14	Allison et al. (1985)
13. Saudi Arabia	70	2400	Dune sand	20	Dincer et al. (1974)
14. Eastern Botswana	447	1220	Sand, sandstone	0.5–6	Carlsson et al. (1989)
15. Southern Cyprus	390	1450	Fine sand	10–94	Kitching et al. (1980)
16. Central Sudan	225	No report	Sandstone	0.2–1.3	Darling et al. (1992)
17. Northeast Sudan	220	No report	Sandstone	<1	Darling et al. (1992)

From Stephens, 1994. With permission.

different methods that were used and, more importantly, the significant variability in geologic setting, soil texture, geographic area, and vegetation. For instance, at four sites (1 through 9 in Table 1) within about a 180-km radius in southern New Mexico and western Texas, natural recharge calculated at the various sites ranged from only about 0.01 to 37 mm/year. In light of the diversity of soil hydraulic properties and vegetation patterns among sites, one possible explanation for the rather small variability is that most diffuse recharge is controlled not so much by the soil properties as by slow infiltration flux rates (e.g., Chapter 3, Figure 1A), during times when there is no ponding or significant plant uptake, such as during winter months.

The nature of variability of diffuse recharge within a site is also important to recognize. At the Sunland Park site (Table 1) in New Mexico, the chloride profiles in alluvium used to calculate recharge were remarkably similar (Figure 4) in three borings located within about 200-m of each other. Significantly greater variability may occur at other sites due to preferential flow and evapotranspiration. Sharma and Craig (1989) showed, for three sites on coastal dunes, consistent chloride profiles within sites of a particular land use, but they identified significant differences in chloride profiles from different land use groups. Among four profiles within a 100-m radius, Allison, et al. (1985) found more than a 10-fold difference in chloride concentration, and hence inferred recharge in soils beneath crop and pastureland receiving about 300 mm/year precipitation. As suggested by Simmers (1989), the degree of spatial variability of recharge is difficult to predict without field measurements, and perhaps an approach to evaluate recharge variability over a watershed may combine field measurements, remote sensing, and geostatistical analysis.

Data to compute recharge for most of the sites in Table 1 were gathered from the upper several meters of land surface, so we can only infer that this flux actually becomes recharge. Some soil water may be extracted by plant roots from deeper parts of the vadose zone. Also, in the relatively shallow part of a desert soil environment where many of the data were collected, water vapor movement due to temperature gradients near land surface can be important in the interpretation of field data to quantify recharge (Knowlton et al., 1989).

To estimate diffuse recharge on a site-specific basis, two or more methods are recommended to address uncertainty. Chemical tracer methods applied to soil water and groundwater are preferred for long-term indicators of net recharge. It is quite possible that at some sites in semi-arid or arid climates, significant recharge only occurs as a result of a few years of relatively high precipitation. Consequently, estimates of recharge calculated from point-in-time measurements may not represent the most significant recharge events, because these significant events may have return periods of tens or hundreds of years.

II. LOCAL RECHARGE

When we consider recharge mechanisms in areas of low precipitation, one of the first to come to mind is infiltration of runoff in ephemeral channels, arroyos, or wadis. The source of infiltration, ponded water, is relatively easy to measure by

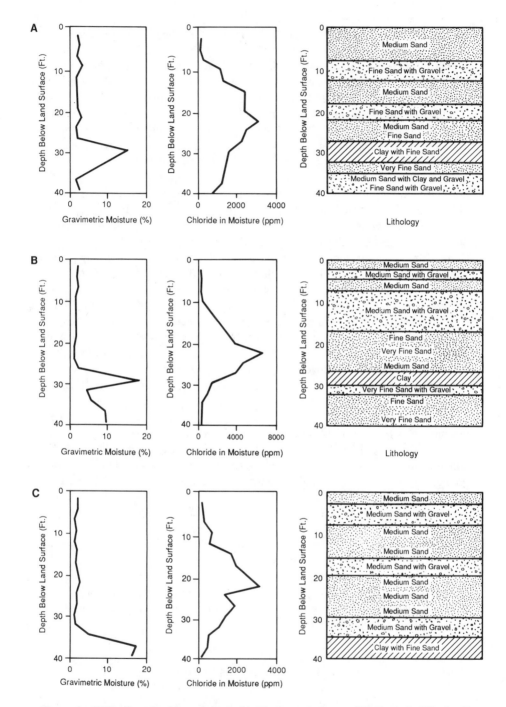

Figure 4   (A) Profiles of moisture content, chloride concentration, and lithology at soil boring B-8, Nu-Mex Landfill, (B) at soil boring B-9, and (C) at soil boring B-10. (From Daniel B. Stephens and Associates, Inc., 1992. With permission.)

gaging, and evidence for deep percolation is readily detected by vadose zone monitoring equipment. The same is generally true for culturally modified recharge such as spreading basins and irrigated fields. We also include in our discussion of local recharge the seepage from water retained in natural depressions, as well as the influence of topography on the spatial variability of recharge, although the latter is more a perturbation of diffuse recharge. Other distinctions between diffuse and local recharge are that (1) local recharge occurs within a more limited part of the watershed, whereas diffuse recharge is likely to occur within most of the watershed; (2) local recharge is more likely to be a two- or three-dimensional process, except in large irrigation projects, whereas diffuse recharge is usually one-dimensional on a large scale; and (3) significant local recharge to a deep water table is more likely to occur seasonally, whereas diffuse recharge to a deep water table is more likely to be slow and distributed more uniformly over long time periods. These general criteria are only intended to be used for gaining a better understanding of recharge mechanisms. However, the physical processes influencing diffuse recharge are also applicable to local recharge.

A. CHANNEL INFILTRATION

In humid climates, most groundwater recharge probably occurs by diffuse percolation of precipitation. This is because the frequent, regular precipitation maintains a high antecedent water content in the soil so that there is little additional water storage capacity; thus, infiltration can be routed quickly through the vadose zone to the phreatic zone. This recharge raises the water table, which leads to increased stream flow. That is, in humid climates flowing perennial streams are typically groundwater discharge areas sustained by diffuse recharge in the basin.

On the other hand, in areas of low precipitation the water table is usually well below the base of the channel; consequently, infiltration of runoff in channels is often the largest source of recharge (Bouwer, 1989). The magnitude of the infiltration depends upon a variety of factors such as vadose-zone hydraulic properties, available storage volume in the vadose zone, channel geometry and wetted perimeter, flow duration and depth, antecedent soil moisture, clogging layers on the channel bottom, and water temperature. Many of these factors are discussed extensively in civil engineering and hydrology textbooks, and we refer you to them for more detail (e.g., Linsley et al., 1982; Bouwer, 1978). This section highlights the important physical processes relevant to natural recharge due to localized infiltration along drainage channels in areas of low precipitation. Many of the examples are from Arizona in the Sonoran Desert and New Mexico in the Chihuahuan Desert.

The Tucson Basin in Arizona is one area where there have been excellent hydrologic investigations to quantify the importance of recharge from ephemeral channels. The basin has an area of about 2600 km^2 and is rimmed by mountains attaining elevations of more than 2740 m. Relatively permeable alluvium from the basin margin fans and ephemeral streams fills the basin. The ephemeral stream channels typically have large width-depth ratios. Burkham (1970) used streamflow gaging measurements to derive an empirical relationship between infiltration rate from the channel and inflow rate to a reach of the channel. He found for the seven major channels that drain the basin that on average about 70% of the inflow to the channels was lost by infiltration. Although the channels flow only about 10 to 20% of the time, infiltration from the channels represented about 43% of the total inflows to the aquifer (Davidson, 1973). Lesser amounts of recharge to the aquifer were contributed by recharge along the mountain fronts, underflow from adjacent basins, and return flows from irrigation and domestic uses.

Renard (1970) presented an excellent analysis of the hydrology of a desert rangeland, the 150-km^2 Walnut Gulch watershed near Tombstone, AZ, southwest of Tucson. The Walnut Gulch watershed is covered mostly with desert shrubs, native grasses, and shrubs. Channels are underlain by thick, permeable sand deposits, and the depth to the regional water table is 45 m. Main channel widths range from about 11 to 65 m at bank full stage. During runoff, water levels in wells near the channel increased as much as 4.5 m in response to runoff, but outside the main channel areas where water may accumulate (as in grassy swales), no downward migrating pulses of soil moisture were noted below the 150 cm depth. The study concludes that channel recharge is the principal mechanism for recharging groundwater. Transmission losses in the watershed (groundwater recharge plus evapotranspiration) were about 4.4 cm of the 30 cm average annual precipitation.

Hydrogeological investigations utilizing isotope geochemistry have also demonstrated the importance of channel recharge. Clark et al. (1987) studied the groundwater in the Najd Region in southern Oman (mean annual precipitation, $P < 100$ mm/year), where high tritium activity and stable isotopes suggested that the shallow phreatic aquifer was recharged by rapid infiltration in the wadis following infrequent storms. In the Nubian sandstone aquifer in Sudan ($P \approx 220$ mm/year), Darling et al. (1992) used stable isotopes and radioisotopes in soil and groundwater to demonstrate that there is no significant recharge to this deep aquifer outside the Nile Valley. Shallow perched aquifers beneath wadis were recharged, but infiltration from the perched aquifers did not significantly recharge the underlying Nubian sandstone.

The process by which channel recharge occurs is most relevant to our analysis and interpretation of hydrogeologic information. There are very few field investigations that detail the pathways of water migration from ephemeral stream channels or from canals to the water table. However, the problem has been addressed by mathematical modeling by Riesenauer (1963), who used a variably saturated finite-difference model to study seepage from an unlined irrigation canal (Figure 5). The most interesting feature of his simulation is the steady-state moisture-content and pressure-head distributions, which reveal incomplete saturation through the vadose zone beneath the edge of the channel even at steady state. Owing to the relatively great depth to the water table and to capillary effects, we infer from this model study that no groundwater mound would rise through the vadose zone to intersect the channel. This would be true at all times, even if the flow duration is sufficiently long for the vadose zone to reach steady-state moisture distribution, as long as the aquifer can transmit the recharge away from the area. For relatively deep water-table conditions, saturated zones do occur beneath the channel, but only to a limited depth. The base of the saturated zone beneath the channel would be regarded as an inverted water table. Unsaturated flow would occur between the inverted water table (0.33

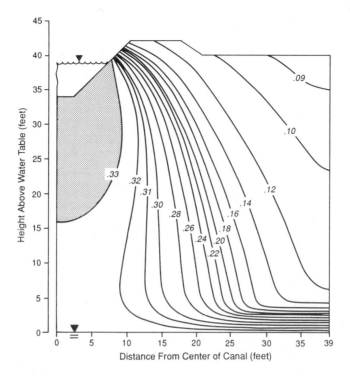

Figure 5 Steady-state moisture content pattern below a 15-ft-wide canal in homogeneous soil. The shaded area indicates where the water content equals the porosity and the soil is fully saturated. (Modified from Riesenauer, 1963. With permission.)

cm^3/cm^3 contour in Figure 5) and the regional water table. Where the water table is relatively shallow, however, there may be complete saturation between the channel and the regional water table.

Transient numerical simulations by Peterson and Wilson (1988) clearly demonstrate the importance of recognizing unsaturated flow when predicting the increase in recharge from stream infiltration that occurs when water tables are lowered by groundwater pumping, especially where a relatively low permeable clogging layer is present on the channel bottom. If the free surface on a groundwater mound rises from the shallow regional water table to intercept the water level in the channel, the stream-aquifer system is hydraulically connected (Figure 6). On the other hand, if there is unsaturated media between the channel and the regional water table, then the system may be hydraulically disconnected. Peterson and Wilson's (1988) simulations showed that even when the unsaturated condition was present, the stream and aquifer may in fact be connected, in the sense that further lowering of the regional water table could increase channel losses. There is some critical depth to the water table, however, where further lowering has no influence on channel losses. At this depth, which depends mostly on soil properties and head in the channel, the aquifer becomes hydraulically disconnected from the stream.

A relevant field investigation of ephemeral stream infiltration was conducted on the Rio Salado north of Socorro, NM (Stephens et al., 1988). The stream channel is

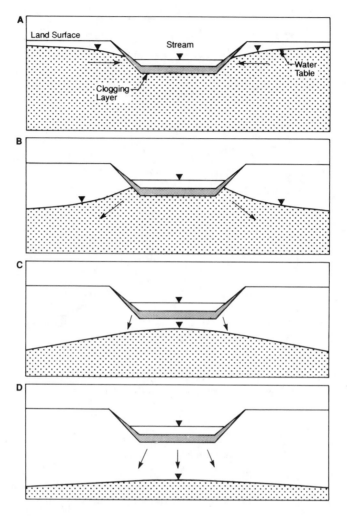

Figure 6   Stream-aquifer relationships for the case of a clogged streambed: (A) connected gaining stream, (B) connected losing stream, (C) disconnected stream with a shallow water table, and (D) disconnected stream with a deep water table. (From Peterson and Wilson, 1988. With permission.)

about 200 m wide, about 1 m deep, and is braided. Mean annual discharge is about 0.4 m³/s. Unconsolidated sand, gravel, and cobbles are the predominant materials beneath the channel. The stream flows for about 1 month each year beginning in late July, depending upon the summer monsoon season. Neutron probe access tubes to measure soil-water content and groundwater monitor wells were installed in the channel and on the banks of the channel. The depth to the water table is about 9 m below the base of the channel. During a period of flow in July 1986, approximately 30% of the channel conveyed water in localized rivulets. The water-table mound height increased about 1 m above the initial position of the regional water table (Figure 7). The moisture content in the neutron probe located about 2 m from the

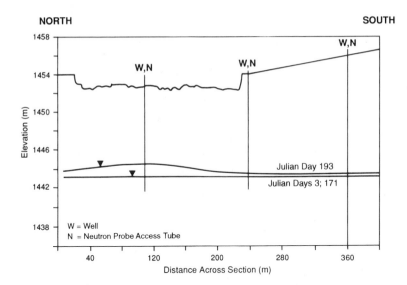

Figure 7 Groundwater mound developed at the lower Rio Salado site near Socorro, NM, during the discharge events of July 1986. (From Stephens et al., 1988. With permission.)

edge of a rivulet in the center of the channel increased to only 17% by volume during runoff, or about 50% of saturation. These field data suggest that the conceptual model of flow unsaturated beneath irrigation canals predicted by Riesenauer (1963) is applicable to this site as well.

Another ephemeral stream, the Rio Puerco located about 10 miles north of the Rio Salado, was also instrumented to study seepage pathways through the vadose zone (Stephens et al., 1988). The Rio Puerco main channel at the site is about 10 m wide and 2 m deep. In contrast to the Rio Salado, the sediments beneath the channel are finer, mostly silt, clay, and fine to coarse sand. The Rio Puerco flows periodically throughout the year for a total of approximately 3 months and has a mean annual discharge of 1.3 m^3/s. The site was instrumented with neutron probe access tubes and monitor wells adjacent to the channel. One of the neutron probe access tubes was drilled at an angle from the bank to measure water content within about 30 cm of the channel base. The depth to the water table is about 1 m below the channel during dry periods.

Figure 8 illustrates the stream stage at this site and water-table response in a monitor well adjacent to the stream. In general, it appears that the groundwater levels fluctuate in phase with the stream stage. However, carefully examine the three labeled periods of stream flow and water table response in spring 1986. The third flow event was separated from the previous event by about 10 days of no flow. The peak stream stage increased from the first to the third of these events, but at the end of the third flow event the peak water level in the monitor well was lower in elevation than at the beginning of the flow sequence. Consequently, it appears that the water table is generally receding when stream stage peaks generally are rising. The neutron logging beneath the channel showed the moisture content was about 37% for the first two events and this decreased to 34% during the third event.

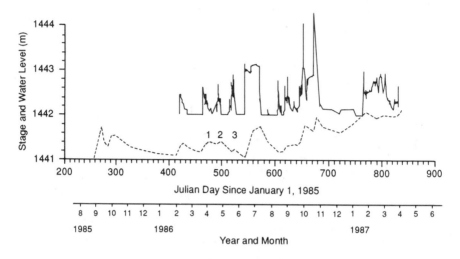

Figure 8   Stream stage and water-level hydrograph beneath the channel at the Rio Puerco site near Socorro, NM. When the stream stage is at 1442 m, there is essentially no stream flow. (From Stephens et al., 1988. With permission.)

Based on these data, the high suspended sediment load, and measurements of relatively low hydraulic conductivity of the channel bottom sediments at the Rio Puerco, Stephens et al. (1988) suggested that the change in groundwater and soil-water conditions was influenced by a clogging layer that may have developed on the channel bottom during the recession of the second runoff event. For these three stream flow events on the Rio Puerco, the stream and aquifer appeared to be hydraulically disconnected, or nearly so. Complete saturation below the channel was achieved only when the stream stage exceeded about 1443 m after about day 550. A vertical cross section shows the moisture content distributions when the stream and aquifer were connected and when they were not connected by a saturated zone (Figure 9). When a clogging layer develops, the flow to the underlying and more permeable vadose zone is restricted so that only part of the pore space needs to be filled to convey the seepage that is allowed through the clogging layer.

From this case study, we conclude that the channel recharge process along a particular reach can be highly variable in both time and space due to the transient development and subsequent erosion of clogging layers. Recharge occurred as unsaturated flow until runoff was sustained for periods of at least a few weeks, when saturated conditions developed beneath the channel.

In spite of the water-table rise due to recharge in the Rio Puerco case study, the shallow depth to groundwater limits the available storage; consequently, the amount of channel recharge would appear to be small relative to diffuse recharge over the basin. To evaluate this, we made rough calculations of the relative magnitudes of channel recharge and diffuse recharge for the Rio Puerco basin. The average annual channel recharge per unit channel length for the Rio Puerco is about 30,000 m³/year-km (Stephens et al., 1988). We roughly estimate the length of the Rio Puerco and its main tributaries at about 1000 km, so channel recharge would be about 3×10^7 m³/year. If we estimate that diffuse recharge is between 1 and 10 mm/year (Table 1),

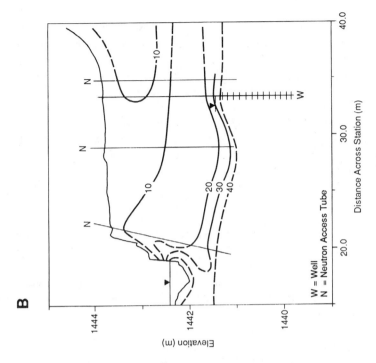

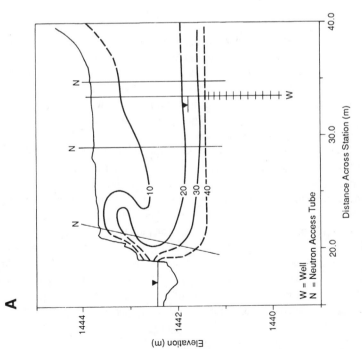

Figure 9 Vertical cross-section at the Rio Puerco site north of Socorro, NM, showing moisture content distribution when below the channel conditions where (A) saturated and (B) unsaturated.

then diffuse recharge for the 18,816-km^2 basin would be 1.9×10^7 to 1.9×10^8 m^3/year. Thus, for this basin it appears that the two recharge processes are approximately of similar magnitude over the long term.

The Rio Puerco case study indicates that, at least in parts of New Mexico and similar sites, stream channel infiltration is a significant source of recharge and the recharge rate is variable in time, owing to the complex relationships between stream flow characteristics and clogging layers. Transient effects of recharge from stream infiltration are also undoubtedly induced by water temperature and viscosity effects. Water temperature differences typical of winter and summer runoff could cause variation in the infiltration rate of roughly a factor of two.

B. DEPRESSION-FOCUSED RECHARGE

Depression-focused recharge occurs where runoff, interflow, or lateral flow of subsurface moisture accumulates within or beneath depressions on the landscape. This type of recharge was recognized on the Canadian prairies where water-table undulations were correlated with potholes and hummocky topography (e.g., Meyboom, 1967). This prairie terrain is underlain mostly by low-permeable glacial till, which enhanced runoff and ponding in the lowlands, where recharge occurred to create local mounds in the groundwater table. Numerical simulations of seepage from circular ponds above the water table illustrate the nature of fluid saturation from the lowland (Riesenauer, 1963). The features of this axisymmetric flow field in the vadose zone are very similar to the two-dimensional (vertical cross section) flow field described earlier in regard to stream channel recharge (Figure 9). Hence, we expect that recharge from the prairie potholes occurs via unsaturated flow through the vadose zone.

Playa lakes are another potential source of depression-focused natural recharge in areas of low precipitation. Playas are nearly flat, relatively barren areas in the lowest topographic parts of undrained desert basins. These areas infrequently hold water derived from rain, snowmelt runoff, and groundwater discharge (e.g., Eakin et al., 1976). Closed desert basins are defined with respect to the surface drainage. Perhaps several hundreds of topographically closed basins occur throughout the basin and range physiographic province in the western and southwestern United States. The hydrogeology of a closed desert basin consists of recharge where fractured and faulted bedrock are exposed in the uplands and where mountain-front streams lose their water along the alluvial fans. In this closed desert basin scenario, the groundwater table intersects the playa lake, and the groundwater discharge is consumed by evapotranspiration near the lake (Figure 10). This conceptual model led Brookings and Thomson (1992) to suggest that playas would be excellent waste disposal sites, because they believed upward-flowing groundwater would prevent subsurface migration of the waste. However, two-dimensional mathematical analyses by Duffy and Al-Hassan (1988) indicated that downward flow of groundwater could occur beneath playa lakes due to convection cells created by brine density effects between the surface-water and groundwater system (Figure 11). Based on their analysis, direct precipitation or surface runoff to these closed basin playa lakes could become recharge to groundwater, if the lake water salinity is sufficiently increased by evaporation. Examples of closed basin playas fed by groundwater

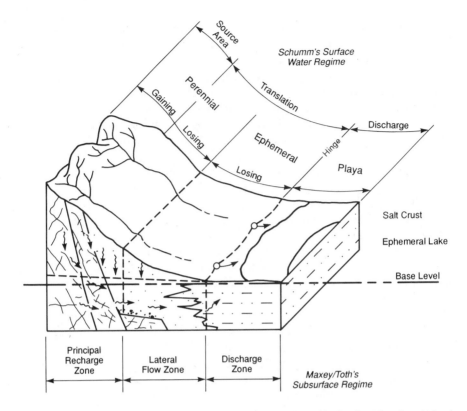

Figure 10   Conceptual drawing of the hydrology of a topographically closed and undrained desert basin. (From Duffy and Al-Hassan, 1988. With permission.)

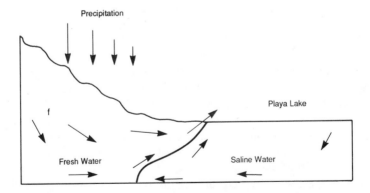

Figure 11   A two-dimensional conceptual model of the hydrogeology of a closed basin, showing natural salinity-induced density effects on groundwater circulation. (Modified from McCleary-Hanagan and Duffy, 1988. With permission.)

discharge are the Devils Golf Course in Death Valley, CA, and Lake Estancia east of Albuquerque, NM, among others.

Most playas, however, are not fed directly by groundwater, inasmuch as the depth to the water table is often tens or hundreds of feet below land surface. Consequently, few playas are actually closed with respect to groundwater flow. Most playas are usually dry throughout most of the year, except when direct precipitation and surface runoff pond on the low permeable playa floor. Eakin et al. (1976) referred to these as dry playas. When water is ponded on the playa, much of the water evaporates, but there is ample opportunity for infiltration to occur and recharge the water table. The amount of recharge would be controlled mostly by the duration of playa ponding, surface area of the pond, desiccation fractures in the clay, and permeability of the playa bottom sediments. For example, in Kuwait (P ≈ 127 mm/year), Bergstrom and Aten (1964) studied wadis (i.e., arroyos, ephemeral streams) and playas that fill with surface water following infrequent torrential rainstorms. They found fresh groundwater localized along the wadis and playa lake basins. However, where the wadi channel was steep or the water table was more than about 30 m deep, no fresh water was present. They also observed that the thickest lenses of fresh water occurred beneath the playas, even though much of the playa areas is underlain by clay pans and cemented subsoils, which retard infiltration. Bergstrom and Aten (1964) thought that although recharge apparently occurs within the playa and along the playa edges, most of the recharge to the regional aquifer system is from the wadis.

In summary, playa lakes represent potential sources of depression-focused recharge. This could occur where dry playa lakes are fed by runoff that ponds and then seeps toward the underlying water table. In closed-basin playa lakes fed by groundwater as well as runoff, groundwater could be recharged by density driven processes, although such recharge would probably be unsuitable for drinking. In either case, playas would represent a poor choice for siting a waste disposal facility.

Multidimensional flow of water within the vadose zone that originates as precipitation over wide areas may produce variability in the spatial distribution of natural recharge. If the recharge distribution is sufficiently focused, localized water-table mounds may be observed, giving the impression that the recharge is from a local source of ponded water. For example, such undulations in the water table were encountered beneath the Sand Hills of Nebraska (Winter, 1986). In this area of dunes and interspersed lakes, water table mounds occurred beneath portions of dunes that contain hummocky topography. At this site there was no support for the common assumption that the water table is a subdued reflection of the topography. McCord and Stephens (1987) raised the issue of how water could accumulate in the interdunal swales to create the underlying groundwater mounds. In contrast to the Canadian prairie discussed earlier in this section, the dunes are expected to be so highly permeable that no surface runoff would be likely to occur in depressions. Also, active dunes are typically uniform in texture and contain few low-permeable soil horizons above which interflow could be anticipated. Therefore, there must be some other mechanism to cause water to accumulate in the lowlands where recharge is focused. Field experiments (McCord and Stephens, 1987) and numerical modeling (McCord et al., 1991), based in part on theoretical work by Zaslavsky and Sinai (1981) and Yeh and Gelhar (1983), demonstrated that anisotropy in unsaturated hydraulic conductivity, discussed in Chapter 3, can cause lateral flow components in material as uniform as sand dunes and therefore can lead to the accumulation of soil moisture in topographic low areas (e.g., Figure 14). This mechanism may play a significant role in controlling the spatial distribution of recharge in hummocky terrain. The importance of state-dependent anisotropy may also explain, at least in part, how recharge may be enhanced where aquifers dip basinward from an outcrop, as we describe in a later section.

C. CULTURALLY MODIFIED RECHARGE

The human desire to deliver water to areas of scarce precipitation has had profound consequences on the amount and distribution of recharge through the vadose zone. In some areas, agricultural, industrial, municipal, and domestic water use and wastewater disposal practices have led to increased recharge far above what would otherwise have occurred under natural conditions. This culturally modified recharge often carries with it undesirable chemicals, which, in some cases, leads to groundwater contamination.

Culturally modified recharge from agricultural sources includes leakage from unlined canals and laterals that convey water to irrigated fields. It also includes the amount of irrigation water applied to the fields in excess of the crop requirements, which eventually percolates through the vadose zone to recharge the shallow aquifer. When an irrigated area is first developed, the water table usually rises, sometimes to sufficiently shallow depths where soils become waterlogged and crop production is reduced due to salt accumulation and lack of aeration to the roots. To alleviate this problem, extensive drain systems must be installed to control the position of the water table. A very good example is in the Mexicali Valley of the Colorado River delta in Mexico, Arizona, and California. In many flood-irrigated fields, it would not be unreasonable to expect that half or more of the water diverted from a stream or pumped for irrigation eventually recharges the aquifer. The amount of recharge would depend upon the field geometry, frequency and duration of irrigation, crop evapotranspiration, and soil hydraulic characteristics. Quantifying the actual irrigation water consumptive use and irrigation return flows can be the center issue in water rights investigations. Excessive recharge not only can lead to drainage problems, but it also could convey agrochemicals to the water table, along with the salt accumulations that are leached from the shallow soils. For example, salt leached by irrigation in western Arizona contributes so much to the salinity of the Colorado River that a desalinization plant has been constructed near Yuma, AZ, to meet U.S. treaty obligations to Mexico for water quality.

Anthropogenic induced changes in vegetation type may produce important modifications in recharge. An excellent study of this nature was conducted in a recharge area known as the Gnangara Mound on the Swan Coastal Plain near Perth, Western Australia (Sharma et al., 1988). The investigation to select the optimal vegetation was undertaken as part of a long-term regional water management plan. Recharge was computed by the water-balance method, a bromide tracer method, chloride mass balance method, and by a numerical model of the Richards equation. Three different vegetation types were investigated: native Banksia woodland, planted pine forest, and typical pasture. Each site contained neutron probe access tubes for periodically monitoring water content profiles. The results showed that, compared to beneath native Banksia woodland, the average recharge rate was much higher (two to four times) beneath a pasture and much lower (less than half) beneath a mature pine plantation. The study also found that recharge decreased with increasing vegetation density and age of pine forest.

At the regional scale, a dramatic example of the importance of vegetation on recharge is the influence of clear-cutting forests to create farmland in Australia. The loss of the forest reduced evapotranspiration and increased recharge to such an extent that water tables rose to near land surface in the lower areas of the catchments, causing salt accumulations in the soil and, ironically, causing loss of arable land, as well as degradation of both surface-water and groundwater quality (Peck and Williamson, 1987).

Industrial facilities often require significant quantities of water for processing, cleaning, or cooling. Clean water stored in unlined or leaky reservoirs and tanks, as well as chemical-laden liquids from waste outfall drains, and unlined sumps, pits, ponds, or lagoons, may seep into the vadose zone and potentially could recharge the aquifer. Considerable efforts are now made by both industries and regulators to prevent such recharge from occurring, owing to the high potential for adverse impacts. In too many cases in the past, groundwater contamination occurred because of misconceptions about the potential for recharge by seepage. Developers or their consultants often mistakenly believed seepage from impoundments would be virtually nil, citing reasons such as the impoundment is self-sealing because impermeable clay particles settle out, or the potential evapotranspiration exceeds mean annual precipitation, or the vadose zone is more than 100 feet thick so any seepage would be absorbed. Careful review of available hydrogeologic data, including on-site vadose zone measurements, would show that such claims are rarely supportable.

A large fraction of the municipal water diversions is available for recharge when the effluent from wastewater treatment systems is discharged to arroyos, playas, losing streams, or surface impoundments situated above the water table. Stormwater runoff in unlined conveyances as well as in storm sewers also may recharge aquifers. Urbanization typically leads to paving and building construction over extensive areas, which limits recharge potential beneath these low-permeability areas. However, runoff from urban areas may be routed to unlined conveyance channels where water, because it is spatially concentrated, has a high potential for infiltration and recharge. In some cities, such as Phoenix, AZ, artificial recharge of storm runoff occurs in retention basins constructed along the ephemeral Salt River and its flood plain. Many cities have ordinances that require developers to control storm runoff from their properties by directing water to retention ponds or to dry wells (i.e., vertical french drains completed above the water table), because infiltration from these facilities has the beneficial effect of enhancing groundwater recharge. Other municipal recharge could occur by leakage from pipes conveying fresh water or waste water. Septic tank seepage and domestic water applied to lawns and gardens in excess of plant water requirements also comprise potential sources of culturally modified recharge.

The same physical processes control both natural and culturally modified recharge. Consequently, methods to quantify and monitor natural recharge are usually applicable to culturally modified recharge as well.

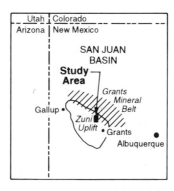

Figure 12   General location map of the recharge site near Prewitt, NM, in the southern San Juan Basin. (From Stephens, 1983. With permission.)

III. CASE STUDY CONSIDERING COMBINED RECHARGE MECHANISMS

The Westwater Canyon sandstone member of the Morrison formation is an important aquifer and uranium source in the southern San Juan Basin in New Mexico and Arizona. The formation dips north at about 3 to 5 degrees and outcrops on the steep slopes of prominent south-facing mesas between Grants and Gallup in New Mexico (Figures 12, 13, and 14). The Westwater Canyon sandstone member is overlain by low-permeable shale in the Brushy Basin member of the Morrison formation (Table 2). These formations are exposed on the walls of mesas, which are capped by the well-jointed Dakota sandstone. Stratigraphically overlying the Dakota is the low permeable Mancos shale, which forms broad east-west trending valleys north of the Dakota outcrop. Numerous dry lakes, such as Casamero Lake and Ambrosia Lake, occur in the valley of the Mancos shale (Figure 14).

Stephens (1983) described groundwater conditions and sources of recharge along Casamero Draw north of Prewitt, NM, which generally seem to be representative of conditions along many sections in this part of the basin. The mesas and sedimentary rock units are incised by south-flowing ephemeral streams in alluvial valleys. Along Casamero Draw, the alluvium is typically less than 33 m thick and is saturated only in about the lower 8 m. The saturated alluvial aquifer is in contact with the Westwater Canyon aquifer in the subsurface, north of the mesa escarpment, and the direction of shallow groundwater flow is south (Figure 15).

An interesting problem in this area is identifying the sources of groundwater recharge to the Westwater Canyon aquifer. The fact that the mean annual precipitation is about 28 cm and lake evaporation is about 114 cm (P/E = 0.24) says little about recharge processes or quantities, as discussed previously in this chapter. The northward slope of the potentiometric surface in the Westwater Canyon aquifer (Figure 16), mapped from deep wells in the southern part of the basin, suggests that recharge does occur in the outcrop area. The northerly flow direction in the Westwater Canyon aquifer generally coincides with the dip direction, which is consistent with a conceptual model whereby recharge occurs due to infiltration on the outcrop. However, the

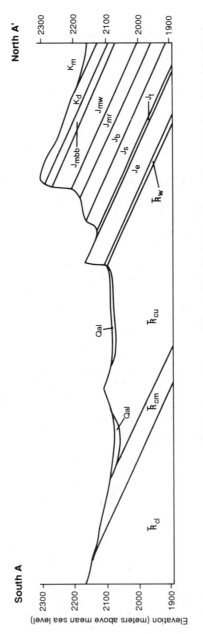

Figure 13   Geologic cross section along Casamero Draw near Prewitt, NM. Westwater Canyon Sandstone Member of the Morrison formation (J_{mw}) outcrops in steep cliffs and is overlain by low-permeable Brushy Basin Mudstone Member (J_{mbb}), Dakota Sandstone (K_d) and Mancos Shale (K_m). Other units include the lower, middle, and upper Chinle Formation (TR_{cl}, TR_{cm}, TR_{cu}), Wingate Sandstone (TR_w), Entrada Sandstone (J_e), Todilto Limestone (J_t), Summerville Formation (J_s), Bluff Sandstone (J_b), Recapture Member of the Morrison Formation (J_{mr}), and Quaternary alluvium (Qal). Section line is shown in Figure 14. (From Stephens, 1983. With permission.)

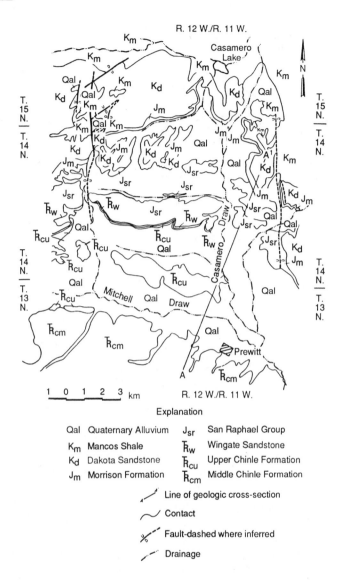

Figure 14   Geologic map near Prewitt, NM. Geology by Cooper and John, 1968. (From Stephens, 1983. With permission.)

outcrops form cliffs inclined more than about 70 degrees from horizontal in places (Figure 13). There would be little contact time or surface area for precipitation to infiltrate into the formation. That precipitation that does infiltrate undoubtedly would have some tendency to migrate laterally, due to the high anisotropy in unsaturated hydraulic conductivity created by the northerly dipping stratification. Because the horizontal distance which water would have to flow in an unsaturated state from the outcrop to the water table within the Westwater is more than about 2 km, it is unlikely that infiltration on the outcrop is a significant source of recharge.

Table 2   General Hydrogeologic Data from Formations in the Southern San Juan Basin

Unit	Lithology	Thickness (m)	Hydraulic conductivity (m s^{-1})
Alluvium	Sand, silt, and clay	30	7.0×10^{-6}
Mancos Shale	Calcareous, shale with minor sandstone layers	76–335	1.0×10^{-10}
Dakota Sandstone	Massive sandstone; some coal, shale, and conglomerate	24–43	1.0×10^{-7}
Morrison Formation			
Brushy Basin	Bentonitic mudstone, minor sandstone	24–37	Unknown
Westwater Canyon	Sandstone, minor siltstone	46–61	2.5×10^{-6} 7.8×10^{-6}
Recapture	Claystone, siltstone, and sandstone	34–43	Unknown
San Rafael Group			
Bluff	Massive sandstone	30–46	1.8×10^{-6}
Summerville Formation	Siltstone and shale	29–43	Unknown
Todilto Limestone	Limestone	6–9	Unknown
Entrada Sandstone	Massive sandstone; lower siltstone, and silty sandstone	53–61	1.1×10^{-6} 3.9×10^{-6}
Wingate	Massive sandstone	11–14	3.5×10^{-7}
Chinle			
Upper	Siltstone, mudstone with minor sandstone	274–305	Unknown
Middle	Sandstone with siltstone and mudstone (includes Sonelsa sandstone)	18–69	2.5×10^{-7}
Lower	Siltstone and mudstone	91–122	Unknown

From Stephens, 1983. With permission.

Another potential source of diffuse recharge is rain and snowmelt infiltration through the Dakota sandstone that overlies the Morrison formation. The Dakota outcrop forms dipslopes that cover extensive areas, at least relative to the small outcrop area of the Westwater Canyon sandstone (Figures 14 and 15). On the mesa tops, preferential flow through the joints and fractures in the Dakota would route water below the piñon and juniper root zones toward the Brushy Basin member mudstone. Unsaturated flow probably would occur through the matrix and fractures of the Brushy Basin before the infiltration reaches the Westwater Canyon sandstone. Some may dismiss this source owing to the low permeability of the Brushy Basin sandstone. But bear in mind that the vertical fluxes likely to be important over extensive areas may only be on the order of 10^{-9} to 10^{-8} cm/s. Because this flux is probably less than the saturated hydraulic conductivity of the Brushy Basin mudstone, this unit will not significantly impede deep water movement. In fact, there is little evidence of springs or perched water associated with the upper contact of the Brush Basin, which, if present, would indicate the Brushy Basin mudstone impedes deep vadose zone flow.

Infiltration of rain and snowmelt runoff in the ephemeral lakes in the Mancos shale valley north of the Westwater Canyon outcrop also represents a potential source of local recharge to underlying formations. Because of the relatively small areas and

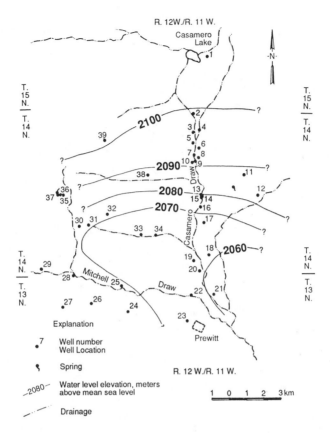

Figure 15   Water-level contours in shallow aquifers near Prewitt, NM. (From Stephens, 1983. With permission.)

ephemeral nature of the lakes, as well as the very low permeability of the thick, underlying shale, the amount of local recharge from the dry lakes is believed to be relatively small on a regional scale.

A fourth source of recharge to the Westwater Canyon sandstone is leakage from the alluvial aquifer in the sandstone subcrop area where the two aquifers connect. Although groundwater in the alluvial aquifer flows south in the direction of topographic slope, opposite to the regional flow and dip direction in the Westwater Canyon sandstone, numerical simulations showed that locally beneath the channel valley along Casamero Draw, the two systems are most likely in hydraulic connection, such that the alluvial aquifer probably recharges the Westwater Canyon in the subsurface (Stephens, 1983) (Figure 17).

In this case study of an arid site in western New Mexico, we have developed a conceptual model that recognizes some of the major types of recharge described already: diffuse recharge on dipslopes, local recharge on outcrops enhanced by anisotropy, local recharge along ephemeral streams, and depression-focused recharge. Additional studies would be required to quantify these processes and evaluate the relative importance of each of the components of recharge. A recommended

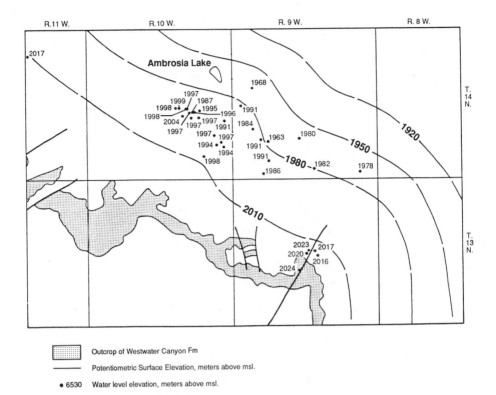

Figure 16   Potentiometric surface in the Westwater Canyon sandstone aquifer in the area overlapping and adjoining the Prewitt site on the east. (From Kelly, et al., 1980. With permission.)

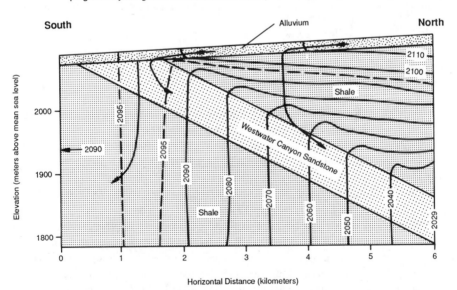

Figure 17   Hydraulic head and stream lines from a steady-state numerical model of flow, showing recharge to the Westwater Canyon aquifer occurring due to seepage from the alluvial aquifer and some leakage through the overlying shale. (From Stephens, 1983. With permission.)

general approach to a particular hydrogeological problem involving recharge mechanisms is to first develop a conceptual model, such as we presented here, that broadly considers multiple possible mechanisms, next apply analytical or numerical models to test identify the most significant of the recharge mechanisms, and then develop a field program to quantify the important mechanisms in order to reduce uncertainty in the solution to the problem.

CHAPTER 5

Characterizing Hydraulic Properties

In Chapter 1, we briefly discussed the physical properties necessary for modeling flow in the vadose zone, and in Chapter 3 we described the important physical processes that need to be considered for accurate prediction of fluid behavior in the vadose zone. In this chapter, we present some of the field and laboratory techniques that are available for obtaining parameters that can be used in quantitative analyses and model simulations of vadose-zone processes. However, before presenting these various methods, we must address the portion of a vadose-zone characterization effort that is often critical to its success: planning.

I. PLANNING SITE CHARACTERIZATION

The following example illustrates the importance of site characterization. During a recent geotechnical investigation at a facility that generated hazardous waste, approximately 400 borings and monitor wells were installed. The information base included stratigraphy, structural geology, geomorphology, and basic hydrogeology. The depth to groundwater contained in an upper water-bearing zone was only about 2 m. This upper water-bearing zone was underlain by a thick clayey till, which in turn was underlain by a major sand aquifer at about 30 m. Subsequent to the main phase of fieldwork, as part of the regulatory compliance program, it was necessary to develop a conceptual model of the vadose zone. Based on the field results, it was uncertain whether the vadose zone was only 2 m thick or whether the upper water-bearing zone was merely perched on the thick clay till. In the latter case, there would be two unsaturated zones, an upper one 2 m thick, and a lower one extending from somewhere within the clay till to the water table in the deep sand aquifer 30 m below land surface. The distinction was critical because regulatory requirements stipulated the development of a characterization and monitoring program of the first regional aquifer. The costs of implementing such a program are substantially different if the top of this unit is 2 m or if it is 30 m deep. Unfortunately, even after a few million dollars for site investigations, the data necessary to resolve this vadose zone issue were not collected. Most of the borings extended only about 7 m below land surface and in none of these or in any of the several deeper borings were samples collected to determine water content and porosity. The presence of unsaturated conditions in the lower part of the clay till, established by moisture contents less than porosity, would have confirmed that the upper water-bearing zone was perched. This simple analysis, at a cost of less than $40 per sample, likely would have provided the information necessary for regulatory purposes. Instead, those planning the investigation apparently failed to fully understand how their data would be used. Consequently, the vadose zone evidently was not incorporated into their preliminary conceptual model that guided the initial site investigation. The lack of adequate planning, unfortunately, is rather common, most likely because many traditional hydrogeologists and civil engineers focus on aquifer hydraulic characteristics and soil mechanical properties, respectively, inasmuch as that is their professional interest. Additionally, their formal training usually does not expose them to vadose-zone processes, nor to many of the methods of vadose-zone site characterization and monitoring.

We briefly outline a five-step process that is usually relevant to planning most subsurface site investigations involving the vadose zone; however, the basic outline could apply to almost any hydrogeological investigation.

1. **State the problem that requires solution**. Also, show why this problem must be addressed. Examples of questions that commonly form the motivation for vadose zone site characterization include, among others:
   - What is the natural recharge rate for the groundwater basin?
   - What is the seepage rate from the landfill or impoundment?
   - Will chemicals found in the soil leach to the water table?
   - Could contaminants in the soil have migrated laterally from an adjacent, off-site source?
   - What is an appropriate and cost-effective method of vadose-zone monitoring?
2. **Develop a preliminary conceptual model of the system**. In this step, one collects and integrates basic information from existing field data or limited preliminary site assessments to establish the hydrogeological framework. The scale of interest and domain for the conceptual model should be selected that is appropriate to address the problem. The conceptual model should address all the physiochemical and biological processes that appear to be relevant and may include computing water budget components from available literature. The types of chemicals present and their state (gas, dissolved, free phase), as well as potential fluid migration pathways should be identified in contaminant transport problems. To facilitate developing the conceptual model, it is often helpful to diagram the physical hydrogeological setting in a two- or three-dimensional view, such as one developed at Los Alamos, NM (Figure 1).
3. **Identify an approach to address the problem**. In this step, develop a methodology to meet the project objective. The methodology should include quantitative tools and procedures for using the tools. For example, you may decide that modeling is the best approach to predict travel time to the groundwater. Then you must decide whether the question can be answered with a simple or a complex model, identify the model or models that will be used, decide whether the model will be one-dimensional or multidimensional, and determine whether the flow field will be steady-state or transient. The approach also sets forth the sequence of tests or cases to be analyzed with the model. The approach need not be a model or mathematical equation; it could also be field instrumentation, geochemical analyses, or laboratory tests, for instance, depending upon the objective and budget.

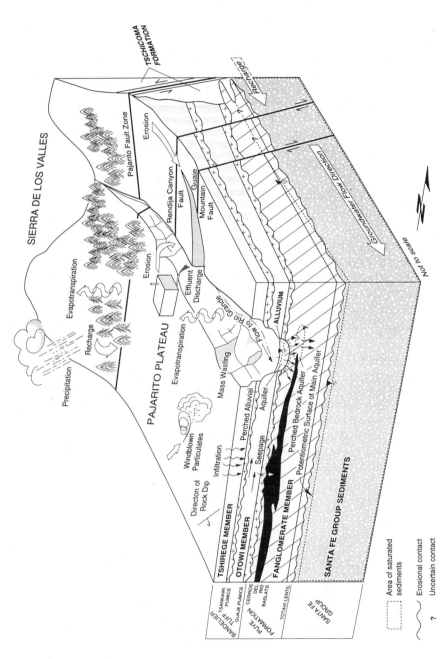

Figure 1 Conceptual geohydrologic model and general relation of major geologic units for OU 1071 on the Pajarito Plateau, Los Alamos, NM. (From Aldrich et al., 1992. With permission.)

4. **Identify the data needed to solve the problem**. In this step, examine the approach to assess data needs. Site characterization of vadose-zone thickness and geology should be a basic element of any field investigation. The extent to which the geology and hydraulic properties are characterized, however, must be guided by the objectives of the investigation. Before collecting field data, ask why the data are necessary. This will force one to think completely through the process of identifying data essential to analyzing the problem as stated in the objective. Site characterization is not an objective in itself.

In the modeling example, consider the hydraulic parameters and coefficients that must be specified as input. Also consider the nature of spatial variability within the domain of interest before selecting targets for characterization. Recognize uncertainty in the parameters that lead to establishing confidence limits to your solution. If the approach requires identifying the probability of a certain outcome, then it is likely you will need not only mean values for hydraulic properties, but also their variance and perhaps their spatial correlation structure. The required type and number of samples depend critically upon the goals and approach, as well as on the heterogeneity of the vadose zone.

5. **Develop a site characterization work plan**. This plan sets forth the goals of the project, identifies what data will be collected, and establishes standard operating procedures describing how the data will be collected. Specify here all drilling methods, sample collection procedures (methods, number, location, depth, etc.), and quality control/quality assurance procedures. Also generally included here is a health and safety plan to assure that the data are collected without risk to the health of workers and others near the work area.

The remainder of this chapter is devoted to a brief description of some common tools that are available for characterizing the hydraulic properties of the vadose zone. The principal properties discussed include saturated hydraulic conductivity, unsaturated hydraulic conductivity, air permeability, and specific water capacity. The descriptions of field and laboratory methods presented here are not intended for use as a standard operating procedure that may be copied into a site characterization work plan. Rather, the methods should be considered as general information on what technology is available, and should serve as a guide when selecting appropriate methods for solving the problem at hand. Planners of hydrogeological site investigations should find the following techniques highly complementary to the suite of tools they have previously applied in traditional soil engineering and aquifer characterization programs.

II. SATURATED HYDRAULIC CONDUCTIVITY

A. LABORATORY METHODS

Because saturated hydraulic conductivity is one of the most widely characterized hydraulic properties of soil, representative values have been published for a wide range of soils (Figure 2). Laboratory methods to measure it have been available since the first published experiments by Henri Darcy in 1856. Soils engineers, petroleum engineers, and agronomists have devoted enormous efforts to establishing drilling,

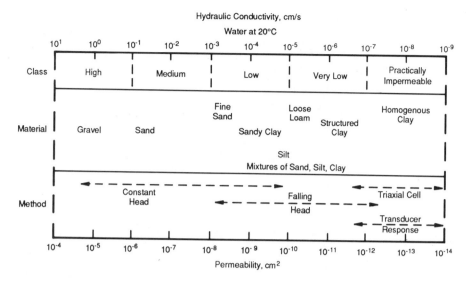

Figure 2 Hydraulic conductivity of various materials at saturation. (From Klute and Dirksen, 1986. With permission.)

core sampling, and testing procedures pertaining to measuring saturated hydraulic conductivity. In fact, there are standards published by the American Society for Testing and Materials (ASTM) for laboratory tests on granular and fine-textured soils (ASTM D-2434-68 and D-5084-90). Laboratory tests involve removing a small sample of soil and testing it in the laboratory under controlled conditions. Because of concern for problems of sample disturbance and representativeness of the relatively small sample, field methods are often preferred. Although field methods may overcome the limitations of the laboratory methods, they often suffer from lack of control on the experimental conditions and other logistical problems. In this section, laboratory methods are summarized first, followed by the field methods. It is common, as well as convenient, to refer to field or laboratory devices to determine saturated hydraulic conductivity simply as permeameters.

There are in general two types of laboratory tests for saturated hydraulic conductivity: constant-head and falling-head permeameters. In a constant-head permeameter, water is introduced into the sample by maintaining inflow and outflow reservoirs at constant positions relative to the sample (Figure 3). The steady flow rate, sample length and cross-sectional area, difference in reservoir elevations, and water temperature are used to calculate hydraulic conductivity according to Darcy's equation (Chapter 1, Equation 12). In a falling-head permeameter, water is introduced to a water-saturated sample by gravity drainage from a standpipe, while the head on the downstream end of the sample remains constant (Figure 4). The elevations of the water level (H) in the standpipe at different times (t), pipe cross-sectional area (a), sample length (L), and area (A) are used to calculate the saturated hydraulic conductivity in the following equation:

$$K_s = \frac{aL}{At} \ln\left(\frac{H_1}{H_2}\right)$$ *falling head permeater*   (1)

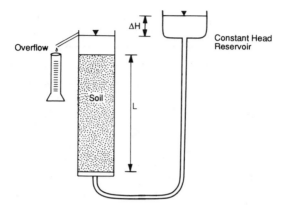

Figure 3   Constant-head permeameter.

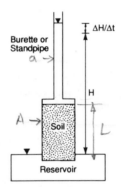

Figure 4   Falling-head permeameter.

According to Klute and Dirksen (1986) the falling-head method is most accurate in the hydraulic conductivity range from about 10^{-3} to less than 10^{-7} cm/s, whereas the constant head permeameter range best applies to soils having a conductivity greater than about 1 cm/s to about 10^{-5} cm/s (Figure 2). For reference, the saturated hydraulic conductivity of sand typically ranges from about 10^{-1} to 10^{-3} cm/s, and clay is less than about 10^{-6} cm/s (Figure 2).

There are many important considerations associated with both the constant-head and falling-head permeameters. Sample disturbance and repacking are always impor-tant and virtually unavoidable to some degree, but this consideration is not so critical for loose, granular, unstructured soils. Undisturbed samples are generally preferred, although if repacking is necessary, samples should be repacked to the field bulk density. Where soils are structured, that is, where they contain small cracks or aggregates, for example, the size of the soil sample should be sufficiently large to incorporate these secondary pore structures. Othman et al. (1993) reviewed proce-dures to test the hydraulic conductivity of soil subjected to freeze-thaw conditions. For compressible soils in particular, such as clay liner material, it is also important to conduct the tests under confined conditions which reproduce the overburden pressures. Soil engineers use fixed wall and flexible wall cells (e.g., triaxial cells) to accomplish this (e.g., Daniel, 1989; Daniel, 1993), as illustrated in Figure 5. Fixed-wall cells are recommended for coarse-textured soils, and flexible cells are recommended for fine-textured soils, and in either permeameter, falling-head or constant-head conditions can be imposed. In all tests, de-aired water is recommended to reduce the potential for entrapping air in the sample. Flushing the samples with carbon dioxide prior to infiltration also may reduce air entrapment. For tests of more than about 2 d duration, the infiltrating water should be sterilized to reduce the potential for accumulation of microorganisms, which can plug parts of the pore space. The chemistry of the water is also important. Soil scientists often use a 0.005 or 0.01 M $CaSO_4$ solution, which minimizes the dispersion of clays, and engineers prefer distilled water. Where the field problem involves a fluid other than water, that fluid should be used to determine the hydraulic conductivity. Fluid chemical interactions with the soil matric can lead to large increases or decreases in hydraulic conductivity relative to that for water (e.g., Shackelford, 1994). And finally, it is important to minimize the hydraulic gradient during the tests. Excessive gradients may violate Darcy's equation, cause high flow velocities to entrain soil particles and enlarge the pore spaces (piping), or induce liquefaction of the sample.

There are a number of advantages to characterizing sites using laboratory analyses of core samples. For instance, after the core is used to measure saturated hydraulic conductivity, other hydraulic and physical properties can be measured on the same sample, such as moisture retention curves, unsaturated hydraulic conductivity, bulk density, porosity, particle size distribution, or other geotechnical indices. Also, the same drilling methods commonly employed by geologists and geotechnical engineers to sample the soils for visual description and classification can be used to collect the cores suitable for testing hydraulic properties. Typically, multiple core samples are collected in rings stacked above each other, so that one sample can be analyzed for chemical concentrations and another for physical properties. Where the field problem is of a local nature and there is significant spatial heterogeneity in the soil, core samples usually afford the best means to quantify small-scale variability in hydraulic properties. For additional guidance refer to appropriate ASTM standards or reference materials such as Klute and Dirksen (1986).

B. FIELD METHODS*

There are several different field techniques to measure the saturated hydraulic conductivity. Here we discuss the air-entry permeameter, borehole permeameter, disc permeameter, sealed double-ring infiltrometer, and air and gas permeameters. During the following descriptions of methods, bear in mind that some permeameters measure conductivity in the vertical direction, whereas others produce results that include components of conductivity in the horizontal direction as well. For additional general information, refer to ASTM Standard D-5126.

* Significant parts of this section are modified from Havlena and Stephens (1991) and are reproduced here with permission of the senior author.

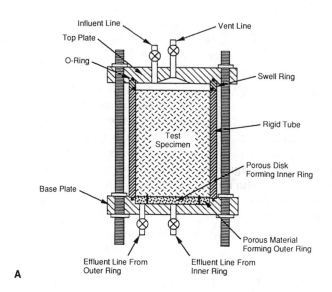

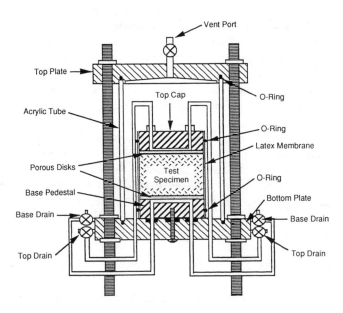

Figure 5 (A) Rigid wall permeameter designed to detect side-wall leakage and (B) flexible wall (triaxial) cell with (C) detail on a system to control applied fluid pressure on the inlet and outlet ends of the sample, as well as side-wall confining pressure. (From Daniel, 1993. With permission.)

1. Air-Entry Permeameter

The air-entry permeameter (Figure 6) was first proposed by Bouwer (1966) as a means to determine the vertical, field-saturated hydraulic conductivity of geologic materials above the water table. Most applications of the air-entry permeameter have

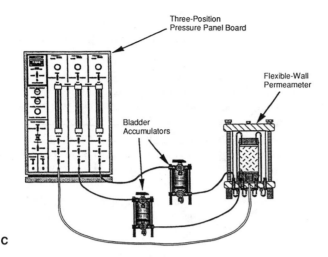

Figure 5 (continued).

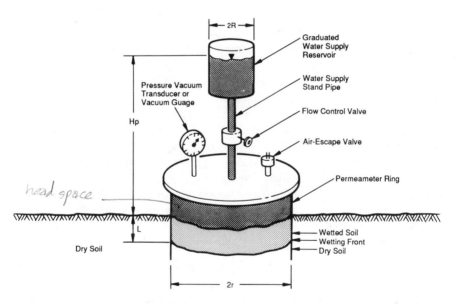

Figure 6 Air-entry permeameter. (From Havlena and Stephens, 1991. With permission.)

generally been limited to fairly homogeneous sands and silty sands. Recent work
(Knight and Haile, 1984; Stephens et al., 1988; Havlena and Stephens, 1991) has
shown the air-entry permeameter produces good results in low-permeability clays
and engineered clay liners. The air-entry permeameter consists of a single ring,
typically 30 cm in diameter and 25 cm deep, which is driven vertically into the
material to be tested. A water supply reservoir is mounted on top of an adjustable
standpipe. To begin the test, the air-entry permeameter reservoir is filled and water
is allowed to infiltrate into the ring. When the water displaces the air between the soil
and top plate, the air-escape valve is closed. The rate of decline of head applied at

the soil surface is then recorded, either through the use of a transducer set within the ring at the soil surface or by observing the decline of water in the graduated supply reservoir (ASTM D-5126). Infiltration continues until the wetting front approaches the bottom of the ring, as indicated usually by a change in soil tension monitored by a tensiometer or the change in electrical conductivity at buried resistance probes, or as inferred by a mass balance approach using an assumed fillable porosity of the soil. The water supply valve is closed, and the soil water is allowed to redistribute. The negative pressure created by the redistributing soil water is monitored by a vacuum gauge or pressure transducer mounted at the top of the ring. When air begins to enter into the saturated soil from below the wetting front, the gauge will have recorded a minimum pressure. The minimum pressure achieved during this portion of the test is used to calculate the air-entry pressure head of the soil, ψ_a. The field saturated hydraulic conductivity, K_{fs}, is calculated using the following equation:

$$K_{fs} = \frac{(R_r / r)^2 L \, (dH / dt)}{(H_P + L - \psi_f)} \tag{2}$$

where R is the radius of the reservoir, r is the radius of the cylinder, L is the depth to the wetting front when dH/dt is measured, dH/dt is the rate of decline in head in the reservoir, and H_P is the depth of ponding when dH/dt is measured. Bouwer (1966) indicated that the actual saturated hydraulic conductivity may be about twice K_{fs}, due to entrapped air. Stephens et al. (1983) reached a similar conclusion and showed that flooding the cylinder with carbon dioxide gas prior to infiltration minimized entrapped air because the carbon dioxide gas is much more soluble in water than air.

The determination of vertical, field-saturated hydraulic conductivity is based in part on the wetting front capillary pressure head, ψ_f. Bouwer (1966) estimated this to be approximately equal to $\psi_a/2$. The wetting front pressure head has been shown to represent the effective capillary driving force during infiltration (Morel-Seytoux and Khanji, 1974) and, as such, is a major component of the overall applied hydraulic gradient. Neuman (1976) showed that the ψ_f can be more accurately calculated from hydraulic conductivity:

$$\psi_f = \frac{1}{K_s} \int_{\psi=\psi_i}^{0} K(\psi) \, d\psi \tag{3}$$

where ψ_i is the initial pressure head in the soil. Stephens et al. (1988) calculated ψ_f with Equation 3 based on laboratory-measured moisture retention (θ-ψ) data for the field-tested soil and van Genuchten's procedure applied to those data to calculate the K-ψ function. This approach was applied in clay soil where the time to wait for drainage to the air-entry pressure would have been impractically long.

2. Borehole Permeameters

The borehole permeameter is a method to determine *in situ* saturated hydraulic conductivity, K_s, of unfissured, homogeneous, isotropic soil and rock above the water table. The method has been variously referred to as a well permeameter, a

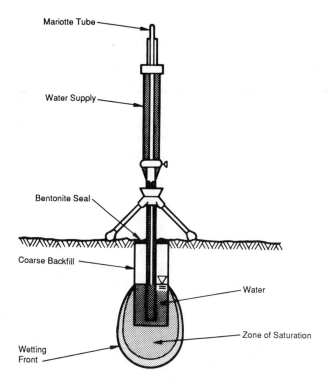

Figure 7 Guelph permeameter. (From Havlena and Stephens, 1991. With permission.)

constant-head borehole infiltration test, a reverse auger-hole method, and a well pump-in test. The borehole is constructed utilizing an auger (or drilling) to the desired depth, and if the soil contains fines, the sides of the borehole must be scraped to remove any smear zone. In caving formations, well screen and a coarse sand or gravel pack must be placed in the borehole. In a typical borehole permeameter, a constant head of water is maintained in the borehole until the infiltration rate is steady. The borehole testing equipment typically consists of a water-supply reservoir and a means of controlling the head of water within a 5- to 15-cm-diameter borehole. The equipment used to maintain a constant level of water in the borehole most commonly consists of either a float-type stock-tank valve installed within the borehole and connected to a remote water supply reservoir (U.S. Bureau of Reclamation, 1974; Stephens, 1992), or a Mariotte device as in the Guelph permeameter (Reynolds and Elrick, 1985) (Figure 7). Most of these testing methods that are used to determine field-saturated hydraulic conductivity may be easily modified to provide values of hydraulic conductivity that are much closer to full water saturation by preflooding the test material with carbon dioxide gas (Stephens et al., 1983).

Because the flow rate from a borehole, Q, can be affected by the depth to the water table, there are two types of equations to compute hydraulic conductivity: deep and shallow water table conditions. Table 1 summarizes some of the important solutions under deep water-table conditions for K_s. According to the U.S. Bureau of Reclamation (1974) deep water-table conditions for borehole permeameter tests exist when the distance from the water surface in the borehole to the water-table, T_u, is at least three times the depth of water in the borehole, H. In a mathematical sense, there is a semi-infinite unsaturated zone when $T_u \geq 3H$. Depending upon the method of analysis, there are some restrictions in the ratio of water depth, H, and borehole radius, r. Most of the steady-state solutions are valid where $H/r > 10$. Analytical solutions for the deep water-table case were derived by Nasberg (1951), Glover (1953), Terletskaya (1954), Cecen (1967), and Reynolds et al. (1983). All these formulas neglected capillarity. However, numerical simulations by Stephens (1979) revealed that when the flow field was at steady state, a finite bulb of saturation enveloped the well bore (Figure 8), the size of which decreased as capillary effects of the soil increased. Stephens (1979) also developed a solution to the borehole permeameter problem for deep water-table conditions which included the effects of unsaturated flow (see also Stephens and Neuman, 1980; Stephens and Neuman, 1982). This early work, which included capillarity, was based on numerical simulations and regression analysis, as were subsequent extensions of this approach (Stephens et al., 1983; Stephens et al., 1987). Stephens' solutions for K_s use actual capillary properties obtained from the K-ψ curve for selected soils as represented by moisture retention models of Brooks and Corey, Burdine, and Mualem (as modified by van Genuchten (1978)). Figure 8 illustrates the relationship between the dimensionless flow rate, C_s, and the dimensionless head of water in the borehole, based on the numerical analysis by Stephens et al., 1987. Philip (1985) and Elrick et al. (1989) developed approximate analytical solutions to compute K_s that take capillarity into account by assuming that $K = K_s \exp(\alpha\psi)$. Reynolds et al., (1985) and Reynolds and Elrick (1985, 1986) suggested that K_s, sorptivity, and the unsaturated hydraulic conductivity could be determined in the field by conducting two tests in a borehole at two different water depths or stages and solving the equation for each test simultaneously. However, Laase (1989) and later others (Amoozegar-Fard, 1989; Salverda and Dane, 1993) concluded that the two-head test may give negative values for K_s in heterogeneous soil.

As defined by the U.S. Bureau of Reclamation (1974), shallow water-table conditions exist if the distance between the water table and the water surface in the borehole, T_u, is less than three times the depth of water in the borehole (H). An equation for K_s under shallow water-table conditions developed by Zanger (1953) for a fully screened well is

$$K_s = \frac{2Q}{C_u r T_u} \tag{4}$$

in which

$$C_u = \frac{2\pi(H/r)}{\sin h^{-1}(H/r)^{-1}} \tag{5}$$

Stephens (1979) conducted numerical simulations to evaluate the reliability of the solutions. Stephens (1992), in summarizing results of field experiments at a shallow water-table site conducted by Rabold (1984) and Mikel (1986), concluded that the reliability of these equations is not as good as for the deep water table case based on the evaluations to date.

Table 1 Equations for Saturated Hydraulic Conductivity by the Borehole Permeameter with a Single-Stage Test

Equation	Comments	Ref.
$K_{fs} = \dfrac{Q}{C_u rH}$	Q is the steady infiltration rate $$C_u = \dfrac{2\pi H_D}{\left[\sin h^{-1}(H_D) - \left(\dfrac{1}{H_D^2+1}\right)^{1/2} + \dfrac{1}{H_D}\right]}$$ $H_D = H/r$ r is borehole radius H is borehole water depth	Glover (1953)
$K_{fs} = \dfrac{Q}{C_s rH}$	$\log_{10} C_s = 0.653 \log_{10} H_D - 0.257 \log_{10} \alpha_v$ $- 0.633 \log_{10} H + 0.021 H_D^{0.5} - 0.313 N^{-0.5}$ $+ 1.456r + 0.453$ α_v (cm^{-1}) and N are Van Genuchten's θ-ψ fitting parameters H and r are in meters (m)	Stephens et al. (1987) (refer also to Figure 6)
$K_{fs} = \dfrac{CQ}{\left(2\pi H^2 + \pi r^2 C + 2\pi \dfrac{H}{\alpha^*}\right)}$	$\alpha^* = K_{fs}/\phi_m$ (L^{-1}) $$\phi_m = \dfrac{CQ}{\left[\alpha^*(2\pi H^2 + \pi rC) + 2\pi H\right]}$$	Elrick et al. (1989)
$K_{fs} = \dfrac{Q}{C_p rH}$	$$C_p = \frac{(H_D^2-1)^{1/2}}{H_D}\left\{\frac{4.117H_D(1-H_D^{-2})}{\ln\left[H_D+(H_D^2-1)^{1/2}\right]-(1-H_D^{-2})^{1/2}} + \frac{4.028+2.517 H_D^{-1}}{A\ln\left[H_D+(H_D^2-1)^{1/2}\right]}\right\}$$ $A = 1/2\alpha r$, with α obtained from the slope of ln K-ψ	Phillip (1985)

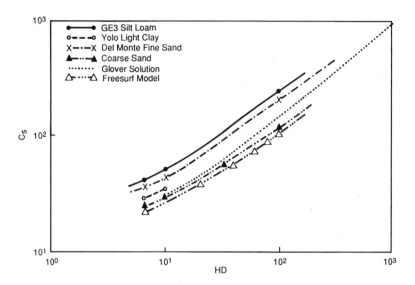

Figure 8   The relationship between C_s and H_D varies with soil texture because of differences in unsaturated hydraulic conductivity-pressure head relationships. Results from Glover's solution and from a numerical model that neglects capillarity (FREESURF) are shown for comparison. (From Stephens, 1979. With permission.)

Applications of the borehole permeameter methods are not extensively published, although some early applications were reported by Robinson and Rohwer (1959) and Winger (1965). Interest in this application continues (e.g., Bell and Schofield, 1990; Jabro and Fritton, 1990), and recently, Boutwell and Tsai (1992) have proposed a falling-head method to measure K_s in a borehole cased with well screen. Unfortunately, in field soils, negative values of K_s may occur with this falling-head method. Even more recently, Philip (1993) derived an approximate solution for a falling-head permeameter in which water is only permitted to flow from the base of the casing; however, preventing water from also flowing up the outside of the casing seems to be a practical limitation.

Borehole permeameters may be used for a wide range of applications and materials, including lithified rock. The greatest advantage of the borehole permeameter method over most other methods is its ability to be used at virtually any depth. This is especially useful for investigating the vertical, spatial variability of hydraulic conductivity in areas where the water table is deep. The test methods that incorporate the effects of capillarity are considered the most accurate, especially in fine-textured materials. However, the methods that ignore capillarity may be used to approximate field-saturated hydraulic conductivity in coarser materials, or where a less rigorous determination is acceptable. Because of the requirement that steady state be attained, borehole permeameter tests may take considerable time to complete, especially in low-permeability materials. Stephens (1979) suggested graphical methods to approximate the steady infiltration rate from early time data, but these have not been thoroughly tested in the field. For additional discussion on borehole permeameters refer to Stephens and Neuman (1982), ASTM Standard D-5126, and Elrick and Reynolds (1992).

3. Disc Permeameter/Tension Infiltrometer

The disc permeameter has recently been developed (Perroux and White, 1988) as a rapid means to determine field-saturated hydraulic conductivity, sorptivity, macroscopic capillary length, and characteristic pore size, as well as unsaturated hydraulic conductivity at low tensions in a variety of materials. The disc permeameter consists of a circular disc, a fine nylon membrane typically about 20 cm in diameter, to which is attached a water supply reservoir and a Mariotte siphon head-control device (Figure 9A). The reservoir is made of clear plastic to allow the user to read the rate of water level decline and calculate infiltration rate. Ankeny et al. (1988) developed a similar device called a tension infiltrometer that includes pressure transducers to monitor applied pressure (−2 to 50 cm) and infiltration rate with a data logger (Figure 9B). The position of the bubble tube base controls the tension of the water above the disc. For the determination of field-saturated hydraulic conductivity, a circular ring is driven a small distance into the soil surface, and the permeameter, without the nylon membrane, is placed on the ring. In this mode, the disc permeameter is similar in operation to a single ring infiltrometer to characterize ponded infiltration. Whether under a tension or positive pressure, the test primarily consists of measuring infiltration rates for a period of a few minutes to usually no more than an hour or two. In the unsaturated mode, it is critical to maintain excellent contact between the disc and the soil. Where there is no ring driven into the soil to constrain the flow from the disc, it is important to recognize that disc permeameters induce a three-dimensional flow field and, strictly speaking, the theoretical developments pertain to isotropic soils.

The solution for hydraulic conductivity at the supply potential, K_0 (Wooding, 1968) requires measuring the steady-state flow rate of water into the soil, Q, for a disk of radius r_0:

$$K_0 = \frac{Q}{4\pi r_0^2} - \frac{4\Phi_m}{\pi r_0} + K_i \pi r_0^2 \tag{6}$$

where Φ_m is the matric flux potential, and K_i is the hydraulic conductivity at the initial water content. Usually K_i is negligible and this last term is often dropped. The matric flux potential could be calculated if one knew the unsaturated hydraulic conductivity and initial pressure head from

$$\Phi_m = \int_{\psi_i}^{\psi_0} K(\psi)\, d\psi \tag{7}$$

Development of disc permeameter theory by Perroux and White (1988) overcomes the need to compute Φ_m by obtaining the capillary properties directly from the early time test data, which give a parameter called sorptivity, S_0:

$$K_0 = \frac{Q}{\pi r_0^2} - \frac{4 b_d S_0^2}{\pi r_0 (\theta_0 - \theta_i)} \tag{8}$$

where θ_i and θ_0 are the initial water content and water content at the supply potential, respectively, and b_d is a dimensionless constant having a value between about 0.5 and

A

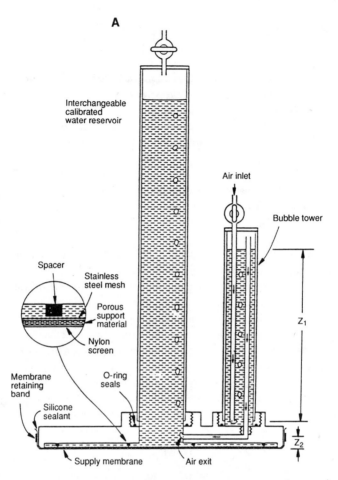

Figure 9 Hydraulic conductivity measurement by the (A) disc permeameter and (B) tension infiltrometer. The distance from the bubble inlet to the water level in the bubble tower, Z_1, and the distance between the bubble inlet in the reservoir to the soil, Z_2, control the pressure head ($\psi = Z_2 - Z_1$) at which infiltration occurs. (From Ankeny et al., 1988. With permission.)

0.79. The water content must be sampled immediately after infiltration within about 5 mm of the soil surface. The term $\mathscr{S}_0$ is obtained as the slope of a straight line fitted through a plot of cumulative infiltration versus the square root of time. Hussen and Warrick (1993) applied the preceding equations and other algebraic equations to field and model generated disc permeameter data. They concluded all the solutions gave consistent predictions of the steady infiltration rate. Hence, we can infer that there would be no practical differences in the calculated K_0 values. A method by Ankeny et al., (1991) obtains K_0 and Φ_m by conducting two steady-state tests at different supply potentials and by solving simultaneous equations.

Disc permeameters are useful in a wide variety of applications, including thickly bedded, heterogeneous sediments, gravelly sands, and lithified rock. Contact between the formation and plate can be achieved using a very thin layer of the loose

B

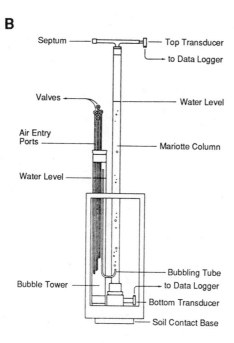

Figure 9 (continued).

field soil, fine silica flour, or diatomaceous earth. One distinct advantage of disc permeameters is that they can operate in soils containing macropores. Because the pressure head along the bottom of the membrane during unsaturated testing is maintained at a slight tension with respect to atmospheric pressure, water does not flow into macropores. Subsequent tests conducted at lower pressure heads may show increases in flow rate attributable to macropores (e.g., Jarvis et al., 1987). Accordingly, soil matrix properties are measured at high tensions, whereas the bulk permeability of the soil-macropore composite is measured during tests at low tensions or during ponded tests for field-saturated hydraulic conductivity. Comparison of the results of the different tests run on the same soil volume allows for an assessment of the contribution of macropores to the overall bulk saturated hydraulic conductivity.

4. Sealed Double-Ring Infiltrometer

Open-top, cylinder infiltrometers have long been used to measure the infiltration rate of natural soils (Daniel, 1989; ASTM D-5126). Their range has recently been extended to allow measurements in lower permeability materials and clay liners by the addition of a top cap over the water reservoir to prevent evaporation and disturbance to the water supply (Daniel and Trautwein, 1986; ASTM D-3385-88). For this reason, closed-top, or sealed, infiltrometers are usually preferred to the open variety.

The sealed double-ring infiltrometer (SDRI) is currently very popular as a means of approximating field-saturated hydraulic conductivity of low permeability materials. In the strictest sense, SDRIs measure the steady infiltration rate, not field-saturated

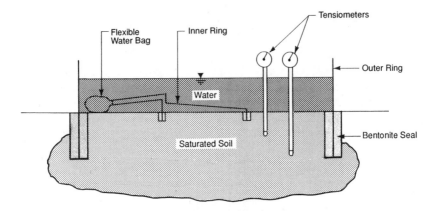

Figure 10   Sealed double-ring infiltrometer. (From Stephens et al., 1991. With permission.)

hydraulic conductivity. The SDRI consists of two concentric rings inserted a shallow distance into the material to be tested (Figure 10). The inner ring is commonly 2 m across, and the outer ring may be 1 m larger. In some installations, such as in bedrock where concrete saws must be used to excavate the trench, the infiltration surface is configured as a square. The rings are set into a bentonite slurry within two concentric trenches approximately 10 to 20 cm deep each. The bentonite provides a bedding for the rings and reduces the potential for preferential flow along the sidewalls.

Water is allowed to infiltrate between the two rings and within the inner ring. The water infiltrating between the two rings provides a barrier that constrains the water from the inner ring to infiltrate vertically downward. This assumption of one-dimensional, vertical infiltration is the main advantage of the double-ring over the single-ring infiltrometer. The inner ring is sealed with a cap to limit evaporation during the often long-term tests. The flow rate through the inner ring is periodically measured by weighing the sealed water reservoir used to supply the inner ring. The steady-state flow rate through the inner ring is used to determine the infiltration rate of the material being tested.

When the SDRI steady infiltration rate data are used to calculate the vertical, field-saturated hydraulic conductivity, one must assume that the flow within the inner ring is essentially one-dimensional downward at a unit hydraulic gradient. The solution for vertical, field-saturated hydraulic conductivity follows directly from

$$K_{fs} = \frac{Q_i}{\pi r_i^2} \tag{9}$$

where Q_i is the steady flow rate into the inner ring and r_i is the radius of the inner ring. In recent modifications, two or more tensiometers set within the soil profile beneath the inner ring allow calculation of the hydraulic gradient. Consequently, Equation 9 would be modified by dividing the right-hand side by the hydraulic head gradient to obtain a more reliable determination of saturated hydraulic conductivity.

Because of the typically complicated and lengthy setup and lack of portability of the equipment, the SDRI is generally limited to lower permeability materials with field-saturated hydraulic conductivities in the 10^{-9} to 10^{-5} cm/s range. At conductivities on the order of 10^{-7} cm/s, which are typical of clay liners, more than several weeks may be required to achieve steady-state and complete the test. With proper excavating and rock-cutting equipment, the SDRI may be used effectively in cobbley soils and lithified rock. A distinct advantage of the SDRI is given by the large volume of soil tested, which provides a more representative sampling of actual field conditions than smaller scale tests. Where portability, ease of use, and low cost are of particular importance, it may be more feasible to use an open-top infiltrometer, for situations where the soil is sufficiently permeable to allow rapid tests where evaporation can be neglected.

5. Air and Gas Permeameters

Air or gas, such as nitrogen, can also be used as the permeating fluid in field permeameters. Three types of air permeameters are discussed. The first involves air injection into a vertical cylinder containing soil. In the second method, air is injected or withdrawn from an interval of a borehole. The third method uses natural, *in situ* pressure measurements, which fluctuate due to barometric pressure. With any of these methods, the time to complete a test is usually quite short, and one avoids problems sometimes associated with introducing water as the permeant such as swelling and mobilizing dissolved soil contaminants. Refer to Chapter 1 for a review of basic concepts of air flow.

Before proceeding to discuss these methods, it is worthwhile to note processes that may affect permeabilities obtained by any of the methods. The rate of air flow is proportional to pressure gradient as well as to the air hydraulic conductivity, K_a. Therefore, air permeability measurements are potentially sensitive to the applied pressure and soil water content. Air in the soil pores at low pressure may slip along the surface of the solids in a phenomenon known as the Klinkenberg effect, which causes the measured air permeability to exceed the intrinsic permeability (Rasmussen et al., 1993):

$$k_{app} = k\left(1 + \frac{b_k}{\overline{P}}\right) \qquad (10)$$

where k_{app} is the apparent measured air permeability (L^2), k is the intrinsic permeability to liquid or high density gas (L^2), b_k is the Klinkenberg slip flow coefficient for the gas-solid system (Pa), and $\overline{P}$ is the mean absolute air pressure on the sample (Pa). This correction is especially important in fine-textured soils that have intrinsic permeabilities less than about 100 millidarcies (hydraulic conductivity $\approx 10^{-4}$ cm/s) (Baehr and Hult, 1991). The importance of the Klinkenberg effect also increases as the soil becomes drier, because it is in the small air-filled pores that the air slippage is most prevalent.

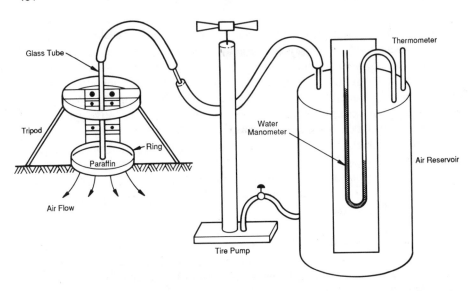

Figure 11   Apparatus for measuring air permeability of surface soil. (From Evans and Kirkham, 1949. With permission.)

Because air and water in soil comprise a two-phase system with air as the nonwetting fluid, the soil-water content will also affect the air permeability (Chapter 1, Figure 10). As water content increases, the air content decreases until, at the air-entry value, the air permeability becomes zero, because the air phase is discontinuous. However, in dry soils where the water content is near the specific retention, the permeability to air will only be about 0.6 to 0.8 less than that of the completely air-dry medium (Weeks, 1978). It is also important to recognize that as the soil wets or dries, clay minerals that absorb or desorb the water may expand or contract. This changes the pore structure and influences the measurement of air permeability. For some practical applications, such as soil vapor extraction in dry climates, water content is not likely to change significantly; consequently, air permeability measurements at *in situ* water content are all that may be necessary for a system design.

a. Vertical Air and Gas Permeameters

Vertical air and gas permeameters utilize a ring emplaced 5 to 30 cm into the soil similar to an air-entry permeameter installation (Figure 11). Air or gas (typically nitrogen or carbon dioxide) is allowed to flow at low pressure into the permeameter ring and through the soil profile. The pressure applied to the soil surface is monitored by a pressure gauge or transducer ported into a manifold on the permeameter ring. Falling pressure and constant pressure type tests are both possible.

Falling pressure type tests require measuring the decline of pressure within the permeameter, as well as the permeameter and fixed supply reservoir geometry. Constant pressure tests require that the permeameter geometry, flow rate, and pressure be known. The temperature of the infiltrating gas and the moisture content of the soil must be known for both type tests. Several different solutions for permeability are applicable (Corey, 1959; Evans and Kirkham, 1949; Groenewoud, 1968; Grover,

1955; Kirkham, 1946; Reeve, 1953) according to the depth of burial of the ring and the type of test (falling or constant pressure). For instance, the equation developed by Evans and Kirkham (1949) for a falling pressure test is

$$k = \frac{\mu v}{A_f P_a t} \ln \frac{y_0}{y_1} \tag{11}$$

where μ is the air viscosity (Pa-s), v is the volume of the air tank (L^3), A_f is a geometry factor that depends on the radius of the cylinder containing the soil and the radius of the air supply tube feeding the reservoir (L) (Figure 11), P_a is atmospheric pressure (Pa), y_0 is the water level in the manometer on the air pressure reservoir at time t_0 (L), y_1 is the manometer reading at time t_1, and t is ($t_1 - t_0$).

Although vertical air and gas permeameters may be used in a variety of materials and conditions, these permeameters have the greatest utility in lower permeability applications where the vertical permeability of the tested material is required. Such applications could include measuring the permeability of an engineered cap overlying radioactive mill tailings, which has been emplaced to prevent the upward release of radon gas. It should also be noted that the application of water low in salinity and divalent cations to soils containing clay minerals may result in dispersion of clays and swelling, with a resulting decrease in the hydraulic conductivity in most situations. Because of this, it is possible to use the ratio of permeability-to-air to permeability-to-water as an indicator of soil stability and the combined effects of swelling, slaking, and dispersion (Reeve, 1953).

The success of vertical air and gas permeability permeameter tests depends largely on the ability to properly emplace the permeameter ring into the soil and seal the ring tightly against the soil without disturbing the soil structure or creating pathways for leakage along the outside wall of the permeameter. The presence of macropores, channels, and soil disturbance can greatly increase the air flow rate and lead to severe overestimates of soil matrix permeability.

b. Borehole-Type Air and Gas Permeameters

Borehole-type air and gas permeameters provide a more horizontal determination of permeability. Both single-hole and cross-hole type tests may be conducted. The testing equipment for a single-hole type test is set up over a borehole of known geometry, and a pressure or vacuum is applied at the surface over the borehole at a constant head or flow rate. The applied pressure (or vacuum) is monitored by a pressure gauge or transducer, and the flow rate is monitored by a variable area or electronic flowmeter. The temperature, density, and viscosity of the gas and the moisture content of the soil also need to be known. The cross-hole type test is similar, except that the pressure (or vacuum) is monitored in a separate borehole at a known radial distance from the source. Within a single borehole, it is also possible to have multiple parts, so one part can be used as the pressure source while one or more are pressure detectors.

The solutions for permeability for borehole-type air and gas permeameter tests most often assume radial, horizontal flow to (or from) the tested borehole. This simplifying assumption allows the use of a one-dimensional, cylindrical solution for steady-state, compressible gas flow (Keller et al., 1973). Rasmussen et al. (1993) used the following equation to obtain the apparent air permeability in the Apache Leap Tuff site in central Arizona where air is injected between two packers in a borehole:

$$k_{app} = \frac{Q\mu_a P_a \ln(L/r)}{L(P^2 - P_a^2)} \qquad (12)$$

where Q is the steady-state volumetric air flow rate at standard temperature and pressure (L^3s^{-1}), μ is air viscosity (Pa-s), P_a is atmospheric pressure (Pa), L is the length of the test interval (L), r is borehole radius (L), P is the measured steady absolute air pressure in the borehole interval (Pa). A more rigorous, multidimensional solution may be used where required. However, for most applications where horizontal permeability is greater than vertical permeability, the horizontal radial flow solution is frequently considered appropriate. It is generally advised to conduct the permeability measurement in the dimension and the scale of interest.

Borehole-type air and gas permeameters may be used to help in the design of soil vapor extraction systems (e.g., Johnson et al., 1990) and as an aid in the interpretation of soil gas sampling data by providing information on the permeability of the tested material to air at the applied vacuum. Multiple observation wells are utilized to determine the radius of influence of a vapor extraction system to design the spacing between extraction wells. In many instances the pressure drop versus time can be analyzed by standard groundwater well hydraulics equations. These permeameters can operate in a wide range of soil types and conditions, including coarse and bouldery alluvium and clays. Because of the tendency of an applied vacuum to improve the surface seal around the top of the borehole, it is best to use a vacuum-type test in applications where higher pressures or flow rates are required.

c. Natural Air Pressure Method

Weeks (1978) developed and tested a field method to determine soil diffusivity to air. Recall that the diffusivity as it applies to a fluid flow problem is the ratio of a permeability term to a fluid storage capacity term. Weeks installed nests of air piezometers in a west Texas site to depths of 38 m and monitored natural soil air pressure in piezometers set to using an inclined water manometer. The soil air pressure fluctuated in response to barometric pressure, with some phase lag and amplitude damping. This information was incorporated into a computer program to analyze the soil air diffusivity at the field water content.

6. Other In Situ Test Methods

Several other methods of determining field-saturated hydraulic conductivity with water are available. These include BAT probe permeameters (Torstensson, 1984), variants of the borehole permeameter method (Stephens, 1979; Daniel, 1989), and ring-type infiltrometers (Daniel, 1984, 1989; Bouwer, 1961; ASTM D-5126).

III. FIELD PERMEAMETER CASE STUDIES

Over the past 17 years or so, I, my former students, and more recently my co-workers have conducted extensive field tests with the borehole permeameter, as well as with other field and laboratory permeameters described previously. These case studies, which highlight some of that work, examine the viability of the different solutions for K_S, as well as the effects of heterogeneity, temperature, and entrapped air. Native soils tested include sand, loam, and clay. Tests presented here for the sand and silt sites were undertaken for research purposes, whereas the test on the clay was a consulting project.

A. FLUVIAL SAND

The borehole permeameter tests in fluvial sand were conducted initially to validate computer-simulated behavior (Stephens, 1979) of infiltration from a bore-hole and the effect of capillarity on K_s predicted by borehole permeameters in uniform soils. The site for the field tests in sand was located approximately 32 km north of Socorro, NM, on the Sevilleta National Wildlife Refuge, adjacent to the flood plain of the Rio Salado. In this semi-arid climate, vegetation is sparse and consists mostly of creosote bush, salt bush, grasses, and a few juniper. Mean annual precipitation is about 20 cm.

The site was underlain by unconsolidated, uniform, fine fluvial sand to a depth of approximately 4.5 m (Byers and Stephens, 1983). The profile was stratified, cross-bedded, and contains a few thin silt layers. There was no developed soil profile. Anisotropy in saturated hydraulic conductivity (ratio of horizontal to vertical) determined from five pairs of oriented core samples (100 cm³) ranged from about 0.8 to 4.0, with a geometric mean of about 1.3 (Stephens et al., 1983). The mean particle diameter was about 0.26 mm, and the uniformity coefficient was 2.4, based on more than 271 core samples. Porosity, calculated from bulk density, ranges from 36 to 46%, and *in situ* water content was 7 to 10%. The depth to the water table was approximately 5 m in the test area.

1. Borehole Permeameter Tests

The first borehole permeameter test at the site began in July 1981 (Stephens et al., 1983). Through July 1983, 27 tests under deep water-table conditions were successfully completed at seven stations within an area of several hundred square meters. Multiple borehole permeameter tests were conducted at each station. During most tests, tensiometers were used to monitor pressure head and the neutron probe was used to monitor water content within about 1.5 m of the borehole. Refer to Stephens et al. (1983) for procedural details. Table 2 summarizes all test conditions and results (Stephens et al., 1983; Laase, 1989). In the calculations of K_s by the Stephens method (Table 1) (Stephens et al., 1987), we used the Mualem/van Genuchten model with $\alpha_v = 0.025$ cm^{-1} and $N = 4.7$; in Philip's method (Philip, 1985) $\alpha = 0.09$ cm^{-1}; and in the Reynolds method (Reynolds et al., 1985) $\alpha = 0.08$ cm^{-1}. These capillary parameters were determined from laboratory measurements of moisture

Table 2   Summary of Borehole Infiltration Test Conditions and Results for Fine Fluvial Sand at the Sevilleta National Wildlife Refuge

| Experiment | Measurement[a] | | | | | | | | | Hydraulic conductivity, 10^{-3} cm/s | | | | Comments[b] |
	H (cm)	R (cm)	H/R	A (cm)	θ_i (cm³/cm³)	Duration (min)	V_{min} (L)	V_{total} (L)	Q_s (L/min)	Glover (1953)	Stephens (1987)	Philip (1985)	Reynolds et al. (1985)	
S1T1	113.0	3.2	34.9	113.0	10.0	241	438	1014	3.1	2.1	2.6	2.8	1.9	1, 2
S2T1	60.2	4.7	12.9	60.2	10.0	232	115	—	2.2	3.6	3.2	4.4	3.4	1, 2
S2T2	91.4	5.1	17.9	91.4	10.0	120	327	737	5.1	4.2	4.5	5.4	4.0	1, 2
S2T3	103.0	4.7	21.9	103.0	10.0	480	418	1772	3.4	2.4	2.7	3.1	2.2	1, 2
S2T4	92.5	5.7	16.2	92.5	10.0	527	360	2342	4.0	3.0	3.3	4.0	3.0	3, 5, 10
S2T5	94.0	8.9	10.6	94.0	8.0	265	534	1328	4.2	2.6	2.8	3.4	2.5	3, 5
S3T1	90.0	6.0	15.0	88.9	10.0	314	347	2005	5.7	4.5	4.8	5.8	4.4	3, 6, 10
S3T2	96.5	8.9	10.8	88.9	12.5	253	487	1746	5.8	3.4	3.7	4.6	3.3	3, 6
S3T3	96.5	8.9	10.8	88.9	12.5	230	487	2214	8.4	5.0	5.4	6.7	5.2	4, 5, 8, 10
S3T4	96.5	8.9	10.8	88.9	13.5	270	470	2176	6.6	3.9	4.3	5.3	4.1	4, 5, 8
S3T6	96.5	8.9	10.8	88.9	13.0	300	479	4617	15.0	8.9	9.7	12.0	9.5	4, 5, 7, 8
S3T7	97.0	8.9	10.9	88.9	7.0	177	592	1802	9.4	5.5	6.0	7.4	5.8	4, 5, 8
S4T2	91.4	5.8	15.9	91.4	11.5	280	333	712	6.9	5.4	5.7	7.0	5.3	7
S4T3	91.4	5.8	15.9	91.4	12.5	255	322	532	1.3	1.0	1.1	1.3	0.81	7
S4T4	22.9	5.8	4.0	22.9	12.5	255	17	125	0.30	1.7	1.1	1.7	1.1	10, 7
S5T3	76.2	5.8	13.3	76.2	13.0	1560	205	8847	8.0	8.3	8.1	10.6	8.2	9
S5T4	76.2	5.8	13.3	76.2	13.0	312	205	1896	4.9	5.1	5.0	6.5	5.0	7, 9
S5T5	76.2	5.8	13.3	76.2	13.0	349	205	2488	6.6	6.9	6.7	8.7	6.8	7, 9
S5T6	76.2	5.8	13.3	76.2	13.0	305	205	2252	6.8	7.0	6.9	9.0	7.0	7, 9
S5T7	76.2	5.8	13.3	76.2	14.0	375	198	2685	6.9	7.2	7.0	9.1	7.1	7, 9
S5T8	76.2	5.8	13.3	76.2	14.5	735	194	4582	9.0	9.4	9.2	11.9	9.3	7, 9
S6T1	155.0	5.8	27.0	122.0	8.8	507	1330	11,200	18.3	6.0	8.4	8.3	7.0	3, 6, 7, 8
S6T3	121.0	5.8	21.0	121.0	9.8	1295	697	10,512	9.1	4.5	5.5	6.1	4.4	3, 6, 7, 9
S6T4	58.0	5.8	10.1	58.0	10.1	1290	121	810	.8	1.2	1.0	1.5	1.0	3, 6, 7, 9
S6T5	57.0	5.8	9.1	57.0	9.7	1275	118	4248	3.0	4.9	4.2	5.9	4.7	3, 6, 7
S6T6	91.4	5.8	15.9	85.3	8.4	2874	370	8342	3.0	2.3	2.5	3.0	2.3	3, 6, 8
S7T1	91.4	15.2	6.0	91.4	7.7	1425	801	23,101	15.7	7.1	8.1	9.7	7.4	6, 7, 11

[a] H = water depth, R = borehole radius, A = screen length, θ_i = initial water content, V_{min} = water volume requirement from USBR (1974), V_{total} = actual water volume used, Q_s = steady infiltration rate.

[b] 1, Test terminated prior to steady-state; 2, gravel-filled borehole; 3, Johnson PVC screen, 0.008-in slots; 4, hacksaw-cut PVC casting with/wire; 5, closed-casing bottom; 6, open-casing bottom; 7, CO_2 injection preceded test; 8, open interval slightly less than H; 9, Q_s long duration result; 10, θ_i estimated from previous tests; 11, Johnson wire wrap screen.

Table 3 Comparison of K_{fs} as Determined by Single-Head and Dual-Head (Reynolds and Elrick, 1986, with permission) Borehole Permeameter Solutions in Sand, Sevilleta Site, New Mexico

		$K_{fs} \times 10^{-3}$ cm/s				
Test number	Experiment number	Glover (1953)	Stephens (1983)	Philip (1985)	Reynolds, et al. (1985)	Reynolds and Elrick (1986)
1	S4T4	1.64	1.14	1.71	1.06	
2	S4T3	1.01	1.09	1.32	8.14 ?	0.73
1	S4T4	1.64	1.14	1.71	1.06	
2	S4T2	5.37	5.78	7.01	5.26	6.73
1	S6T4	1.26	1.01	1.54	0.97	
2	S6T3	4.50	5.56	6.07	4.40	7.61
1	S6T5	4.85	4.24	5.91	4.69	
2	S6T3	4.50	5.56	6.07	4.40	4.14

From Stephens, 1992. With permission.

retention curves and the calculated slopes of relative unsaturated hydraulic conductivity versus pressure head from van Genuchten (1978). Laase (1989) found no statistically significant difference among the three methods for the 27 tests.

Laase (1989) reexamined the multiple permeameter tests conducted in a single borehole (Stephens et al., 1983) to evaluate the method of Reynolds and Elrick (1985, 1986) which uses simultaneous equations to solve for K_s as well as for capillary terms. From the tests listed in Table 2, pairs of tests with different heads in the same borehole were used in the analysis of simultaneous equations proposed by Reynolds and Elrick (1986) (Table 3). For comparison, Table 3 also shows K_s values interpreted from results of a single test. There is very good agreement among all methods for experiments S6T3 and S6T5. However, the two-head test approach produced inconsistent results for the other three pairs of experiments. This may be due to the effects of poorly defined initial moisture content conditions, and inherent limitations in the two-head test approach (Laase, 1989), or heterogeneity. For instance, in tests at Site 6 (S6T3, S6T4, S6T5, S6T6) there are two silty zones several centimeters thick near the lower part of the borehole in which the saturated hydraulic conductivity is about 2×10^{-3} cm/s based on laboratory permeameter tests; in the rest of the profile, the saturated hydraulic conductivity is about ten-fold greater. This type of minor heterogeneity could be all that is necessary to produce poor, or even negative, values of K_s results with the two-head test suggested by Reynolds and Elrick (1985, 1986). Consequently, we strongly recommend that this test be avoided.

2. Comparison to Different Permeameters

At this same site we also determined saturated hydraulic conductivity by ponding experiments, air-entry permeameters, and laboratory core samples as summarized in Table 4 (Stephens et al., 1983; Byers and Stephens, 1983). There is some suggestion that the mean K_s depends on sample size, inasmuch as the largest values of K_s were measured in the 100-cm^3 cores and the smallest values were from the ponding experiments where water infiltrated over an area of about 9 m^2. Nevertheless, the difference among the methods is less than a factor of two. Saturated hydraulic conductivities by the borehole permeameter (Table 2) on the order of 8×10^{-3} cm/s

Table 4   Summary of Hydraulic Conductivity Tests at the Site of
Instantaneous Profile Test, Sevilleta, NM

Type of test	Number of tests performed	Geometric mean, $K_s \times 10^{-3}$ cm/s
Ponding tests	2	7.5
Air-entry permeameter tests	24	8.4
Cores	271	14.0

From Stephens, 1992. With permission.

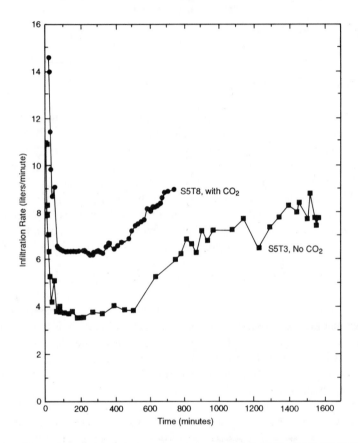

Figure 12   Flow rate from borehole permeameter corrected to 20°C. (From Stephens, 1992. With permission.)

are quite consistent with K_s values by other methods (Table 4), especially for those borehole tests that were relatively long in duration or used carbon dioxide gas to minimize entrapped air.

Figure 12 illustrates the differences in infiltration rate with and without the carbon dioxide pretreatment. Carbon dioxide causes less air entrapment, as indicated by the greater water content near the borehole (Figures 13 and 14). The increase in infiltration rate after several hundred minutes is attributed to decreasing water temperature and its effect on allowing the entrapped air bubbles to dissolve (Stephens et al., 1983). In each case, the zone of maximum saturation is limited to a bulb-shaped

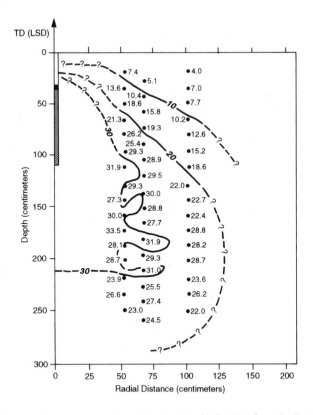

Figure 13 Moisture-content distribution after 400 min of infiltration from the borehole at S5T3 (no CO_2). (From Stephens, 1992. With permission.)

zone surrounding the borehole, even at steady-state infiltration. This field behavior was predicted earlier by computer simulations (Stephens, 1979).

B. LOAM

Borehole infiltration tests were conducted in a loam soil beneath a playa on M-mountain west of Socorro, NM. The purpose of the research tests was to study the behavior of infiltration from a borehole in moderately permeable soils and to compare solutions for K_s in the field with laboratory results. Figure 15 illustrates the variability in particle size and initial moisture content. Together these two parameters often reveal a great deal about the site stratigraphy. Here there is essentially a two-layer profile. Below about 200 cm depth the soil is slightly coarser and less moist than in the upper part. Also, at depths less than 200 cm the uniformity coefficient is about 50, whereas at greater depths the uniformity coefficient is about 15. Porosity of the loam in the upper 200 cm is about 50% (Herst, 1986).

In the borehole, the depth of water was maintained constant at 76 cm using a carburetor valve and styrofoam floats. The borehole radius was about 5 cm, and the hole was cased and screened with polyvinyl chloride (PVC) pipe. Prior to infiltration, carbon dioxide gas was injected into the borehole. The infiltration rate at steady state

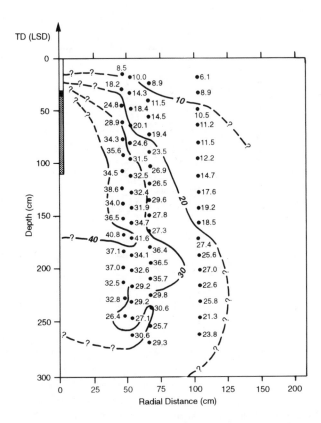

Figure 14   Water-content distribution after 400 min of infiltration from the borehole at S5T8 (with CO_2). The 40% water content contour approximately outlines the zone of complete saturation. (From Stephens, 1992. With permission.)

was 0.12 L/min, after about 12,000 min (8.3 d) had elapsed and more than 3800 liters of water had infiltrated (Figure 16). This volume is approximately 12 times greater than one would estimate from the water volume estimation method proposed by the U.S. Bureau of Reclamation (1974).

Based upon the Mualem/van Genuchten model in the Stephens equation (Table 1), with $\alpha_v = 0.05$ cm^{-1}, $N = 1.1$, the saturated hydraulic conductivity was 1.2×10^{-4} cm/s. Glover's solution, which neglects capillarity, yielded K_s of 1.3×10^{-4} cm/s (Herst, 1986). For most practical problems, these results are quite consistent.

Soil samples were collected in thin-walled shelby tubes (3.6 cm radius) by pushing the samples to a depth of 173 cm with a drill rig. The shelby tubes were converted to constant–head permeameters to determine K_s without further disturbance by extrusion. The K_s of the cores ranged from 2.1×10^{-4} to 1.3×10^{-3} cm/s, using low-salinity water that was virtually the same as that used in the field experiment. The borehole methods produced significantly lower K_s values. We believe that the borehole results are more reasonable for this soil texture (Figure 2) and that possibly the laboratory samples were disturbed. It is also possible that borehole

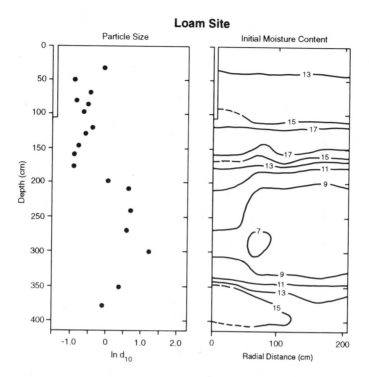

Figure 15   Particle size (μm) and initial moisture content (cm³/cm³) from neutron logging at the M-mountain loam site. (From Stephens, 1992. With permission.)

smearing during hand-auguring and clogging by suspended sediment could have reduced infiltration, thereby causing the borehole test to underestimate K_s. Some difference in K_s between field and laboratory data may be attributed to heterogeneity. Evidence for this is suggested by the successive changes in slope on the cumulative infiltration plot, which may reflect different textures encountered by the advancing wetting front. Regardless of the cause in this case, where fine-textured layers are encountered, extraordinary care should be applied to minimize the effects of smearing along the borehole wall. A wire brush specifically sized for the hole diameter may be a convenient tool to attempt to remove the smear layer. In many fine-textured soils, the effects of smearing may be very difficult to avoid.

The moisture content was monitored by neutron logging in access tubes located at distances of 45, 75, 100, 155, and 200 cm from the borehole (Figure 17). Beneath the borehole the downward propagation of the wetting front was impeded by the coarse, more uniform, lower layer at a depth of about 200 cm, resulting in some lateral flow components that could have contributed to the low K_s values obtained by the borehole permeameter. Note that the entire area monitored for water content is unsaturated, and that the saturated bulb extends less than 45 cm from the borehole. As predicted by computer simulations (Stephens, 1979), and a rough comparison with tests at the sand site (Figure 14), the zone of maximum saturation decreases in size as the effect of capillarity increases.

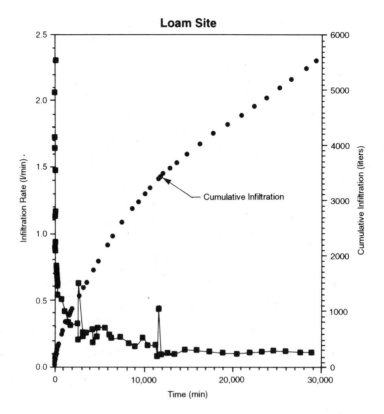

Figure 16   Infiltration rate and cumulative infiltration at the M-mountain loam site. (From Stephens, 1992. With permission.)

C. CLAY SITE

The clay site is located at the Imperial Valley Facility, a landfill site in southern California. The purpose of the site characterization was to quantify the effectiveness of the native materials to mitigate potential seepage through the vadose zone. The Imperial Valley Facility is underlain by a thick sequence of predominantly fine-textured lacustrine sediments. Field tests for saturated hydraulic conductivity were conducted within each of five major stratigraphic units located above the water table; we focus on the results of tests conducted in units 2, 3, and 4 (referred to as units QL_2, QL_3, and QL_4, respectively). A stratigraphic column and brief geologic descriptions of each unit are given in Figure 18.

Unit QL_2 at the test site consisted of very thinly laminated clay and silty clay. The median particle diameter ranged from 1 to 4 μm, as determined by hydrometer analysis. Many near-vertical fractures and desiccation cracks were observed within the test horizons; however, most fractures were filled with silt. The thickness of unit QL_2 at this site was approximately 2 m. The basal part of QL_2 included a thinly bedded silty sand, 1.5 to 20 cm thick.

Unit QL_3 was comprised of a stiff, thickly bedded clay. This unit appeared to be the least permeable on the basis of field observation. The median particle diameter was determined by hydrometer analysis to be between 5 and 10 μm. The test site contained intersecting, subparallel, near-vertical fractures up to several centimeters wide, which were mostly filled with silty sand. Unit QL_3 was approximately 2 m thick.

Unit QL_4 consisted of alternating beds of silt, clayey silt, and stiff clay, each approximately 13 to 20 cm thick. The median particle diameter was determined by hydrometer analysis to be 0.6 μm for the clay, and 13 μm for the silt. Many intersecting, near-vertical fractures were observed; the fractures within the clay beds were filled with clayey silt, whereas the fractures within the silt beds were often filled with clay. Unit QL_4 was approximately 3 m thick. The underlying unit, QL_5, is composed of unfractured fine sand to sandy silt.

Methods for determining saturated hydraulic conductivity to characterize the native clay included air-entry permeameters, slim-diameter (Guelph) borehole permeameters, large-scale borehole permeameters in the field, and falling-head permeameter laboratory analysis of core samples. The results of these techniques are compared and discussed next. Refer to Stephens et al. (1988) for additional detail.

1. Slim-Diameter Borehole Permeameter Tests

The slim-diameter Guelph permeameter (Soil Moisture Equipment Corporation, Santa Barbara, CA) is essentially a Mariotte siphon designed to maintain a constant head of water 5 to 25 cm deep within a 5-cm-diameter borehole (Figure 7). The annular space around the screen was backfilled with 20-mesh silica sand. The rate of water flow into the borehole was determined by measuring the water level decline in the supply reservoir. The temperature of the infiltrating water was also recorded.

The method of analysis for the Guelph permeameter using the two-head test is described in detail by Reynolds and Elrick (1986). To apply the methods of Glover (1953) and Stephens et al. (1987), the steady flow rate and depth of water obtained from the last stage of the two-head test were used. The capillary parameters in the Stephens solution are given in Table 5.

2. Large-Diameter Borehole Permeameter Tests

Three large-diameter, constant-head borehole permeameters also were used in the study. These permeameters were used to determine the saturated hydraulic conductivity of unit QL_2 (borehole permeameter, BH-3), unit QL_4 (borehole permeameter, BH-2), and the composite saturated hydraulic conductivity of the sand and clay of units QL_2 and QL_3. The borehole permeameters were constructed of 10-cm-diameter, 0.015 cm (0.006 in) slot PVC well screen set in 15-cm-diameter (BH-2) and 18.6-cm-diameter (BH-3) boreholes. The annular space around the well screen was backfilled with 20-mesh silica sand. The head of water within the boreholes was maintained at a constant level using a downhole float valve. Infiltration rates were determined by measuring the water-level decline in the water reservoir. The temperature of the infiltrating water was also recorded. Cumulative infiltration curves from the three large-diameter borehole permeameter tests are shown in Figures 19, 20, and 21, respectively. The straight line segments in the cumulative infiltration curves suggest

Loam Site

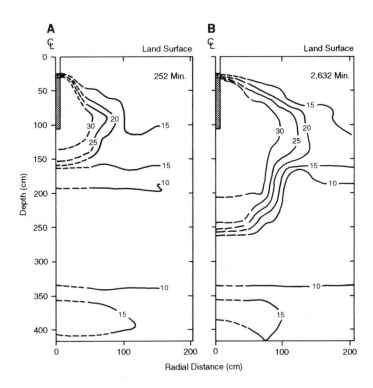

Figure 17   (A,B) Moisture content distributions during borehole infiltration at the M-mountain loam site. (From Stephens, 1992, with permission.) Moisture content profile at (C) $t = 9827$ min and (D) $t = 27,627$ min. (From Herst, 1986. With permission.)

that QL_2 and QL_4 are influenced by fissures. For all three tests, the time to reach steady infiltration ranged from 500 to nearly 2000 h.

3. Comparison of Results of Different Permeameters

Test conditions for the borehole permeameter tests are summarized in Table 6, and the values of saturated hydraulic conductivity from the tests are summarized in Table 7. As indicated in Table 7, the methods of Glover (1953), Reynolds and Elrick (1986), and Stephens et al. (1983) (Table 1) produced results which differ by a factor of about three in unit QL_2 and QL_3 clays and unit QL_4 silt and clay zones. Some of the differences in results from the various methods are attributed to inherent differences in the conceptual models used to develop the mathematical solutions. For example, Glover's method ignores capillary effects. Although the Reynolds and Elrick (1986) and Stephens et al. (1983) solutions do incorporate capillary effects, the Reynolds and Elrick approach is based primarily on analytical methods, whereas the Stephens et al. (1983) approach is based primarily on numerical simulation under

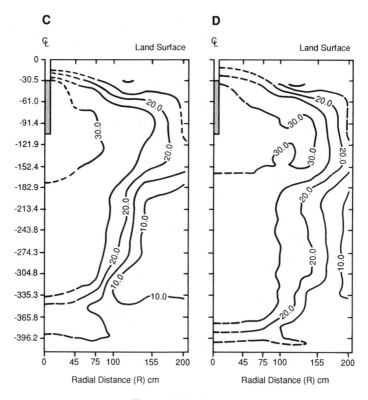

Figure 17 (continued).

variably saturated flow conditions. The results obtained by the Stephens et al. (1987) solution, when applied to the second-stage head and flow rates of the slim-diameter borehole permeameter, agree favorably with air-entry permeameter (AEP) test results (Table 7). The two-head method of analysis proposed by Reynolds and Elrick (1986) produced inconsistent results. For example in the unit QL_3 clay, where the air-entry permeameter seemed effective, the two-head method of analysis yielded negative values of saturated hydraulic conductivity. This may be attributed to a lack of achieving steady-state in the first stage. In summary, results from the clay tests indicate that methods that incorporate capillarity yield smaller K_s values than methods that ignore capillarity. Moreover, when capillarity is considered, the K_s values agree more closely with air-entry permeameter and laboratory core results.

4. Wetting Front Behavior

Upon completion of many of the field tests, the soil was excavated to reveal the wetting front position. Two samples representative of the observed behavior are shown in Figure 22. In the test in QL_5 sandy silt, the wetting front was nearly symmetric, although there was some tendency for greater lateral movement than vertical. This behavior is consistent with that expected for mildly anisotropic, fine-textured, uniform

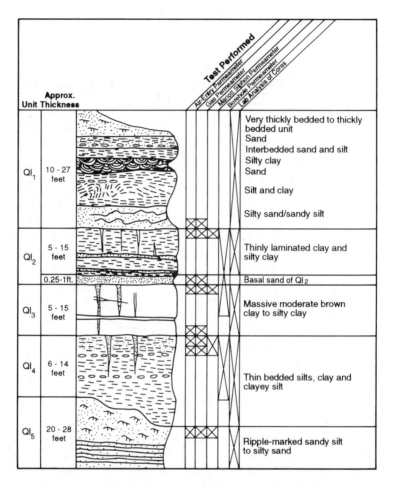

Figure 18   Stratigraphy and field tests conducted at the clay site, Westmoreland, CA. (From Stephens et al., 1988. With permission.)

Table 5   Summary of Mualem/van Genuchten Parameters, Westmoreland, CA

	α_v (cm^{-1})	N	M	θ_s (% vol)	θ_r (% vol)
QL$_2$	3.15×10^{-3}	1.65	3.94×10^{-1}	50.4	26.9
QL$_3$	4.20×10^{-3}	1.66	3.98×10^{-1}	41.2	19.1
QL$_4$ composite	2.19×10^{-3}	1.85	4.59×10^{-1}	44.6	21.1

From Stephens, 1992, with permission.

soil (Stephens, 1979). In the QL$_4$ clay test, fractures clearly inhibited the lateral movement of moisture and may have reduced the flow from the borehole in comparison to what would have occurred in an unfissured clay. In spite of the influence of the fractures within a short distance of the borehole, values of K_s seem to be reasonable for clay. These and other tests at the site suggest that, while some of the theoretical assumptions are clearly violated, in some cases heterogeneity may have only a small effect on the borehole permeameter values of K_s.

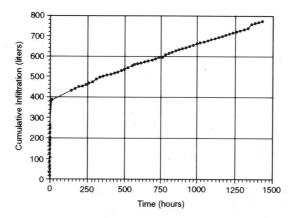

Figure 19   Cumulative infiltration curve at the clay site for large diameter borehole test in QL₂. The linear relationship suggests fractures or macropores dominate over capillary effects throughout infiltration. (From Stephens et al., 1988. With permission.)

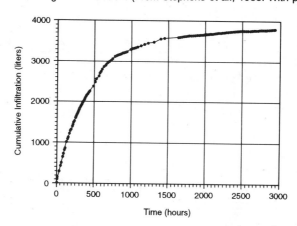

Figure 20   Cumulative infiltration curve at the clay site in QL₂ and QL₃. (From Stephens et al., 1988. With permission.)

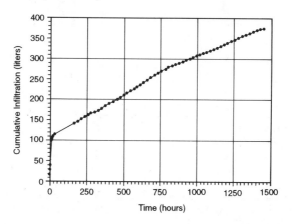

Figure 21   Cumulative infiltration curve at the clay site in QL₄ showing the effect of fractures. (From Stephens et al., 1988. With permission.)

Table 6   Summary of Conditions for Selected Borehole Permeameter Tests, Westmoreland, CA

	Slim-diameter						Large-diameter			
Unit	H_1 (cm)	H_2 (cm)	r (cm)	Q_1 (10^{-2} cm^3)	Q_2 (10^{-2} cm^3)	T (°C)	H (cm)	r (cm)	Q (10^{-2} cm^3)	T (°C)
QL$_2$ clay	15	25	2.54	0.14	0.28	32	129.5	9.3	6.31	29
QL$_2$ and QL$_3$ (composite)	—	—	—	—	—	—	139.0	7.62	4.21	29
QL$_3$ clay	5	15	2.54	0.38	0.14	33	—	—	—	—
QL$_4$ composite	19	32	2.54	0.36	0.56	33	129.5	7.62	4.21	29

From Stephens, 1992. With permission.

Table 7   Saturated Hydraulic Conductivity in 10^{-7} cm/s at the Clay Site, Westmoreland, CA

	Lab cores	AEP	D = 5.4 cm			D = 10 cm	
			Glover	Reynolds and Elrick[a]	Stephens[b]	Glover	Stephens
QL$_2$	0.1–0.5	0.8–1.4	11.0	9.1	3.5	11.0	9.3
QL$_3$	0.3–1.0	0.3	11.0	U[c]	3.2	—	—
QL$_2$ sand/QL$_3$	—	—	—	—	—	7.3	5.7
QL$_4$	—	—[t]	15.0	3.2	4.6	8.3	5.6

[a] QL$_2$: C_1 = 1.62, C_2 = 2.03; QL$_3$: C_1 = 0.9, C_2 = 1.6; QL$_4$: C_1 = 1.8, C_2 = 2.15.

[b] Capillary factors can be found in Table 4 of Chapter 4.

[c] U, Unsuccessful.

From Stephens, 1992. With permission.

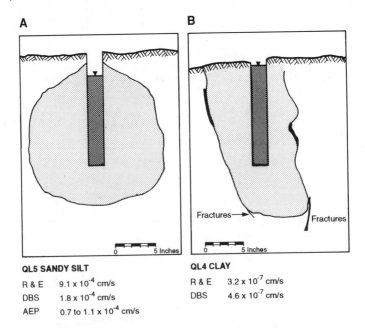

A

B

QL5 SANDY SILT

R & E   9.1 x 10^{-4} cm/s

DBS   1.8 x 10^{-4} cm/s

AEP   0.7 to 1.1 x 10^{-4} cm/s

QL4 CLAY

R & E   3.2 x 10^{-7} cm/s

DBS   4.6 x 10^{-7} cm/s

Figure 22   Wetting front position observed at the end of infiltration with Guelph permeameter showing a test (A) in unfissured silt and (B) in fractured clay. (From Stephens et al., 1988. With permission.)

IV. UNSATURATED HYDRAULIC CONDUCTIVITY*

The purpose of this section is to review methods to measure unsaturated hydraulic conductivity in the field and laboratory. Methods to estimate unsaturated hydraulic conductivity are also presented. These estimation methods are invaluable when the task at hand does not require direct measurement of soil properties or when other data are available which can be used to calculate hydraulic conductivity. For additional detail on many of the methods discussed here, refer to Dirksen (1991). The section concludes with a case study in which unsaturated hydraulic conductivity is determined by laboratory, field, and estimation techniques.

A. LABORATORY METHODS

In comparison to the field methods, there are a larger number of laboratory methods for determining unsaturated hydraulic conductivity. These laboratory methods have been grouped into two general categories: steady-state flow methods and transient flow methods. In most cases, the transient techniques will offer comparable accuracy with considerably less time to complete testing.

1. Steady-State Methods

$$q_i = -K(\theta)_{ij}\left[\frac{\partial \psi}{\partial x_i} + \frac{\partial z}{\partial x_i}\right]$$

z pos. upward

Steady, unsaturated flow can be introduced in horizontal or vertical laboratory columns under constant head or constant flux conditions. By whatever method this is achieved, the unsaturated hydraulic conductivity is calculated from Darcy's equation as simply the ratio of steady flow rate per unit cross-sectional area to the hydraulic gradient (Chapter 1, Equation 17a). The hydraulic conductivity is associated with the mean water content and pressure head established in the column.

A constant head of water at less than atmospheric pressure can be applied to the top and bottom of a vertical soil column when the ends are fitted with porous plates (Figure 23). These plates fit tightly against the soil and have high negative air-entry pressures so that the plate remains saturated as water flows through it under a tension. To supply the water from the soil column under a constant tension, Mariotte siphons are commonly used. A hanging water column is commonly used for collecting drainage at constant pressure. The flow rate can be calculated from transient water level measurements in a calibrated reservoir or burette. Tensiometers are often installed through the wall of the soil column at two different positions to measure soil-water pressure head for computing both the hydraulic gradient and the mean pressure head at steady state. A series of steady-state tests is run to obtain a sequence of conductivity measurements, beginning with a nearly saturated column and ending with a low water content. The range of measurements ideally represents conditions likely to be encountered in the field. Because of the sequence of the tests, it is important to recognize that the unsaturated hydraulic

---

* Significant parts of this section are modified from Stephens (1993) and are reproduced here with permission of the American Society for Testing and Materials.

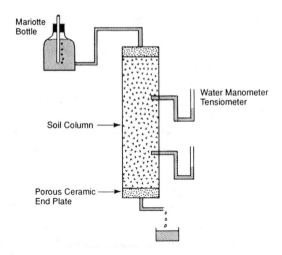

Figure 23   Measuring unsaturated hydraulic conductivity using the column method for establishing steady infiltration.

conductivity is associated with pressure head on the drainage cycle, owing to hysteresis in the K-ψ relationship.

Steady state can also be achieved by applying water at a constant rate until the inflow and outflow, as well as the pressure in the soil column, are constant. The easiest means to apply water at a constant rate is with a peristaltic pump. The water is diffused uniformly across the vertical soil sample surface through a porous end plate or a network of small tubes. Water is allowed to freely drain from the sample to the atmosphere at the lower end of the column. Under this steady flow condition, and if the column length is much greater than the height of the capillary fringe for the soil, the hydraulic gradient will be close to unity in most of the column. Such a test is called the long column method. If the gradient is unity, then the hydraulic conductivity is equal to the steady applied flux. To associate this value of hydraulic conductivity with water status, either a tensiometer can be used to measure pressure head, or a gamma-ray device (Chapter 6) can be used to measure *in situ* water content, for example, but other methods may be equally appropriate. As with the constant-head methods, a series of tests must be conducted to calculate conductivity over a range of moisture conditions. By applying many of these same concepts, Olsen et al. (1993) recently have developed a means to measure unsaturated hydraulic conductivity in soil cores in a modified triaxial test cell. Such a test allows for evaluating the effects of over burden and confining pressure on hydraulic conductivity of partially saturated soil.

Although the calculation of conductivity is very simple, the steady-state methods are tedious to conduct. Moreover, the entire process is very slow to complete, especially for low-water-content conditions, due to the long time requirements to reach steady state. During this equilibration time, care must be taken to overcome potential problems such as bacterial growth, air entrapment, evaporation, dissolution, or shrinking and swelling.

2. Transient Methods

To reduce the time to obtain useful unsaturated hydraulic conductivity data, various transient methods have been proposed. In general, the time savings is at the expense of somewhat more sophisticated apparatus, increased experimental data collection requirements, and more complex mathematical analysis. Here, five transient techniques are discussed: the instantaneous profile method, the Bruce-Klute method, the pressure-plate method, the one-step outflow method, and the ultracentrifuge method.

a. Instantaneous Profile Method

The instantaneous profile method, apparently first developed by Richards and Weeks (1953) and presented later by Watson (1966), uses disturbed or undisturbed soil cores that are subjected to known infiltration or drainage rates. By applying the equation of one-dimensional water mass conservation between two locations along the column, we find that

$$q_{a,t} = q_{b,t} - \frac{1}{\Delta t} \int_b^a \Delta\theta \, dz \tag{13}$$

where $q_{a,t}$ and $q_{b,t}$ are the Darcy velocities at depths a and b ($a>b$) at time t; $t = (t_1 + t_2)/2$; $\Delta t = (t_1 - t_2)$, the time interval between measurements of the change in water content at depth a; and $\Delta\theta = (\theta_1 - \theta_2)$, the change in water content between times t_1 and t_2 at depth a.

Measuring the change in the water content on the right-hand side of Equation 13 is readily accomplished by sequential water content measurements with equipment such as a gamma ray device (see Chapter 6, Section II.A.1). Alternatively, this term can be determined by measuring pressure head and computing water content from soil-water retention data for the soil. If the Darcy velocity, or soil-water flux density, is determined at point b, then the unsaturated hydraulic conductivity at location a is obtained by substituting $q_a = -K(\overline{\psi}) \, \nabla H$ and solving Equation 13 for hydraulic conductivity at the mean pressure head, or mean water content, in the depth interval. The hydraulic gradient, ∇H, is measured with soil-water potential sensors such as tensiometers.

There are various different approaches to obtain the first term on the right-hand side of Equation 13, $q_{b,t}$. One approach is to pump or allow water to flow water into a sample at a known rate. Another is to conduct the experiment in such a way that $q_{b,t} = 0$. Such a condition would occur if somewhere in the column during redistribution there would be a plane of zero flux separating zones where flow is in opposite directions, or if the top of a vertically oriented column were capped to prevent evaporation from a column draining from a state of complete saturation.

This laboratory method has the advantage in its versatility. For instance, the boundary conditions can be steady or transient, data can be obtained for either the wetting or drying cycle, and one does not need to wait for steady state of equilibrium.

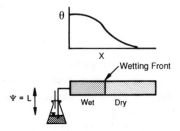

Figure 24   Apparatus for measuring soil-water diffusivity by horizontal infiltration. The water is introduced under a slight tension, ψ. (From Bruce and Klute, 1956. With permission.)

Additionally, the material tested can be heterogeneous, although more extensive instrumentation is required to obtain the properties of individual layers.

The range of soil-water status over which conductivity is obtained depends upon the method of pressure head instrumentation. For example, with tensiometers, the hydraulic gradient can only be determined if the pressure head exceeds about –0.8 bars. However, Hamilton et al. (1981) applied this method in a clay soil by measuring soil-water potential as low as –80 bars using thermocouple psychrometers. They obtained hydraulic conductivities from 10^{-7} cm/s at saturation to nearly 10^{-12} cm/s. The time to complete the test on samples with a length of 11.4 cm was only about 20 d.

b. Bruce-Klute Method

This method utilizes dry soil packed into a thin column that is oriented horizontally (Bruce and Klute, 1956). Water is introduced to one end at a small but constant tension using, for example, a Mariotte siphon (Figure 24). The only data requirements are water-content measurements, either at one location over time or at many locations along the column at the same time. Both approaches produce a set of $\theta(x,t)$ data. Water-content distributions at an instant in time can be obtained by using a specially designed column composed of segments such that, at the desired time, the column is instantly sectioned and water content can be determined gravimetrically from the soil in each segment. A nondestructive method utilizes the gamma beam attenuation method to measure moisture content repeatedly at one point or at discrete locations within a relatively short time span.

The mathematical model is based on the diffusivity form of the unsaturated flow equation:

$$\nabla \cdot D(\theta)\nabla\theta = \frac{\partial\theta}{\partial t} \tag{14}$$

where

$$D(\theta) = \frac{K(\psi)}{C(\psi)} \tag{15}$$

The two independent variables measured during the test (x,t) are required to generate the Boltzman variable $\lambda(\theta)$:

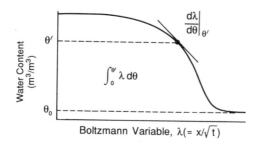

Figure 25 Analysis of horizontal infiltration data to determine soil-water diffusivity by the Bruce-Klute method.

$$\lambda(\theta) = \frac{x}{\sqrt{t}} \tag{16}$$

where x is the horizontal distance from the source to a particular value of θ at time t. Application of the Boltzman variable $\lambda(\theta)$ linearizes Equation 14 to give the solution for diffusivity at water content θ' from the following equation:

$$D(\theta') = -\frac{1}{2}\left(\frac{d\lambda}{d\theta}\right)_{\theta=\theta'} \int_{\theta_i}^{\theta'} \lambda(\theta)\, d\theta \tag{17}$$

As indicated in Figure 25, the terms in Equation 15 can be determined graphically by integration of the area beneath the λ-θ plot up to the water content θ' where D is to be calculated, and $d\lambda/d\theta$ is simply the inverse slope of the λ-θ curve at θ'. To compute the hydraulic conductivity, the specific moisture capacity needs to be determined from an independent analysis of the moisture retention curve, and then Equation 15 is applied to calculate conductivity. The analysis is easily programmed in order to calculate soil water diffusivity and conductivity at any water content.

The Bruce-Klute method is a valuable laboratory tool for obtaining hydraulic conductivity under wetting conditions. In almost all published applications of this method, the tested soils are repacked, and therefore disturbed. Nevertheless, there is nothing to preclude the extension to undisturbed samples if care is taken to preserve soil pore size distributions and structures during sample preparation. The graphical method of analysis is generally most useful for obtaining conductivities at relatively low water content, but it produces results that are less reliable near saturation, because of the large slope of the λ-θ curve there. Where the conductivity is calculated from the diffusivity using moisture retention data from a separate soil sample, there could be a concern about bias introduced from the possible variability in properties between the two samples.

c. Pressure-Plate Method

The pressure-plate method is a third method to obtain conductivity in the laboratory (e.g., Gardner, 1955). The principle here is to force water out of a soil sample under pressure, and examine the rate of water outflow over time. The testing

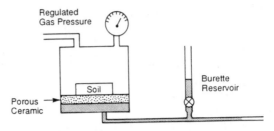

Figure 26 Volumetric pressure-plate extractor to measure unsaturated hydraulic conductivity. Rate of outflow (or inflow) from sample due to an applied pressure change is measured by the advance (or retreat) of the meniscus in the thin-horizontal glass tube.

assembly consists of a tightly sealed metal chamber that contains inside near its base a saturated porous ceramic plate with a large negative air-entry value (Figure 26). A thin soil sample is firmly placed on the ceramic plate to ensure good contact. Then the chamber is closed and pressurized, and nitrogen gas is applied to the chamber space above the porous plate. Usually the test is initiated with the sample near saturation. A small increment of gas pressure in excess of atmospheric pressure is applied and maintained constant, as the volume of water forced from the sample chamber is measured over time. The rate of outflow is conveniently measured in a horizontal glass tube and burette assembly. When the outflow from that pressure increment ceases, another small increment of pressure is applied, and so on, until the pressure-head range of interest is covered. With proper equipment design, the apparatus can be used to rewet the soil and therefore to examine hysteretic behavior.

Although the pressure-plate method is relatively easy to apply, even in the dry range, it does not seem to be widely used in practice. This stems from several problems related to the theoretical development and logistics. For example, the mathematical model is based upon a form of Equation 14 that assumes that during a pressure increment, the diffusivity is constant, even though the moisture content is actually changing. At high soil permeability, flow impedance through the ceramic plate can affect results unless this is taken into account (e.g., Kunze and Kirkham, 1962). Furthermore, shrinkage of the soil during drainage may cause loss of contact with the plate.

d. One-Step Outflow Method

This variation on the pressure outflow method apparently was proposed by Doering (1965) and later by Passioura (1976). This approach forces water from the soil in response to a large step increase of either known applied pressure above the sample or known tension below the sample. One-step outflow data can be collected by draining the sample under a tension using a hanging water column apparatus or by applying a positive pressure in a pressure cell. The rate of outflow is plotted and evaluated. The analysis produces soil-water diffusivity as in Equation 14, but it is assumed that diffusivity is an exponential function of water content, instead of a constant over the increment of applied pressure. Constantz and Herkelrath (1984) designed an outflow cell and procedure for conducting tests under controlled temperature conditions in order to analyze the effect of high temperatures on unsaturated hydraulic conductivity. Kool et al. (1985) developed a convenient computer code (ONESTEP) to analyze the outflow data using the Richards equation along with a nonlinear least-squares technique to adjust unsaturated flow parameters to obtain a best fit to measured outflow. Unfortunately, this approach may lead to nonunique results (e.g., Toorman et al., 1990). There are presently a number of researchers attempting to overcome this limitation of the one-step outflow method, which, in most respects, is a very practical method.

e. Centrifuge Method

This last laboratory method has only recently started to gain popularity, and therefore it has not been extensively tested (e.g., Nimmo, 1987). The principal is to establish a steady-state flow through a sample while it is in a spinning centrifuge and to compute hydraulic conductivity using a form of Darcy's equation that takes into account both matric potential and centrifugal forces. At sufficiently high rotational speeds, the gradients due to matric potential are negligible compared to those imposed by the rotation. Therefore, hydraulic conductivity can be determined from the following equation:

$$K(\psi) = \frac{q}{\rho s_r^2 r} f \tag{18}$$

where q is the flux density into the sample (LT^{-1}), ρ is the fluid density (M/L^3), s_r is the rotation speed (T^{-1}), r is the radius (L) from the axis of rotation, and f is a conversion factor with units of force per unit volume.

Conca and Wright (1990) have developed an ultracentrifuge apparatus that is capable of measuring conductivities as low as about a 10^{-10} cm/s. Not only does the centrifuge method provide conductivities in the very low water content range, but a steady flow field is achieved in a matter of hours, and tests can be completed in a week or two, at an extraordinary savings of time compared to all other methods. Conca and Wright (1992) reported favorable agreement between unsaturated hydraulic conductivity measured by the ultracentrifuge technique and conductivity estimated by the van Genuchten/Mualem procedure (Section IV.C.3.2) from measured soil-water retention data on the same soil. Problems with the technique are few and lie primarily with issues centering on the representativeness of the small sample size (maximum 40 cm^3) and on the compression and change of pore structures during centrifuging. Additionally, at this time the application of the ultracentrifuge method is not widely used because of the great expense of the ultracentrifuge equipment — $80,000 to $100,000 for the commercially available ones. Roughly 2 weeks are required to obtain K-ψ for a single sample, so one ultracentrifuge would only analyze about 25 samples per year. To overcome this, modifications to the ultracentrifuge reportedly allow for multiple samples in the rotor.

B. FIELD METHODS

Field methods are often preferred over laboratory methods because they typically are more representative of bulk or average properties in a heterogeneous soil.

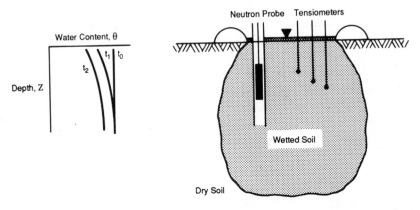

Figure 27 Instantaneous profile test for field analysis of unsaturated hydraulic conductivity.

Additionally, field methods seem to offer greater reliability because laboratory samples are sometimes disturbed to the extent that the value of conductivity is not representative of the field soil sampled. This summary of field methods includes the instantaneous profile method, flux control methods, the flow net method, the bore-hole point source method, and the air permeameter method.

1. Instantaneous Profile Method

The instantaneous profile method, based on the transient drainage of a soil, was described as a laboratory technique and Equation 13 in Section IV.A.2. A square plot, approximately 3 to 10 m on a side, is prepared at a level site and a berm is made on the perimeter. The smaller area is more appropriate for coarse soils in which capillary effects that cause lateral flow are relatively small. If significant lateral flow components cannot be avoided, then sheet pile or trenches filled with impermeable material need to surround the plot. Duplicate tensiometer nests and a neutron probe access tube are emplaced inside the plot, near the center where the flow field is likely to be unaffected by lateral flow beyond the perimeter of an unbounded plot. The bermed area is filled with water until the profile is saturated to the depth of interest. Usually this depth is within 2 to 3 m of the surface, but for clay soils the practical depth of testing may be much less. During infiltration, the tensiometers and neutron probe measurements show when the soil has reached maximum saturation. Either the infiltration rate at constant ponding when the soil is saturated, or the rate of decline in ponded depth upon cessation of water application, may be used to estimate the field saturated hydraulic conductivity if lateral seepage is negligible; otherwise, a small ring can be placed in the center of the berm and the infiltration rate inside it can be used for determining the saturated hydraulic conductivity, provided that the head inside the ring is equal to that outside.

To obtain the unsaturated hydraulic conductivity, the water supply is shut off, the plot is covered to prevent evaporation, and pressure head and water content are measured as the profile drains (Figure 27). The transient data are used to calculate unsaturated hydraulic conductivity at some depth below the top of the soil profile, L, according to the following equation:

$$K(\overline{\theta}) = \frac{\int_0^L (\partial\theta / \partial t)\, dz}{dH / dz} \tag{19}$$

At discrete depths, simultaneously, the hydraulic gradient is calculated from tensiometric data, and the rate of change in moisture content is calculated from the slope of the moisture content versus time plot. The hydraulic conductivity calculated from Equation 19 is associated with the mean water content or pressure head at a particular depth. The analysis progresses from wet to dry conditions to obtain discrete values of hydraulic conductivity over a range of saturations. The measurements are most frequent immediately after infiltration stops and become less frequent with time. The drainage monitoring usually continues until the rate of decrease in moisture content is insignificant. For sandy soils, the test may require only several days to complete. For finer textured materials, tests may require weeks or months to wet and drain the profile. In layered soils, care must be taken in plot preparation to assure that flow is one-dimensional across the layers; nevertheless, conductivity can be calculated at each depth. If tensiometers are used to compute the gradient, the practical lower limit of hydraulic conductivity will correspond to about -850 cm of water pressure head. However, in practice the hydraulic gradient is usually near one so that the determination of pressure head is not always critical to the analysis, at least in relatively uniform soils. If the gradient is one, Equation 13 becomes:

$$K(\overline{\theta}) = \int_0^L \frac{\partial\theta}{\partial t}\, dz \tag{20}$$

Consequently, the hydraulic conductivity often can be estimated from moisture content data alone.

The instantaneous profile method is probably the most widely used of the field techniques. Advantages of the method are that it produces results over a large scale, it uses conventional instrumentation, and the analysis is straightforward. Its disadvantages are the significant time required to complete the tests, the practical limitation to relatively shallow depths and permeable soils, the potential for channeling outside the instrumentation tubing, and the inherent analytical and measurement inaccuracies and uncertainties (e.g., Flühler et al., 1976). For further details on this procedure, including a sample calculation set, refer to Hillel (1980).

2. Constant-Flux Methods

There are two general types of field tests to determine unsaturated hydraulic conductivity that are accomplished by applying water to soil at a constant flux until the profile is at steady state. The methods differ in the manner in which water is introduced: one uses a resistant crust and the other a sprinkler system. The rate of application for both tests is less than that required to cause ponding of the soil. The smaller the application rate, the smaller the hydraulic conductivity. To calculate hydraulic conductivity, simply divide the flux by the hydraulic gradient. The gradient is determined by tensiometers or is estimated to be unity. In contrast to the instantaneous

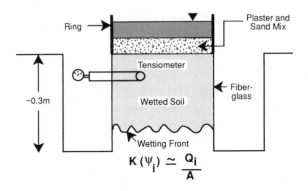

$$K(\Psi_i) \simeq \frac{Q_i}{A}$$

Figure 28   Measuring unsaturated hydraulic conductivity in the field by the crust method applied to a carved soil pedestal.

profile method, the constant-flux methods determine the hydraulic conductivity during wetting rather than drying conditions.

a. Crust Method

The first constant-flux method is referred to as the crust method (Bouma et al., 1974). Here, a pedestal of soil is carved with a nominal diameter and depth of about 0.3 m (Figure 28). The side walls of the soil pedestal are sealed with a metal cylinder or other impermeable material such as a fiberglass coating. The top surface of the soil pedestal is coated with a layer of porous material, such as a mix of plaster of Paris and sand prepared as a slurry, which is less permeable than the underlying soil. This impeding layer forms a crust, which, when a constant head of water exists above it, allows water to infiltrate the soil pedestal at a constant rate. The infiltration rate can be controlled by mixing different proportions of plaster of Paris and sand.

The soil-water pressure head is monitored by installing a tensiometer in the soil beneath the crust. The vertical hydraulic conductivity at the measured pressure head is simply equal to the steady infiltration rate, inasmuch as the hydraulic gradient is assumed to be unity. To obtain a range of values, the test must be run at different rates of infiltration that can be controlled by varying either the crust composition or thickness.

The crust method is not widely used in practice. Because each point on the of K-θ curve requires a new crust and test, the method is very time-consuming to obtain conductivity over a wide range. There also is concern that the flow field may be complex due to the potential for unstable flow created by the low-permeability crust, in which the flow below the fine layer occurs as fingers that occupy only part of the sample cross-sectional area. For layered soils, multiple tensiometers are required to determine both the hydraulic gradient across the layers and average pressure head in the layers. This method is most useful for the wet range, due mostly to the long time to reach steady state.

b. Sprinkler Method

Sprinklers are also used to control water flux to soil (Figure 29). Hydraulic conductivity is determined by the same analysis as described above for the crust

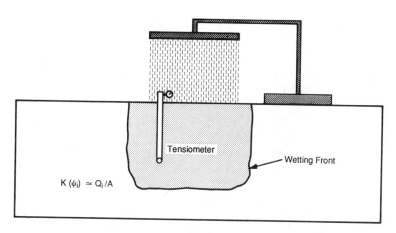

Figure 29   Measuring unsaturated hydraulic conductivity by the sprinkler infiltrometer method.

method (e.g., Youngs, 1964). An apparatus called a sprinkler infiltrometer is commonly used for agricultural and rangeland studies. It or any similar device can be utilized to apply water uniformly over the soil at a constant rate, with an outer buffer area beneath the sprinkler to inhibit lateral flow. For most investigations, the area of application is roughly 3 m². Tensiometers can be used to determine the gradient and mean pressure head. The sprinkler approach overcomes some of the logistical and operational difficulties, as well as perhaps the unstable flow issue, that affect the crust method, but the other limitations remain.

Jeppson et al. (1975) developed a numerical inverse procedure to determine unsaturated hydraulic conductivity from transient sprinkler infiltration data. Their method allows the flow field to be axisymmetric, that is, unconstrained with respect to horizontal flow. However, the porous media must be homogeneous. Instrumentation requirements include tensiometers for measuring pressure head and either neutron probe or gamma ray methods for measuring water content *in situ*. Except for tests in the original article, in which vegetated sites were tested to the 0.3 m depth, there do not appear to be other examples of applications of this method.

3. Flow Net Method

A relatively recent method was proposed to determine the unsaturated hydraulic conductivity by mapping the hydraulic head fields near constant head sources (Stephens, 1985). These sources could be surface impoundments, ditches, or even water-filled boreholes that produce steady-state flow fields above a water table. The principal of the method is based on observations and theory of multidimensional flow in which the water content decreases with increasing distance of flow along a stream tube emanating from the water source. The hydraulic head field is mapped by installing pressure-head sensors such as tensiometers at a sufficient number of locations to contour the measurements in a vertical slice through the flow field. Several flow tubes, arbitrarily selected, are divided into segments that coincide with the equipotential surfaces. The segment of the stream tube closest to the source must be saturated. The hydraulic conductivity of soil at other saturations along the stream tube can be calculated from

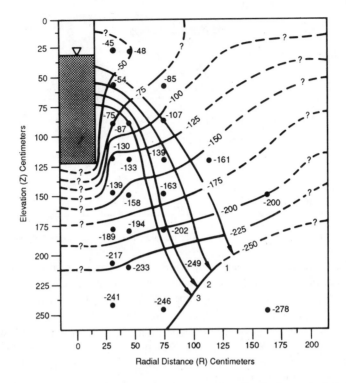

Figure 30   Application of the flow net method for unsaturated hydraulic conductivity for steady infiltration from a borehole. Contours of total hydraulic head in centimeters above land surface datum. A few stream lines are shown with arrows, bold numbers refer to stream tubes, and solid circles show tensiometer locations. (From Stephens, 1985. With permission.)

$$K_{\mathrm{i}} = \frac{K_{S,1} J_1 A_1}{J_{\mathrm{i}} A_{\mathrm{i}}} \tag{21}$$

where the subscript 1 refers to the stream tube segment closest to the source where saturation occurs, subscript i indicates stream tube segments between equipotential lines, J is hydraulic gradient, and A is cross-sectional area of the stream tube in the center of the segment where K_{i} is calculated (e.g., Figure 30). In this method, only the relative hydraulic conductivity (ratio of unsaturated to saturated hydraulic conductivity, K_{i}/K_S) can be obtained, unless the saturated hydraulic conductivity is determined independently (such as with a borehole permeameter).

The flow net method has been applied to sandy and loamy soils with results that compare favorably with other methods (e.g., Larson and Stephens, 1985). The flow net method, which is most useful in the wet range of conductivities, may produce conductivities that are influenced by anisotropy where the flow lines are not in the principal directions. Uncertainties in the graphical flow net procedure can lead to significant uncertainties in the calculated conductivity. Additionally, installation of perhaps more than two dozen or more tensiometers can be labor intensive. Although this method shows promise, it requires further testing before it can be recommended for standard practice.

C. ESTIMATING UNSATURATED HYDRAULIC CONDUCTIVITY

It should be obvious now that most of the field and laboratory methods to characterize unsaturated hydraulic conductivity are either tedious, time-consuming, or have other logistical difficulties. This has led to considerable research in developing ways to estimate unsaturated hydraulic conductivity in lieu of measuring it in the laboratory or field. There are two general approaches for estimation. The first is an empirical approach and the other is to calculate conductivity from moisture retention data.

1. Empirical Approach

Unsaturated hydraulic conductivity can be estimated by assuming that the soil follows a particular model or relationship between conductivity and pressure or moisture content. For example, an equation commonly used, especially in conjunction with analytical solutions, is that developed by Gardner (1955):

$$K(\psi) = K_0' \, \exp(\alpha\psi) \tag{22}$$

where K_0 is the conductivity extrapolated to zero pressure head. In some cases, K_0 is assumed equal to K_s, but usually it is greater than this. For most soils, α ranges from about 0.1 cm^{-1} for sands to 0.005 cm^{-1} for clays.

Brooks and Corey (1964) proposed another commonly used equation:

$$K_r = \left(\psi / \psi_{cr}\right)^{-\lambda p} \tag{23}$$

where ψ_{cr} is the critical pressure head or bubbling pressure head, $\psi \leq \psi_{cr}$, and λ is the pore size distribution index. These two parameters can be calculated graphically from moisture retention curves, or they can be estimated from textural characteristics (e.g., McCuen et al., 1981). In the latter approach, one only needs to know the relative properties of sand, silt, and clay. Then representative parameters are selected, based on statistical analysis of many previously measured values for similar soils (Figure 31).

Clapp and Hornberger (1978) proposed a power function model:

$$K_r = S^{2B+3} \tag{24}$$

where $S = \theta/\theta_{sat}$, and B is an empirical parameter determined for each distinct soil textural classification. Representative values of b were derived by statistical analysis of many data sets of texture and measured moisture retention. To estimate conductivity, only an analysis of soil texture is needed. A similar power-function model was proposed by Mualem (1978), based in part on statistical analyses from a number of soils:

$$K_r\left(S_e\right) = S_e^{3.0+0.015w} \tag{25}$$

where w is evaluated from the moisture retention curve for the soil. In Equation 25, S_e is the effective saturation defined according to

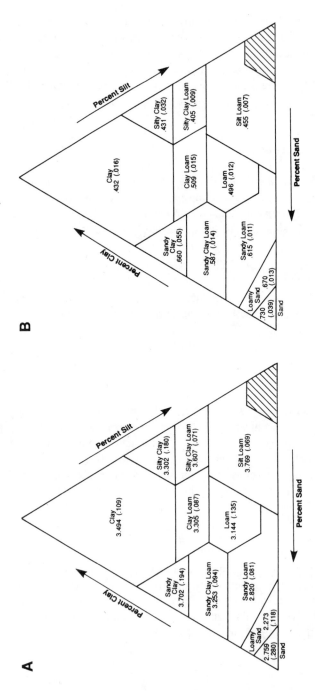

Figure 31 Mean and (standard error) of Brooks and Corey parameters: (A) ln ψ_b and (B) $\lambda^{1/2}$. (From McCuen et al., 1981. With permission.)

$$S_e = \frac{\theta - \theta_r}{\theta_s - \theta_r} \qquad (26)$$

where θ_r is the residual water content and θ_s is porosity. These and other methods are frequently used in lieu of field or laboratory testing, chiefly because they utilize common and easily obtainable parameters.

2. Calculation From Water Retention Data

Calculating hydraulic conductivity (K-ψ) from moisture retention (θ-ψ) curves is perhaps the most popular of all means in current use. The popularity stems from the ease of computation, but the method is also attractive because it is semiquantitative, in that measured properties (porosity, saturated hydraulic conductivity, and moisture retention) are used in the calculation. We forego a discussion of the theoretical development here, such as that in Mualem (1986), and highlight the procedure to calculate the unsaturated hydraulic conductivity.

Statistical models of the pore size distribution have long been applied to develop a mathematical tool for calculating conductivity from moisture retention data (e.g., Childs and Collis-George, 1950; Burdine, 1953; Mualem, 1976). Based upon the Burdine and Mualem models of the porous medium, van Genuchten (1978; 1980) developed a convenient, closed-form analytical solution for calculating conductivity that is based on the following equation, which must be fitted to the measured moisture retention curve:

$$S_e = \left[1 + \left|\alpha_v \psi\right|^N\right]^{-m} \qquad (27)$$

where α_v and N are fitting parameters, and S_e is presented in Equation 26. For the Burdine model, $m = 1 - 2/N$, and for the Mualem model, $m = 1 - 1/N$. A nonlinear least-squares computer routine (SOHYP; van Genuchten, 1978) automatically finds α_v and N by determining the best fit of Equation 27 to observed moisture retention (θ-ψ) data. When residual moisture content, θ_r (Equation 26), is not measured, it too can be obtained by the automatic curve fitting routine. Stephens et al. (1987) compiled a list of the fitting parameters obtained from soil-water characteristic curves for imbibition reported in the literature (Table 8). These fitting parameters are used to compute relative unsaturated hydraulic conductivity from the following equation for Mualem's model:

$$K_r(\psi) = \frac{\left\{1 - \left|\alpha_v \psi\right|^{N-1}\left[1 + \left|\alpha_v \psi\right|^N\right]^{-m}\right\}^2}{\left[1 + \left|\alpha_v \psi\right|^N\right]^{m/2}} \qquad (28)$$

and for Burdine's model:

Table 8   Unsaturated Flow Parameters from van Genuchten's (1978,
           with permission) Three-Parameter Solution of Mualem's
           (1976, with permission) Model Using Imbibition Data

Soil type	Catalog number	α_v (m^{-1})	N	θ_r
Silt "Columbia"	2001	0.015511	1.7676	0.1369
Silt Mont Cenis (limon Silteaux)	2002	0.013647	1.3234	0.0000
Silt of Nave-Yaar	2003	0.072010	2.1969	0.3979
Rideau clay loam	3101	0.069118	2.0604	0.2863
Yolo light clay	3102	0.027000	1.6000	0.1800
Caribou silt loam	3301	0.047125	1.6981	0.2956
Grenville silt loam	3302	0.030702	1.2878	0.0326
Ida silt loam (>15 cm)	3305	0.040000	1.2700	0.0000
Isa silt loam (0–15 cm)	3306	0.089975	1.1768	0.0000
Touched silt loam	3308	0.027302	3.5385	0.0993
Silt loam G.E. 3	3310	0.004233	2.0594	0.1313
Gilat loam	3402	0.017000	2.3000	0.0846
Guelph loam	3407	0.073566	1.7844	0.2193
Rubicon sandy loam	3501	0.052321	1.8570	0.1388
Loam Sand-Hamra Sharon	4004	0.018695	5.1537	0.1997
Plainfield sand (210–250 μm)	4101	0.045177	3.9979	0.0102
Plainfield sand (177–210 μm)	4102	0.038611	4.0409	0.0099
Plainfield sand (149–177 μm)	4103	0.032170	4.0570	0.0069
Plainfield sand (125–149 μm)	4104	0.024903	5.8327	0.0283
Plainfield sand (104–125 μm)	4105	0.022127	4.4446	0.0148
Sand	4106	0.094490	2.0422	0.0000
Sand	4107	0.060000	2.6400	0.0400
Del Monte fine sand	4108	0.016254	4.3600	0.0505
Oakley sand	4112	0.095194	2.0136	0.0255
G.E. 3 sand	4115	0.035965	4.4892	0.0409
Crab Creek sand	4117	0.118896	2.4506	0.0000
Sinai sand	4122	0.023803	5.3076	0.0326
Sand (50–500 μm)	4124	0.019116	4.6747	0.0693
Gravelly sand G.E. 9	4135	0.015048	2.8391	0.0793
Fine sand G.E. 2	4136	0.007192	3.8937	0.0608
Plainfield sand (0–25 cm)	4146	0.033730	3.8518	0.1133
Plainfield sand (25–60 cm)	4147	0.031813	4.1948	0.0724
Aggregated glass bead	5003	0.039748	6.4676	0.0983
Monodispersed glass bead	5004	0.036049	7.6171	0.0363

From Stephens et al., 1987. With permission.

$$K_r(\psi) = \frac{\left\{ 1 - |\alpha_v \psi|^{N-2} \left[1 + |\alpha_v \psi|^N \right]^{-m} \right\}}{\left[1 + |\alpha_v \psi|^N \right]^{2m}} \tag{29}$$

The methods to estimate unsaturated hydraulic conductivity generally have produced good results in nonstructured soils, especially in the wet range (e.g., Stephens and Rehfeldt, 1985). However, there is little experience to establish the reliability of the predictions in the dry range. This includes the pressure-head range below about –800 cm where the Laplace capillary model is not valid because water in this range is held mostly as films on the particle surfaces rather than as fillings of the pore throats. In our practical experience, unless θ_r is measured and prescribed as model input, the model may predict negative values for θ_r. Inasmuch as this is physically impossible, one cannot have much confidence in the calculated conductivities under this condition. On the other hand, θ_r in these models is generally recognized as a fitting parameter, rather than a true physical property (e.g., Ward et al., 1983).

V. SPECIFIC MOISTURE CAPACITY AND MOISTURE RETENTION

The specific moisture capacity is defined in Chapter 1 by Equation 11 as the slope of the soil moisture characteristic curve at a given pressure head. In the Richards equation of unsaturated flow, specific moisture capacity represents the volume of water released from or taken into storage per unit change in fluid pressure head per unit bulk volume of soil. For most agricultural applications, the soil-water characteristic curve, and hence specific moisture capacity, has been most often determined from the drainage cycle. However, for infiltration problems, we need to consider the wetting cycle as well. Differences in the main drainage and wetting curves can be significant, especially for sandy soils, due to hysteresis, as described in Chapter 1.

To obtain this important property of the vadose zone, laboratory methods are most often chosen. These is discussed first, and then we discuss field techniques. In all cases, one needs to calculate the specific moisture capacity directly from the measured soil-water characteristic curve. Although inspection of the Richards equation implies that the specific moisture capacity is required input for solving transient, boundary-value problems, numerical models usually compute this parameter internally from tabular input of the soil-water characteristic curves. For these reasons, this chapter actually presents methods to measure the soil-water characteristic curve. For detailed procedures and discussions of these methods, refer to the excellent references by Klute (1986) and Reeve and Carter (1991).

A. LABORATORY METHODS

There are essentially two types of laboratory methods for measuring soil-water characteristic curves. The first is the hanging water column method, in which the soil water in the sample is subjected to a tension, and the second is the family of pressure-plate techniques. In both techniques it is important to remember that the soil sample is confined to a ring, but in most apparatus the sample is free to shrink or swell, inasmuch as there is no way to reproduce the overburden pressure in these conventional methods. However, the laboratory procedures are designed to minimize shrink and swell effects.

1. Water Column Method

The hanging water column apparatus (Vomocil, 1965) also referred to as the Haines apparatus, is illustrated in Figure 32. The key element of the apparatus is the Büchner funnel, a type of glassware that contains a fritted-glass, porous plate. The pore size of the plate is so fine that after the plate is saturated, the water tension must decrease to –0.1 or –0.3 bars before the air will displace water from the pores of the plate. As described in Chapter 1, this threshold pressure is called the air-entry

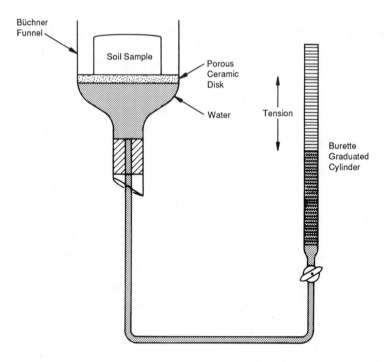

Figure 32   Hanging water column (Büchner funnel) apparatus.

pressure, and it is a characteristic of the pore size distribution of the porous plate. In the hanging water column apparatus, the porous plate is saturated and the tubing below is filled with de-aired water, as illustrated in Figure 32. The soil sample is placed in the Büchner funnel on the upper side of the porous plate. Initially, the tubing and burette assembly are positioned so that the water level in the burette is at the elevation of the top of the porous plate. The testing begins when the soil sample is tension saturated and at equilibrium. During the test, the top of the Büchner funnel is covered to minimize evaporation, and the room temperature is kept constant. To begin testing, the stopcock of the burette is closed and the burette is lowered an arbitrary distance, usually 10 to 30 cm, depending on the soil texture. The stopcock is then opened. At this instant, the soil sample is subjected to a mean negative pressure head equal to the distance between the center of the sample and the water level in the burette. Because of the difference between the atmospheric pressure above the sample and the tension below it, there is a hydraulic gradient created across the sample. Consequently, water flows from the soil, across the porous plate and is stored in the burette. When the fluid level in the burette stops rising, the total amount of water that left the sample is recorded from the burette volume readings. Additionally, the distance between the center of the soil sample and water level in the burette is recorded as a measure of the pressure head in equilibrium with the water still held in the soil. This process is repeated for drainage and wetting cycles as needed. Upon completion, the final water content of the sample is determined. The water content associated with all other previous tensions achieved in the test is calculated by adding to the final water content the incremental volumes of water drained or imbibed from each step. In this way, one obtains pairs of water-content and pressure-head data.

The hanging water column method is practical for relatively wet conditions to pressures as low as about –0.3 bars, depending upon the air-entry value of the porous plate in the Büchner funnel. The method is simple and the apparatus is available from glassware catalogs. Only one sample is tested for each apparatus, but some modifications called porous suction plates and sand suction tables accommodate at least several samples (Reeve and Carter, 1991). When large numbers of points are obtained on a sample, the record keeping and calculation can become quite tedious. In addition, uniform temperature control is essential in the laboratory.

2. Pressure Cell Method

In the pressure cell method, we include apparatus referred to as Tempe cells (Figure 33A), pressure plate apparatus, and pressure membrane apparatus (Figure 33B). The methods share the same concept and equipment configuration, but differ in the composition of the porous plate material that supports the sample and in the strength of materials comprising the pressure chamber. Tempe cells have ceramic plates that remain saturated to approximately –2 bars; the pressure plate apparatus uses ceramic plates with a range to about –15 bars, and pressure membranes of cellulose acetate are useful to about –150 bars. Most commercially available pressure plate and pressure membrane apparatus are large enough in diameter to hold at least several samples, while the Tempe cell accommodates only a single sample.

The basic assembly contains a pressure chamber comprised of a rigid cylinder fitted tightly on the top by a lid or removable plate. Attached to the bottom of the pressure chamber is another removable or fixed end plate. Above the bottom plate is a water-saturated porous disc (ceramic plate, cellulose acetate, or visking membrane), which is seated tightly against the wall of the cylinder by a rubber gasket or O-ring. A wet-soil sample in a sample ring is placed on the porous plate and the top and bottom plates are tightly attached. A nitrogen gas is applied at constant pressure to the sample chamber through a fitting in the cylinder wall. Water is forced out of the sample through the porous disc and into a collection tube, which extends through the wall or base of the cell. Steady pressure is applied until water outflow ceases, typically for periods of 24 h to 7 d or longer. The moisture content still retained in the soil sample is presumed to be in equilibrium with the applied pressure; hence, after the equilibration period, we have obtained only one point on the soil-water characteristic curve. At each equilibration step, the water content can be measured by removing the sample from the apparatus and oven-drying. Such frequent weighing is time-consuming, may disturb the soil, and may cause loss of contact between soil and plate. Where there is only one sample in the pressure cell, usually the pressure is increased stepwise and, without opening the cylinder to the atmosphere, the water content is calculated from the volume of water flowing out from each pressure increment. At the end of the sequence of pressure increments, the final water content is measured by oven-drying. Then the water contents at previous increments are computed by sequentially adding the volume of water released to the volume of water obtained by oven-drying in the final step.

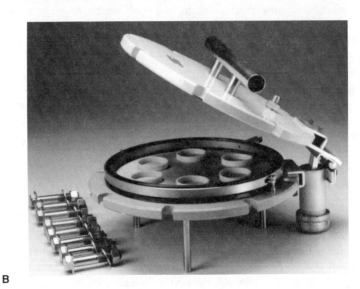

Figure 33   Water retention apparatus: (A) Tempe cells and (B) pressure membrane apparatus. (Courtesy of Soil Moisture Equipment Corp.)

Among the limitations of the pressure plate methods, we note that most equipment is designed to evaluate only the drainage cycle. However, the full hysteretic nature can be measured with a volumetric pressure-plate extractor, an apparatus manufactured by Soilmoisture Equipment Co. (Santa Barbara, CA). Experience has shown that the pressure-plate and pressure-membrane apparatus tend to lead to overestimated values of water content at dry conditions, often at applied pressures greater than about 10 bars. The reason for this is that true equilibrium conditions may take considerably longer to reach than several days; a few months may actually be required. To overcome this problem, we use a laboratory-type psychrometer to determine the actual soil-water potential at the final water content. Considerable care also needs to be exercised in laboratory testing to insure good contact between the sample and porous plate by, for example, applying a thin diatomaceous earth paste and adding lead weights to the top of the soil sample rings.

B. FIELD METHODS

There really are few formal field methods to determine the soil-water characteristic curve. The general procedure is simply to simultaneously measure *in situ*, at the same location, water content and pressure head over a range of conditions and to have an understanding of whether the soil has reached this condition by wetting or drying. The range of moisture conditions can be achieved in any manner of methods. For example, during the instantaneous profile test to measure unsaturated hydraulic conductivity, there usually are both neutron probe access tubes and tensiometers that can measure water content and pressure head, respectively, at a specific depth and time. Stephens (1985) used the same instrumentation within the steady-state moisture field surrounding a water-filled borehole to determine an *in situ* soil-water characteristic curve for the wetting cycle.

One of the inherent problems with field methods is spatial heterogeneity. That is, there are often significant differences in soil texture between locations where water content and pressure head are measured. In developing the moisture retention curve for heterogeneous soils where the spatial correlation length is small, we need to recognize that different instruments produce data that may represent measurements over different regions of the vadose zone. For example, the scale of the tensiometer measurement is on the order of less than a few centimeters, whereas the sphere of influence of the neutron probe measurement may extend over a volume 100 times or so larger.

VI. UNSATURATED HYDRAULIC CONDUCTIVITY CASE STUDY*

The general objective of this case study on water retention and unsaturated hydraulic conductivity determinations is to demonstrate the reliability of the unsaturated hydraulic conductivity estimation model. This example compares unsaturated hydraulic conductivities calculated from moisture retention data for two untilled, uniform sands using the van Genuchten (1980) procedure and the Burdine and Mualem model (Equations 28 and 29) with hydraulic conductivities obtained by (1) field measurements from the instantaneous profile method, (2) one-step outflow experiments (Section IV.A.2.d), and (3) horizontal imbibition (Bruce-Klute) analyses (Section IV.A.2.b).

A. APPROACH

The study area is located on the Sevilleta National Wildlife Refuge along the south bank of the Rio Salado, a braided ephemeral stream about 15 miles north of Socorro, NM. This is the same are where we conducted the borehole permeameter

---

* Reprinted in part from Stephens (1992) with permission from the University of California at Riverside.

Table 9   Summary of Index Soil Properties from
           Laboratory Tests on Cores at the Sevilleta
           Site, NM

	d_{10} (mm)	d_{50} (mm)	C_u	K_s (cm/s)	$\sigma^2_{\ell n\ K}$
Fluvial sand (90 samples)	0.10	0.23	2.7	0.013	0.084
Dune sand (45 samples)	0.16	0.27	1.9	0.012	0.011

From Stephens, 1992. With permission.

tests described earlier in Section III.A. The uncultivated site is sparsely vegetated with salt bush, mesquite, and grasses. The two prominent soils include fluvial sand, representing ancient flood plain deposits of the Rio Salado, and dune sand derived from sandstone outcrops of the Santa Fe Group. The fluvial sand is about 5 m thick. The thickness of the dune sand is unknown, but it exceeds 6.5 m. Table 9 summarizes the textural index parameters (d_{10}, d_{50}, C_u) and saturated hydraulic conductivity, K_s, data that were obtained from transects within each of the soils (Leavitt, 1986). The unsaturated hydraulic conductivity was determined for both fluvial sand and dune sand as described later.

1. Fluvial Sand

The unsaturated hydraulic conductivity of fluvial sand was calculated using moisture retention data obtained from small core samples and from simultaneous measurements of the pressure head, ψ, and volumetric moisture content, θ, collected during a pond drainage experiment. The *in situ* data collected during drainage were also used to measure the unsaturated hydraulic conductivity by the instantaneous profile method, as described in Section IV.B.1.

In preparation for the instantaneous profile experiment to determine K-ψ in the field, an area approximately 400 cm × 400 cm was leveled and diked. Three neutron access tubes were installed near the center of the impoundment to a depth of about 300 cm. A neutron probe (Model 3222 depth moisture gage, Troxler Electronics Inc., Research Triangle Park, NC) was calibrated against gravimetric water content determinations of samples during drilling for the neutron access tubes, and from simultaneous neutron probe readings and gravimetric samples taken outside the probe at the same depth during drainage (Stephens et al., 1983). Mercury manometer tensiometers were installed at about 30-cm increments to a depth of about 240 cm. Water pumped from a shallow alluvial well point was used to create ponded conditions for about 7 h until the water content and pressure head stabilized. At this time the water content averaged about 34%, and the pressure head ranged from 0 to –4 cm. During drainage, the rate of water level decline was measured to calculate the average infiltration rate, 6.0×10^{-3} cm/s at 20°C. The plot was covered with plastic to prevent evaporation during subsequent readings of the tensiometers and neutron probe.

After the instantaneous profile experiment was completed, 100-cm^3 stainless steel core samples, collected from an auger hole near the center of the impounded area, were trimmed, capped, and placed in a padded box for transport. In the laboratory, the samples from the 33, 56, 107, and 122 cm depths were placed in a Büchner funnel filled with tap water and allowed to set overnight. The θ-ψ relationship for desorption was obtained by the hanging column method as described in Section V.A.1.

2. Dune Sand

The unsaturated hydraulic conductivity of the dune sand was calculated from moisture retention data determined from 100-cm³ core samples in a hanging water column apparatus. Continuous core samples of dune sand were collected from depths between 0 and 46 cm (McCord and Stephens, 1989). These samples were thoroughly saturated in the hanging water column apparatus, and then incrementally drained to a pressure head of about −190 cm.

The unsaturated hydraulic conductivity was also determined (Frederick, 1988) on repacked core samples of dune sand by the one-step outflow method (Section IV.A.2.d). To obtain the outflow measurements, the core sample was placed on a Büchner funnel, a tension was applied by quickly lowering the burette, and the cumulative mass of outflow water was measured with a balance. From the fitted parameters obtained from ONESTEP, the unsaturated hydraulic conductivity was calculated using Equation 28.

The laboratory method of Bruce and Klute (1956) (Equation 17) was used to measure the unsaturated hydraulic conductivity by analyzing infiltration into a horizontal column packed with air-dry dune sand (Heermann, 1986). At the end of imbibition, the water distribution along the horizontal column was obtained by quickly extruding the sand from the column, sectioning the column into 1- to 2-cm lengths, and oven-drying the samples. The soil-water diffusivity was computed, and subsequently the unsaturated hydraulic conductivity was calculated by multiplying the diffusivity by the specific moisture capacity. The specific moisture capacity was determined from moisture retention curves for core samples placed on the hanging water column apparatus.

B. RESULTS AND DISCUSSION

1. Fluvial Sand

The moisture retention curves fitted to Equation 27, and the conductivities calculated with Equation 28, for the four cores of fluvial sand are shown in Figure 34. Figure 35A shows the pairs of moisture-content and pressure-head measurements that were collected *in situ* during the instantaneous profile test, as well as the fits of Equation 27 to the field water-retention data. The pore structure parameters obtained from Equation 27 are given in Table 10. Figure 35B also shows the conductivities calculated from the water retention data by applying the SOHYP code. These calculated conductivities are compared to the field results obtained with the instantaneous profile method (Figure 36).

The agreement between all methods is generally good (Figure 36). Near saturation, the calculated conductivities exceed those from the instantaneous profile test.

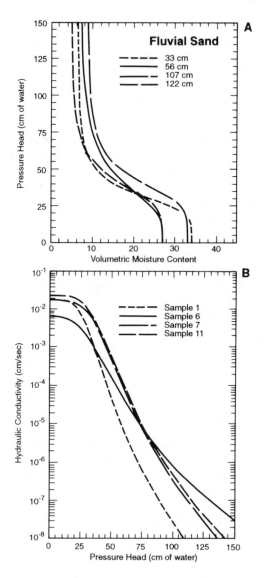

Figure 34   (A) Soil-water retention curves and (B) calculated conductivity using the Mualem
model and SOHYP code on fluvial sand core samples from the Sevilleta site in New
Mexico. (From Stephens, 1992. With permission.)

This is because the saturated hydraulic conductivity determined from the ponded
infiltration rate is about two to three times smaller than the saturated hydraulic
conductivity from the core samples (Table 10). That the laboratory-derived K_s
exceeds the field value is not common, especially where macropores exist. However,
at this site we observed little evidence of macropores. We attribute this result to
entrapped air during the field experiment and possibly to heterogeneity in the
conductivity, inasmuch as the sand did contain thin layers of silt in places.

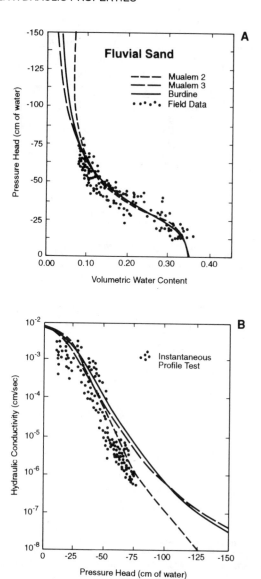

Figure 35   (A) Field-measured soil-water retention data (from tensiometers and neutron probes) and nonlinear least-squares fits with SOHYP. (B) Field-measured and calculated hydraulic conductivity. (From Stephens, 1992. With permission.)

Note from Figure 35A that the Mualem and Burdine models both fit the field moisture retention data almost equally well. However, the calculated conductivities depart significantly from the field measurements in the dryer range (Figure 35B). When SOHYP was used to fit θ_r, the Mualem and Burdine models predicted values of 0.027 and 0.0, respectively. The best agreement between calculated and field measured data is obtained when θ_r in the SOHYP code is fixed at a value of 0.06,

Table 10 Comparison of Calculated Pore Structure Parameters for Fluvial Sand, Derived from Soil Cores (Laboratory) and Instantaneous Profile Test Data (Field), Using Mualem's Model

	Cores (depth, cm)					Field data pairs
	33	56	107	122	Mean	
α(cm^{-1})	0.032	0.027	0.025	0.026	0.028	0.032
N	4.621	3.560	4.710	4.551	4.361	3.230
θ_r	0.031	0.065	0.085	0.045	0.064	0.027
θ_s	0.343	0.271	0.331	0.270	0.304	0.34
K_s ($\times 10^{-2}$ cm/s)	1.8	0.66	1.8	2.4	1.67	0.60

From Stephens, 1992. With permission.

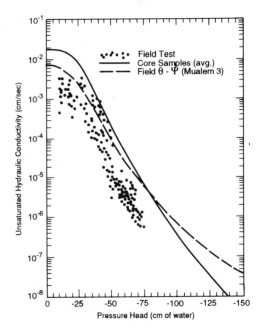

Figure 36 Comparison of hydraulic conductivity calculated from laboratory core-derived and field soil-water retention curves to hydraulic conductivity measured in the field by the instantaneous profile. (From Stephens, 1992. With permission.)

approximately equal to the 15 bar water content, or where $d\theta/d\psi \approx 0$ (Stephens and Rehfeldt, 1985).

The calculated K_r-ψ curves using Mualem's two-parameter model are shown in Figure 37. Differences between the calculated values are less than 50% only at pressure heads greater than about −50 cm of water. At $\psi = -100$ cm the calculated relative hydraulic conductivities for $0.01 < \theta_r < 0.10$ range over more than two orders of magnitude. In Figure 37 there is a clear trend for the slope of the K_r-ψ curve to decrease with decreasing θ_r (and a decreasing N parameter). Thus, for this sand, θ_r can have a significant influence on predicted values of the unsaturated hydraulic conductivity.

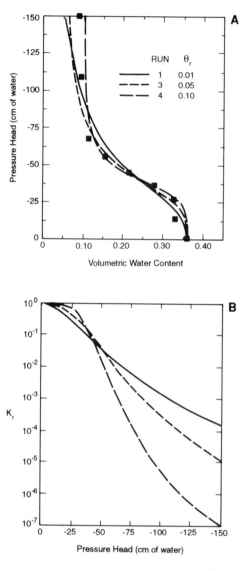

Figure 37   Sensitivity of calculated hydraulic conductivity to residual water content in uniform sand. (From Stephens, 1992. With permission.)

Results of Mualem's three-parameter (α_v, N, θ_r) model fit to the measured θ-ψ data are shown in Figure 38A for three different fixed values of θ_s = 0.332, 0.362, and 0.402. The first value of θ_s was determined by extrapolation of the curve-fitted moisture retention data at pressure heads less than zero, the second θ_s was from the actual value at $\psi = 0$, and the third value was set equal to the porosity, which was calculated from dry bulk density and particle density data. At water contents less than 0.30 the agreement between the curves is excellent. The final fitted θ_r values (0.089,

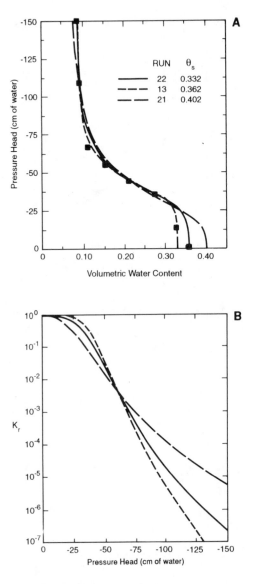

Figure 38   Sensitivity of calculated hydraulic conductivity to saturated water content in a uniform sand. (From Stephens, 1992. With permission.)

0.085, 0.067, respectively) also agree fairly well, and are within the range of water contents measured in the field at large negative pressure heads.

The calculated K_r-ψ curves for these three cases are shown in Figure 38B. The calculated K_r values differ by a factor of about six at $\psi = -30$ cm, even though the θ-ψ curves are nearly identical in this region. There is more than an order of magnitude difference in the calculated relative hydraulic conductivity, K_r, at ψ greater than about –90 cm of water. Figure 38B suggests that θ_r has a strong influence on calculated K_r values in the dry range, in spite of the fact that there is no apparent difference in fit to measured θ-ψ data at water contents less than about 0.30.

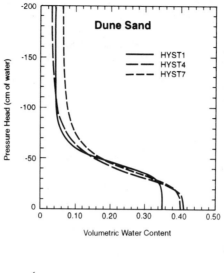

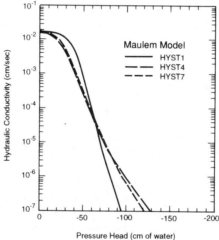

Figure 39 Moisture retention curve and calculated conductivity using the Mualem model and SOHYP code on dune sand core samples. (From Stephens, 1992. With permission.)

2. Dune Sand

Three of the seven cores were selected for representing the moisture retention characteristics and calculated conductivities of dune sand (Figure 39). The calculated mean conductivity of the core samples is compared with conductivities obtained by the one-step outflow and Bruce-Klute methods (Figure 40; see also Table 11). The calculated conductivities based on Equation 27 are in reasonably good agreement with the Bruce-Klute measurements. In the range of data shown, conductivities obtained with these two methods differ by less than 10-fold. These differences are probably due in part to hysteresis, inasmuch as the calculated conductivities are obtained from drainage data, whereas the Bruce-Klute values are derived from an

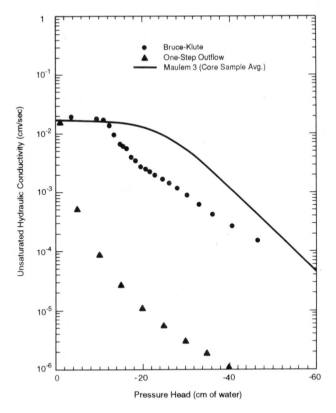

Figure 40 Comparison of calculated conductivity with laboratory measured values for dune sand. (From Stephens, 1992. With permission.)

imbibition experiment. However, the conductivities obtained from the one-step outflow method are significantly smaller than those from the other methods (Figure 40). For reasons that are not clear at present, the very steep K-ψ slope near saturation predicted by the one-step outflow method does not seem to be reasonable for dune sand. The same pore size model (Mualem) was used to formulate ONESTEP and to calculate the conductivity from moisture retention curves Equation 27. The reason for the discrepancy in the ONESTEP results could be due to limitations in the theory, numerical problems in the code, or laboratory-experimental errors. Similar results have been obtained repeatedly for other sandy and loamy soils (Parsons, 1988).

C. CONCLUSIONS

Based upon extensive analyses with fluvial sand and dune sand, as well as many other soils, the SOHYP code to calculate conductivity from water retention data is versatile, convenient, and sufficiently accurate for practical purposes. However, the reliability of the method may depend upon physically reasonable estimates of θ_r and θ_s, and an adequate θ-ψ database. For unstructured sandy soils, improved accuracy in K-ψ predicted with van Genuchten's procedure may be obtained by using a θ_r

value that is based on laboratory-measured water contents at values of ψ that are much greater than those occurring under field conditions. A good visual fit to θ-ψ or data with the Mualem and Burdine models does not guarantee an accurate K-ψ prediction over the entire range of field conditions. A few field- or laboratory-measured values of K-ψ may be necessary to substantiate the K-ψ prediction, particularly under dry conditions. In the dry range ($\psi < -1,000$ cm), none of the pore structure models that are based on capillary theory have been thoroughly validated. Unsaturated hydraulic conductivity measured by the one-step outflow method produced results that were inconsistent with other methods and were unreasonable for the soil tested.

CHAPTER 6

Vadose Zone Monitoring

Through vadose zone monitoring, we obtain information on the status of fluid migration in the vadose zone. Vadose zone monitoring includes methods to measure pressure head and moisture content, as well as methods to sample pore liquids and soil gas. Some of these have been in use for many decades, primarily in applications to agricultural problems. Vadose zone monitoring is also an integral part of many field investigations by hydrologists interested in seepage and natural recharge processes.

More recently, vadose zone monitoring has become an important element of environmental assessments and remediation. In 1980, the U.S. Environmental Protection Agency (EPA) developed regulations under the Resource Conservation and Recovery Act (RCRA) that required vadose zone monitoring at land treatment facilities (USEPA, 1980). In 1988 the EPA proposed to require, on a case-by-case basis, vadose zone monitoring at hazardous waste landfills, surface impoundments, and waste piles, if such monitoring would aid in the early detection of migrating contaminants (Durant et al., 1993). EPA's final determination on vadose zone monitoring requirements on RCRA facilities was expected in 1993 (Durant et al., 1993). States also incorporate vadose zone monitoring as considerations in monitoring systems at underground storage tanks (e.g., in California, (Bonkowski and Rinehart, 1986)) and municipal landfills (e.g., New Mexico Environment Department, 1992). An evaluation of the effectiveness of a soil cleanup action under the Comprehensive Environmental Response Compensation and Liability Act (CERCLA), for example, also can necessitate some form of vadose monitoring. Because of these relatively new regulations and the potential cost savings over deep groundwater monitoring, vadose zone monitoring is gaining increasing attention as a means of monitoring contaminant migration from waste sites. Vadose zone monitoring will likely continue to gain acceptance as more of the techniques prove their utility in the field, and as more effective techniques continue to be developed commercially. The following sections summarize some of the most important vadose zone monitoring devices that are relevant to a wide range of hydrologic and environmental problems.

I. FLUID POTENTIAL

Fluid potential is an important measure of the energy status of soil water. In the vadose zone, spatial differences in the potential determine the direction of soil-water movement. A single sensor also can signal a change in potential energy, or pressure head, for example, from infiltration of rainfall or from seepage from an impoundment. The instruments to be considered include direct methods (tensiometers, psychrometers, and piezometers) and indirect methods (electrical resistance blocks, heat dissipation sensors, and the filter paper method).

A. DIRECT METHODS

In direct methods, the sensor measures the fluid potential in the soil water. Usually these methods are more accurate than the indirect methods, which require measuring some geophysical property that is related to fluid potential through a calibration curve.

1. Tensiometers

A tensiometer is a hydraulic device that measures the matric potential. Gardner et al. (1922) first proposed the concept, and Richards (1928) first applied the concept for developing a tensiometer that is still used today. The device consists of a porous cup, usually composed of a moderately permeable ceramic, that remains saturated under significant tensions (Figure 1). The cup is attached to a water-filled tube that is sealed on the upper end with a removable cap. The purpose of the removable cap is to allow for filling of the tensiometer with water and for purging of air accumulations. A manometer installed through the top part of the water-filled tube measures the pressure of the water in the tensiometer. When the tensiometer is inserted into the soil, the soil imbibes water from the tensiometer, and as this occurs, the water tension in the tensiometer increases until the tensiometer fluid pressure is in equilibrium with soil water outside the cup. The tensiometer pressure measurement can be affected by soil gas pressure, temperature, and overburden pressure (Cassell and Klute, 1986). Typically, the soil air pressure is assumed to be atmospheric pressure, or zero gauge pressure; if this is not the case, then separate measurements of soil air pressure are required to calculate the matric potential. Solutes in the soil are free to pass through the ceramic cup, so the tensiometer measurement does not include the osmotic pressure.

To facilitate rapid equilibrium and accurate results, it is important to install the tensiometer cup with a good contact with the formation. Usually the tensiometer is pushed into the bottom of a slightly undersized hole, or it can be installed in an oversized borehole with a slurry of native soil or silica flour poured around the cup and a seal in the annular space above the slurry to prevent channeling of ponded water.

The principal differences among tensiometers are attributable to the types of manometers used to measure pressure (Figure 1). For the Bourdon or vacuum gauge tensiometer (Figure 1A), the pressure head can be calculated simply as

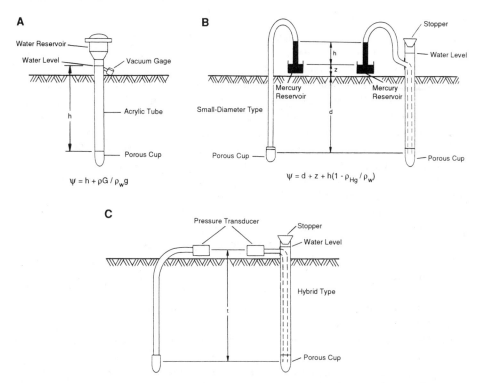

Figure 1 Tensiometer systems using the following pressure sensors: (A) Bourdon gage, (B) mercury manometer, and (C) pressure transducer. (Modified from Stannard, 1990. With permission.)

$$\psi = \frac{P_g}{\rho g} + h \qquad (1)$$

where P_g is the gauge pressure, ρ is the fluid density, g is the gravity constant, and h is the distance between the center of the cup and the manometer connection to the fluid column. However, the Bourdon gauge is a dial-type gauge, with commercially available models having divisions of 2 cbars, or about 20 cm of water. Although this level of accuracy is acceptable to determine irrigation scheduling or to detect relatively large changes in soil-water potential, this type of manometer is not generally recommended to quantify the hydraulic head gradient. For increased accuracy, to within 1 to 2 cm, mercury manometers are preferred (Cassell and Klute, 1986) (Figure 1B). In mercury manometer tensiometers, clear plastic or nylon manometer tubing extends vertically from the mercury reservoir along a scale to read the height of mercury rise above the reservoir. Such tubing material tends to allow air to gradually diffuse into the lines and slow instrument response time. The tensiometer manometer tubing is also prone to thermal expansion. To overcome these difficulties, research grade tensiometer tubing is composed of copper and glass. Thermal expansion can also affect the tensiometer tube, so where highly accurate measurements are required, glass tensiometer tubes are recommended. Some commercially available mercury manometer tensiometers share a single reservoir of mercury to simplify installation of vertically nested tensiometers. In spite of the simplicity and accuracy of mercury manometers, there are potentially significant environmental considerations that should be recognized in handling, spilling, or disposing of mercury.

Transducers are another means of measuring the fluid pressure inside the tensiometer (Figure 1C). But depending upon the type of transducer and recording equipment, this system is relatively more expensive than the others. On the other hand, transducer tensiometer systems are usually convenient, amenable to automation, and do not create potential environmental problems. In one commercially available system, a transducer is attached to the outside of the upper part of the tensiometer tube. Alternatively, the transducer can be specially fitted inside and at the base of the tensiometer column. The latter design allows for installations deeper than about 10 m, which is near the maximum length a water column can remain continuous under a tension. For automatically monitoring a bank of tensiometers with a single transducer, the tensiometer manometer tubing can be connected to a switch called a scanning valve. With the scanning valve system, a data logger is programmed to read the pressure, store the measurement, and switch to the next tensiometer connection. Up to 22 tensiometers can be connected to a single scanning valve, but roughly 10 min may be required for each measurement to reach equilibrium. Although the method is conceptually very attractive, in practice the system requires rather frequent maintenance to remove air from the lines and scanning valve assembly.

Marthaler (1982) designed another means to measure pressure in the tensiometer — a portable transducer system in which the transducer is attached to a hypodermic needle that is inserted through a septum rubber stopper capping the tensiometer (Figure 2). The transducer rapidly records the partial vacuum in the air space above the water column. After the measurement stabilizes and is recorded, the syringe is retracted through the septum and is available for insertion later into the same or another tensiometer. One of the difficulties of this method can be the development of anomalous pressure due to thermal expansion or contraction of the air space at the top of the tensiometer tubing. Heating caused by solar radiation through the clear plastic top can cause pressure buildup if the conductance of the tensiometer cup is not sufficient to allow the excess pressure to quickly dissipate (Stein, 1990).

All tensiometers share a variety of problems. Care must be taken to use de-aired water in the tensiometer and maintain tight fittings to minimize air accumulation in the system. Air also enters the tensiometer by diffusion through the tubing and ceramic cup. Small amounts of air are tolerable, but as the air accumulates, the response time of the system decreases due to the greater compressibility of the air relative to water. Consequently, air must periodically be removed from the system. The most significant limitation is that tensiometers cannot measure soil-water potentials less than about −0.8 bars, because near −1 bar at sea level, the water in the tensiometer would vaporize. For most irrigated soils as well as for uncultivated coarse, uniform textured soils this is not a severe limitation. In fact, tensiometers have functioned quite well even in semi-arid climates in sand dunes where the moisture content is only about 5% (Stephens and Knowlton, 1986). However, temperature variations may seriously affect tensiometer data. Unless the soil-water potential is corrected for diurnal variations in soil temperature, tensiometer measurements in tensiometers exposed at the land surface may only be available once every

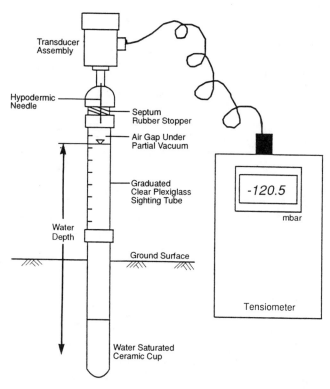

Figure 2 Pressure head measurement using the tensiometer system (Soil Measurement Systems, Inc., Tucson, AZ). *Note:* $\psi = (C_1 \times 9$ tensiometer reading $+ C_2) +$ water level, where C_1 and C_2 are conversion/correction values. (From Stark, 1992. With permission.)

24 h (Nyhan and Drennon, 1989). Freezing conditions can create severe problems for some systems unless antifreeze is used as the solution in the tensiometer (McKim et al., 1976). Be aware that the tensiometer cup is only a few centimeters long, and as consequence, the measurement represents only a very small portion of the soil.

Tensiometers find extensive application in agricultural management, as well as in research. Due to some of the limitations already described, however, tensiometers are not widely used as seepage detectors to monitor waste facilities. The primary factor limiting tensiometer use for detection monitoring is that tensiometers cannot be installed in dry soils where *in situ* soil-water potential is less than about –0.8 bars. Before designing an extensive vadose zone monitoring system that includes tensiometers or other soil-water potential sensors, an effort must be made to estimate the *in situ* fluid potential, in order to select the correct instrumentation. One way to estimate the *in situ* soil-water potential is from measured water content of soil samples and a laboratory-derived moisture retention curve and backgroundwater content.

2. Psychrometers

A soil psychrometer measures the relative humidity within the soil atmosphere from the difference between the wet bulb and dry bulb temperatures. The lower the relative humidity, the faster will be the rate of evaporation, and the lower will be the temperature of an evaporating liquid relative to the dry bulb temperature. Relative humidity, the ratio of the vapor pressure of the soil water (P_v) and the saturated vapor pressure at the field temperature (P_o), is related to the soil-water potential by

$$\psi = \frac{RT}{Mg} \ln\left(P_v / P_o\right) \tag{2}$$

where ψ is the potential energy per unit weight of liquid (i.e., pressure head), g is the gravitational constant, R is the ideal gas constant, M is the molecular weight of water, and T is the Kelvin temperature of the liquid phase. This derivation assumes that water vapor in the soil atmosphere is in equilibrium with the potential energy of the soil water. In contrast to tensiometers, the pressure head determined from psychrometers includes both the matric (capillary plus adsorbed water) and osmotic (solute) potential components, as presented in Chapter 1 in Equation 2. This is because the air-water interface acts as a semipermeable membrane that prevents salts from passing from the dissolved phase into the vapor. The effect of increasing the dissolved salt concentration in the liquid is to lower the vapor pressure of the soil water, and therefore to decrease the soil-water potential. The contribution of osmotic potential to the soil-water potential is usually small and is generally ignored.

Figure 3 illustrates the general features of a Spanner (1951) psychrometer, which is the common design used in field applications. The most important features of the psychrometer are two thermocouples, one for measuring the wet bulb temperature and another for the dry bulb temperature. The thermocouples are very thin wires of dissimilar metal, such as copper and constantan, or Chromel and constantan. If both ends of a thermocouple are joined together and subjected to different temperatures, a current is produced by the so-called Seebeck effect, which allows one to determine temperature. One thermocouple, embedded in insulating material such as Teflon,* serves as a heat sink and the reference electrode to determine the in situ dry bulb temperature. The other thermocouple is located such that its junction is located within the central part of the protective ceramic or stainless steel screen. This thermocouple serves two functions in the process of determining the soil water potential. First, by passing a small current through the thermocouple wire in a particular direction, the junction is cooled to below the dew point by the Peltier effect until water condenses on the junction. Second, after the current is discontinued, the water drop warms and begins to evaporate. During evaporation at the wet bulb temperature, the water drop temperature, and therefore output voltage, decreases very little. When the water drop has evaporated, the temperature quickly returns to the ambient dry bulb temperature. The difference in output voltage between the wet bulb and dry bulb temperatures is compared to a calibration curve to estimate the relative humidity (Briscoe, 1986).

Another approach determines the soil-water potential from the dew point temperature (Neumann and Thurtell, 1972; Campbell et al., 1973; Brunini and Thurtell, 1982). In this mode, the sensor is called a thermocouple hygrometer. Although the same sensor can be used for psychrometric or hygrometric methods, the electronic control and microvolt meter readout device are more complex with the hygrometric approach. The hygrometer thermocouple junction is cooled to the dew point by the

---

* Registered Trademark of E. I. du Pont de Nemours and Company, Inc., Wilmington, DE.

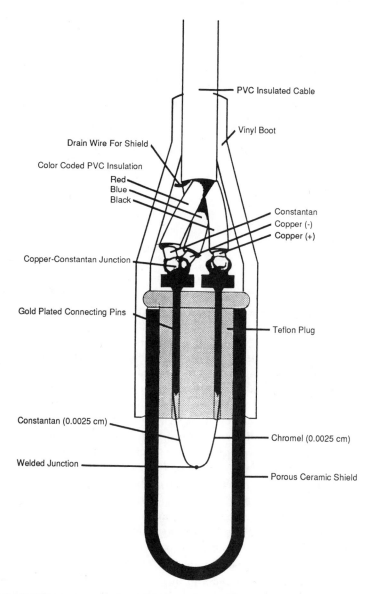

Figure 3 Peltier psychrometer/hygrometer with porous ceramic thermocouple shield. (From Briscoe, 1986. With permission.)

Peltier effect, and the voltage is measured either with the same thermocouple junction (Campbell et al., 1973) or with a second thermocouple junction (Neumann and Thurtell, 1972). Once the dew point is reached, the measurement of temperature depression is continuous, unlike the psychrometric method in which the temperature depression is interpreted from a transient decline of temperature of the evaporating water drop. For this reason, the hygrometric method is considered more accurate (Briscoe, 1986). To illustrate the different responses, Figure 4 compares the output from a psychrometer and a hygrometer. Another reason for greater expected accuracy

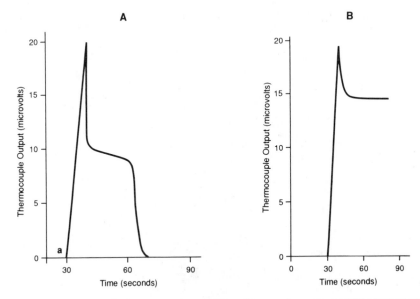

Figure 4   Comparison of a thermocouple output from (A) a psychrometer and (B) a hygrometer. (From Briscoe, 1986. With permission.)

with dew point hygrometers is that, in the psychrometric method, the longer cooling period required to produce the water drop may affect the soil-water environment (Marshall and Holmes, 1979). However, Rawlins and Campbell (1986) indicated that with commercially available devices the accuracy is comparable, although technology exists to improve the accuracy of the dew point system.

Psychrometer calibration is accomplished by using the sensor to determine the output voltage (temperature depression) from a series of sodium or potassium chloride solutions of known different concentrations. These aqueous solutions have known osmotic potentials likely to be encountered in the field (e.g., Lang, 1967). This calibration curve is used to compute the *in situ* soil-water potential from the measured field output voltage. One of the potentially serious problems with all psychrometers is that the calibration can change over time due to microbial growth on the thermocouple wires (Merrill and Rawlins, 1972) or corrosion (Daniel et al., 1981).

To appreciate some of the other limitations of psychrometers, it is important to recognize that over the common range of field conditions, say –0.1 to –15 bars, the soil relative humidity will change very little, from 99.99 to 98.9%, respectively. Thus, even at potentials low enough where most plants permanently wilt, the soil relative humidity is still about 99%. Consequently, psychrometers are used only for dry conditions, usually where soil-water potential is in the range of about –2 to –70 bars or even lower. To accurately determine the field soil-water potential requires extreme care to minimize temperature gradients within the sensor, inasmuch as temperatures must be measured to less than about 0.001°C. Within about 30 cm of the surface, soil temperature fluctuations are often so large that reliable measurements with psychrometers are virtually impossible to obtain. Brown and Chambers

(1987) and Rawlins and Campbell (1986) presented detailed discussions on sources of errors and procedures for using thermocouple psychrometers.

B. INDIRECT METHODS

Tensiometers and psychrometers provide the only direct measurements of soil-water potential in the field. Indirect methods, in which the soil-water potential is determined from a measurement of water content, are also commonly used in the field. The indirect methods include electrical resistance blocks, heat dissipation sensors, and the filter paper method.

All indirect methods share the same general operational principles. When the sensor is placed in firm contact with the soil, water flows into or out of the sensor due to the hydraulic gradient between the potential in the sensor and the soil-water potential, until there is equilibrium. The water content of the sensor depends on the pore size distribution of the sensor as well as the soil-water potential. To relate the measured water content of the sensor to the soil-water potential requires calibration in a pressure-plate apparatus or an independent laboratory determination of the soil-water characteristic curve for the porous sensor. Because the soil-water characteristic curve is hysteretic, the sensor calibration curve for matric potential will also be hysteretic. In practice, this means a water content measurement is not uniquely associated with a single pressure head value, unless one could determine the prior wetting or drying path. The basic differences between the indirect methods for determining soil-water potential are attributed to the approach used to determine the water content of the porous material comprising the sensor. For any of the indirect methods, the accuracy of the potential measurement will be limited to such an extent that the best use of these methods is for general monitoring of soil-water status, rather than for quantifying hydraulic gradient. Additionally, the precision is inherently limited because of hysteresis effects.

1. Electrical Resistance Blocks

Electrical resistance blocks consist of porous gypsum, nylon, or fiberglass within which is embedded an electrode with leads connected to a Wheatstone bridge to measure resistance (Figure 5). The principle of the method is simple. As the water content of the resistance block decreases, the electrical conductivity of the block decreases. The method has been widely used for more than 50 years in agricultural applications (Bouyoucos and Mick, 1940) and is a rather inexpensive method to determine soil-water potential *in situ*. Where frequent measurements are required, electrical resistance units can readily be controlled by a data logger.

There are important differences among the three resistance blocks that affect their response characteristics. Gypsum facilitates passage of electrical current owing to the calcium and sulfate ions that are at saturation within the block. As a result, changes in soil salinity will be masked. The advantage here is that changes in resistance can be more confidently attributed to changes in soil-water potential. Unfortunately, the solubility of the gypsum blocks limits their longevity to perhaps a couple years in wet

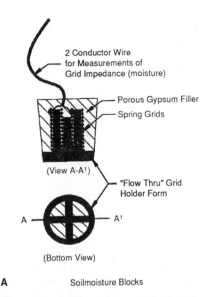

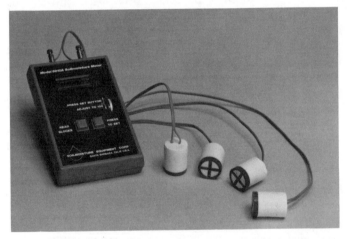

Figure 5   Gypsum block. (Courtesy of Soilmoisture Equipment Corp.)

soils. Daniel et al., (1992) found that wet gypsum blocks disintegrated when repeatedly installed and retrieved. In contrast, nylon and fiberglass blocks are not soluble and can exhibit increases in electrical conductivity because of the increases in ion concentration in the soil solution. Therefore, the resistance of the nylon and fiberglass blocks is influenced by changes in both water content and salinity of fluid within the sensor's pores. Although the nylon and fiberglass blocks are not readily soluble, Campbell and Gee (1986) reported that in some cases these units may fail in 2 to 3 years because of corrosion of the welds. In designing a monitoring system, it is also important to be aware of the different instrument sensitivities that arise because of the different pore size distributions of these blocks. According to Campbell and Gee (1986), gypsum blocks are most sensitive in the dry range at potentials

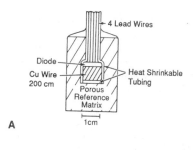

A

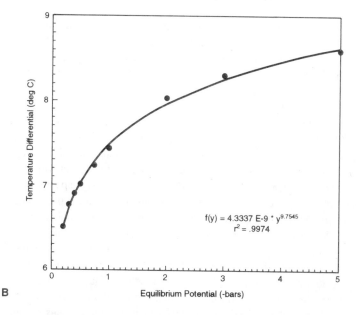

B

Figure 6 (A) Cross-sectional sketch of the Phene heat-dissipation sensor (modified from Phene et al., 1971a, 1971b, with permission), and (B) heat-dissipation probe calibration.

below about –0.3 bars, whereas nylon and fiberglass blocks are most sensitive in the range of 0 to –1 bars.

2. Heat Dissipation Sensors

Heat dissipation sensors are comprised of a porous material, ceramic in the commercially available ones, that surrounds a heating element and thermal detector (Phene et al., 1971) (Figure 6A). The principle, apparently first proposed by Shaw and Baver (1939), is based on the dependence of the thermal diffusivity (thermal conductivity/heat capacity) on moisture content of the porous material in the sensor. A constant current is passed into a resistive heating element along the center axis of the porous material, and during the heating phase, the rate of temperature increase is determined with a thermocouple (Type-T) located at the center of the heating element (J. Belski, 1995, Campbell Scientific, Inc., personal communication). As the moisture content of the porous material increases, the matric potential increases, the thermal

Table 1   Comparison of Calibration Tests Conducted on Four Vadose-Zone
           Monitoring Probes in a Kaolin Clay

| Type of monitoring probe | Range over which probe exhibited maximum sensitivity of response | |
	Volume water content (%)	Soil suction (bars)
Gypsum block	20–50	0–15
Fiberglass res. device	40–55	0–4
Heat dissipation sensor	30–50	0–10
Resistivity probe	20–35	5–15

From Daniel, et al., 1992. With permission.

diffusivity increases, and the temperature differential (heated minus ambient) decreases. The output voltage is related to the potential of the soil by a laboratory calibration curve. The heat-dissipation sensor is insensitive to changes in soil salinity. Over the range of 0 to –1 bar, Campbell and Gee (1986) reported that the precision of the heat-dissipation sensor is approximately ±0.1 bars, and that while the sensor loses sensitivity below this range, it can be used for dry soils. Table 1 compares the range of sensitivity of the heat dissipation sensor with electrical resistance blocks and the portable salinity probe (Section IV.D of this chapter).

Figure 6B shows an example of a heat dissipation sensor calibration curve we obtained in our laboratory. The sensors were embedded in contact with soil material on a pressure plate, and the electrical leads were passed through special fittings in the wall of the pressure plate connecting to a data logger and power pack. The pressure was applied incrementally to the interior of the pressure plate, and as the soil equilibrated with each increment, the output voltage was measured with the data logger. The same setup is used for calibrating electrical resistance blocks. The performance characteristics may differ for each heat-dissipation sensor, so it is important to obtain a calibration curve for each sensor (Daniel et al., 1992).

As an example of a successful large-scale field installation, we note the investigation by Montazer et al. (1988) who placed heat dissipation sensors, along with thermocouple psychrometers, in backfilled boreholes to depths of 387 m at Yucca Mountain, NV. *In situ* potentials ranged from –2 to –15 bars. Over 2 years, 7 of the 10 heat-dissipation sensors eventually became inoperative.

3. Filter Paper Method

The filter paper method is by far the simplest and least expensive of all the indirect methods to measure soil-water potential over a wide range of water contents. Greacen et al. (1987) reported calibrations in the range of about –0.16 to –900 bars. Commercially available filter paper, such as that commonly used in chemistry laboratories, was probably first recognized as a means to measure *in situ* soil-water potential by Gardner (1937). If dry filter paper is buried in soil, it will absorb water, the amount of which depends on the soil-water potential as well as on the water retention characteristics of the filter paper. When the paper water content reaches equilibrium, the paper can be removed from the soil, weighed, and oven-dried and reweighed. The soil-water potential is indirectly determined from the water content and the calibration curve that may be obtained by measuring the soil-water characteristic curve for a stack of filter papers using any of the methods suggested by Campbell and Gee (1986).

Recommended standards for practice have very recently been developed for the filter paper method (ASTM D-5298-92). Greacen et al. (1987) suggested equipment to install filter papers at depth and concluded that this is a very practical way to measure soil-water potential in the range of about –0.5 to –1 bars, where tensiometers are likely to be ineffective. Greacen et al. (1987) concluded that the accuracy of the filter paper method is sufficient to calculate the *in situ* hydraulic gradient. These authors, in addition to Campbell and Gee (1986), pointed out the sources of error and uncertainty in applying the technique. Principal difficulties lie in determining when the paper is in equilibrium with the soil-water potential; a few days to a week may be necessary, with longer times required in very dry soil. Owing to the long equilibration time, the method would be most applicable to determining the moisture status of soil that is at steady-state or very slowly varying. Another reported source of error is attributed to temperature difference between the soil and filter paper, which can be very significant if greater than about 0.001°C, especially in wet soils (Campbell and Gee, 1986). In spite of the simplicity of the method and its low cost, the filter paper method is not widely used. It is labor-intensive for field applications and cannot be automated. Consequently, the filter paper method is not suitable for time-series monitoring at hazardous waste facilities, for instance.

4. Electro-Optical Switch

A related recent development is an electro-optical switch to measure soil-water potential with porous nylon filter disks (Cary et al., 1991). The principle is based on the air-entry value associated with the particular pore size (Chapter 1, Figure 4). At saturation, the nylon filter disk becomes translucent to the transmission of infrared light. At field water contents corresponding to soil-water potentials less than the air-entry value, the disk will dewater; consequently, the light transmission characteristics decrease abruptly from the tension-saturated condition. The soil-water potential can be inferred from the millivolt output change from an electro-optical switch. If the soil contains a number of disks, each having a different air-entry value, then *in situ* soil-water potential could be obtained over a range of conditions. This technique has not been thoroughly tested under field conditions, but seems very promising for automated monitoring purposes.

C. PIEZOMETERS

In some portions of the vadose zone, it is possible that saturated zones occur where fluid pressure is greater than atmospheric. These parts of the vadose zone could include inverted water tables beneath losing streams or perched aquifers, for instance. The simplest method to measure *in situ* positive pressure is with piezometers, although mercury manometer and transducer-type tensiometers can also function under limited positive pressures. The soil mechanics literature describes the many different types of pore pressure piezometers that are commonly in use (e.g., Russell, 1981), and some piezometers with porous stone tips are capable of measuring negative as well as positive fluid pressure. Groundwater hydrologists conventionally construct cased monitor wells to measure water levels as indicators of fluid pressure, so we do not elaborate further on their design. However, we must point out that the monitor wells will not record negative fluid pressure. Consequently, care must be exercised first in placing monitor wells in zones likely to be saturated and second in interpreting time-series data where both positive and negative pressure occur in the soil.

II. MOISTURE CONTENT

Measuring soil moisture content is perhaps the most common form of vadose-zone monitoring, and there is a wide range of methods available. In this section we discuss gravimetric methods, subsurface geophysical methods, and surface geophysical methods. Indirect methods discussed in the previous section can also be used to determine water content simply by calibrating the sensors to water content instead of potential (e.g., Carlson and El Salem, 1987); therefore, the indirect methods, such as electrical resistance blocks, are not discussed again here.

A. GRAVIMETRIC METHOD

The gravimetric method represents the most direct method of measuring water content. The gravimetric method derives its name from the fact that moisture content is determined by weighing. Equations used for determining moisture content, presented in Chapter 1, illustrate two approaches that are commonly in use to express water content. The first is volumetric water content and the second is the mass-based or gravimetric water content. The volumetric moisture content represents the water volume in a known volume of soil, whereas the gravimetric water content represents the mass of water in a known dry mass of soil. When the volumetric water content is determined by a weighing procedure, it too is considered a gravimetric method, but should not be confused with the gravimetric, or mass-based, water content. For typical soils, the volumetric water content (% cm^3/cm^3) will be roughly 20 to 50% larger than the gravimetric water content (%g/g).

To determine moisture content by either approach, a sample of soil is collected from a particular depth by hand auger, power auger, or drilling rig. If the sample is undisturbed and collected so that the bulk volume of soil can be quantified, such as in a known volume of a ring sample (100-cm^3 and 250-cm^3 sizes are common), then the volumetric water content can be determined. Undisturbed samples are usually obtained by carefully pushing a sampling ring into the soil by hand, hydraulic press, or the weight of a drill rig. Undisturbed blocks of soil can also be carved and removed from surface soils or test pits. On the other hand, if the soil sample is disturbed by the collection process so that the *in situ* bulk volume is not known, then only the gravimetric water content can be determined. The gravimetric water content is easier to determine because much less care is required in the collection procedure. For example, gravimetric water content can be determined from approximately 25- to

100-g samples collected from either a hand auger, blade-type or bucket-type tools, drill cuttings pulled off solid- or hollow-stem flight augers, or from split-spoon samples driven into the soil by a 63.5-kg (140lb) hammer on a drill rig.

In all cases, whether the sample is disturbed or undisturbed, great care must be taken to prevent the sample from losing or gaining moisture before it is analyzed. For most standard procedures, samples are placed in a container if they are not in sample rings already; then they are capped, double wrapped in plastic, tightly taped, and placed in a cooler out of direct sunlight for transport. Samples are sometimes dipped in hot paraffin wax as an additional step in preserving soil moisture. Usually, undisturbed soil samples are also tested for hydraulic properties such as conductivity and moisture retention, so the core samples must be carefully packed in the cooler to prevent disturbing the pore structures during shipping.

In the laboratory, the testing process is almost identical for gravimetric or volumetric water content determinations. After removing the field wrapping, the samples are placed in a tared dish and weighed. Then the samples are dried at a temperature of about 105°C in either a conventional convection oven (for 24 h), forced draft oven (for 10 h) or household microwave oven (for 6 to 20 min) (Gardner, 1986). The samples are removed from the oven, and after cooling in a desiccator, they are reweighed to determine the loss of water. This loss on a mass basis, or on a volume basis if converted by dividing by the water density, is directly used to compute the gravimetric or volumetric water content, respectively. Using high-quality laboratory balances, it should be possible to measure the water content with an accuracy between 0.1 and 1.0% (Gardner, 1986). One source of error is attributed to excess heating, which can remove water bound within soil mineral structures, especially clays. Another source of error caused by heating at higher than recommended temperatures is that organic matter that contains water can oxidize. The net effect of overheating is overestimating the water content.

B. SUBSURFACE GEOPHYSICAL METHODS

For many applications, it is important to monitor moisture-content profiles over time. This necessitates a nondestructive measurement that can be made repeatedly. Geophysical techniques are well suited to this purpose, both subsurface and surface techniques. This section presents a review of the subsurface geophysical methods. The first subsurface geophysical technique we discuss is time domain reflectometry, a nonnuclear system of buried probes connected to surface electronics. The next subsurface techniques include several geophysical tools that can be lowered into a borehole to conveniently determine moisture content *in situ*, including the neutron probe, gamma ray apparatus, frequency domain reflectometry, and electrical resistance tomography.

1. Time-Domain Reflectometry (TDR)

In the last decade, there has been a rapidly growing interest in nonradioactive methods to determine soil water content using geophysical properties of soil. One of these methods, time domain reflectometry (TDR), is based on measuring the dielectric constant of soil. Topp et al. (1980), who first proposed the method for soil-water investigations, showed that the dielectric constant of soil is dependent primarily upon the water content through a nearly universal calibration equation that is very insensitive to soil type. Most soil minerals have a dielectric constant of less than 5, whereas water has a dielectric constant of about 78. Thus, water content variations should be easily detectable from measured variations in dielectric constant. However, at very dry water contents, the water adsorbed onto soil particle surfaces may take on a crystalline structure, which leads to a much lower dielectric content. In cold regions the decrease in dielectric constant with freezing allows TDR to distinguish the advance of the freezing front or permafrost depth. Organic liquids typically have lower dielectric constants than water, so one potentially could use TDR to monitor contaminated soil, especially if there are zones of separate phase organic liquids.

To measure the dielectric constant *in situ* using TDR requires determining the rate of travel of electromagnetic energy through the soil. One approach directs a pulse of energy produced from a wave generator along two parallel metallic conductor rods pushed into the soil (Figure 7). The soil between and surrounding the rods is the dielectric medium. The initial approach by Topp et al. (1980) used two coaxial conductor cables with soil infilling the annular space between the coaxial conductors. In either case, the velocity of travel along the transmission lines is described by the following equation:

$$v = \frac{c}{\sqrt{K_{ad}}} \tag{3}$$

where c is the propagation velocity of an electromagnetic wave in free space (3×10^8 m/s), and K_{ad} is the apparent dielectric constant (Topp et al., 1980). Note that soil salinity does not affect transmission velocity; salinity does, however, influence the amplitude of the wave, so that TDR can also be useful to assess soil-water quality. The TDR equipment is usually either a cable tester (e.g., Tektronix model 1502) or a specially designed commercial TDR unit (IRAMS, Foundation Instruments, Ottawa, Canada; TRASE, Soilmoisture Equipment Corporation, Santa Barbara, CA). The TDR source introduces a voltage pulse, and an oscilloscope maps the travel time for the wave to propagate along the transmission lines and reflect back to the source. A typical output trace from the TDR unit is shown in Figure 8. The velocity is determined by dividing twice the length of the transmission line by the travel time shown in Figure 8: $v = 2\ell/t$. From this velocity, one can calculate the apparent dielectric constant from Equation 3 and then use the generic calibration equation (Topp et al., 1980) to compute the water content:

$$\theta = -0.053 + 0.029\,K_{ad} - \left(5.5 \times 10^{-4}\right)K_{ad}^2 + \left(4.3 \times 10^{-6}\right)K_{ad}^3 \tag{4}$$

This calibration equation reportedly has an accuracy of ± 0.02 cm^3/cm^3.

The orientation of the transmission lines in the field influences the resolution of the measurements of water content. The zone of influence surrounding the transmission line mostly includes the soil between the parallel wave guides but some soil

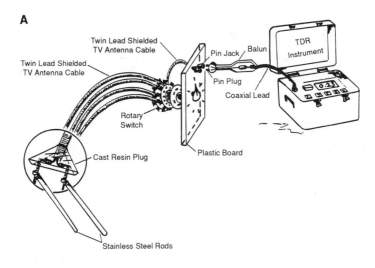

A

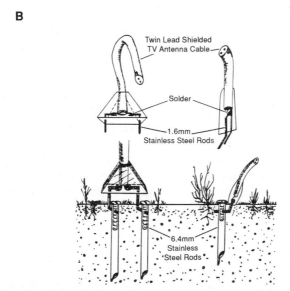

B

Figure 7 Time-domain reflectometry (TDR) equipment: (A) schematically given details of connections, components, and switching used for long-term, in-field installations; (B) connection between the soil TDR line and the shielded television antenna lead. (From Topp, 1987. With permission.)

outside as well, so that the electromagnetic field is an oval-shaped cylinder oriented in the direction of wave propagation. Consequently, the measured water content will represent some average along the waveguides in a space dependent upon the distance between the wave guides. Topp (1987) reported that the sample size is a cylinder having a diameter 1.4 times the distance between the waveguides. Typical installations are stainless steel rods 6 to 12 mm in diameter, 15 to 100 cm long, spaced about 50 mm apart. As the waveguide length and separation distance increase, the resolution of the measurement decreases. Zegelin et al., (1989) developed TDR probes

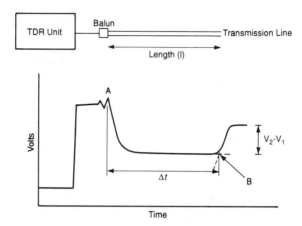

Figure 8   (A) Essential components for measurement of soil-water content by TDR; (B) ideal-
ized TDR output trace showing how the propagation time is determined. (From Topp
et al., 1980. With permisson.)

consisting of three and four wires instead of two. They found that because this design
produced improved clarity in the trace, they obtained improved accuracy in both
water content and electrical conductivity in samples up to 200 mm in diameter.
Regardless of the number of wires, if the waveguides are inserted horizontally in to
the soil, then it is possible to determine water content at discrete depths. If the
waveguides are inserted vertically, then the measured water content will represent a
vertical average. With the vertical emplacement it is not possible to delineate spatial
variations in water content due to soil heterogeneity or soil water movement, unless
clusters of waveguide sets having different lengths are set to overlap to different
depths (Figure 9). Another approach to obtain vertically discrete intervals of water
content is reported to introduce impedances at discrete locations along the transmission line. For any installation, care must be taken to maintain good contact between
the transmission line and the soil.

2. Neutron Probe

Neutron probe methods for measuring soil moisture were first developed in the
late 1940s and commercial devices were available by the mid-1950s (Marshall and
Holmes, 1979). The neutron probe is a geophysical logging tool that is lowered into
a cased borehole on a cable that connects the neutron source with electronic readout
equipment on the surface (Figure 10). The principle is based upon the neutron
thermalization process, in which a radioactive source emits high-energy neutrons
into the soil where, through collisions primarily with hydrogen atoms, the energy of
the neutrons is reduced to lower (thermal) energy levels (Gardner, 1987). Higher
molecular weight elements such as oxygen also slow the neutrons, but far fewer
collisions with hydrogen are required to slow the reaction to thermal energy levels.
The source of the high-energy neutrons in most commercially available neutron
probes is a radioactive americium (atomic weight, 241; half-life, 458 years) and
beryllium mix. The americium actually emits an alpha particle (helium nucleus),

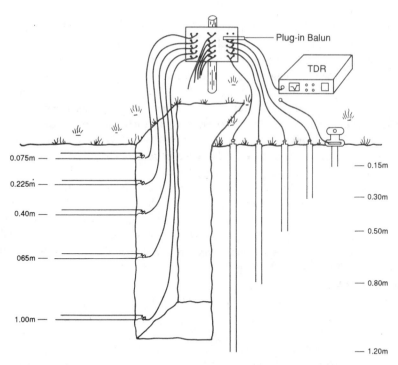

Figure 9   Diagrammatic representation of the TDR transmission line installations. (From Topp, 1987. With permission.)

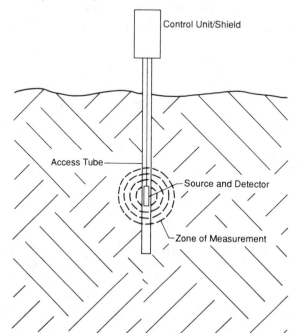

Figure 10   Schematic diagram of water content measurement by the neutron depth probe method. (From ASTM D-5220-92, 1992. With permission.)

which bombards the beryllium atoms that in turn emit a neutron. The CPN 503DR probe (CPN Co., Martinez, CA) produces 110,000 fast neutrons per second at an average energy of 4.5 MeV. The fast neutrons have an average speed of about 1600 km/s, whereas the slow neutrons have an average speed of about 2.7 km/s (Hillel, 1980). The neutrons are emitted more or less radially and form a sphere around the source within which the fast neutrons are attenuated. The size of the sphere of influence, and therefore the sample size, varies inversely with the moisture content. For dry soils the diameter of the sphere is approximately 70 cm, and at saturation the diameter is only about 16 cm; this diameter is unaffected by the strength of the radioactive source (Gardner, 1987). The detector of thermal neutrons commonly consists of a boron trifluoride gas, which, when encountering a thermal neutron, emits an alpha particle as it decomposes to lithium-7 (Goodspeed, 1981); this causes an electrical pulse to be released within the detector. In other words, the number of electrical pulses counted by the detector is proportional to the number of thermal neutrons it encountered. If hydrogen is the only variable affecting the density of slow or thermal neutrons, then a calibration curve can be developed to relate count rate to water content.

Neutron probes with americium-beryllium sources emit alpha particles along with low levels of gamma radiation. The alpha particles have low penetrating power and therefore are easily confined by shields in the container. Other sources of fast neutrons, such as radium-226/beryllium, emit about 60 times more gamma radiation by comparison (Goodspeed, 1981). Consequently, the radium-beryllium source is considered to present more concern for radiation exposure to the operator. For all neutron probes, proper care must always be exercised to limit radiation exposure. Gee et al. (1976) discussed other environmental concerns with operating a neutron probe.

Because the radioactive source decays over time, although very slowly, and because the electronic signal may drift, the field or laboratory measured count rate is divided by the count rate in a known standard. The count ratio is relatively insensitive to these effects, so the neutron probe calibration is commonly expressed as the measured count ratio versus actual moisture content. The standard count is usually obtained in a water-filled tank or in the polyethylene shield that surrounds the source when it is retracted within the probe box.

The calibration of a neutron probe can be done in the field or in the laboratory. Actually, the field approach is more convenient, because it simply involves collecting soil core samples for water content analysis during boring of the hole for installing the access tube. After construction of the neutron probe access tube, the hole is logged to determine the count rate corresponding to the water content at the depths of the gravimetric samples. Most neutron probes have microprocessors that convert the count ratio to water content using a factory calibration curve. The factory calibration curve can usually be modified by reprogramming the slope and intercept if the measured count ratio versus *in situ* water content relationship is linear. Otherwise, one could read out the water content obtained by factory calibration and correct it using another regression equation (e.g., Figure 11). In the laboratory, the neutron probe calibration is developed by packing a tank with field soil recompacted to the field density and water content. An access tube is installed in the tank and the count

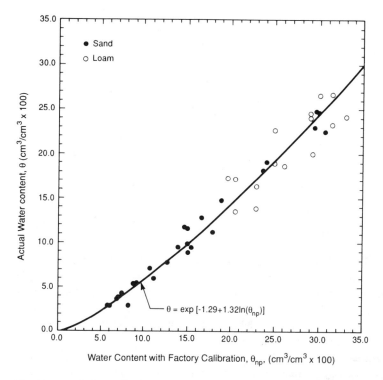

Figure 11   Neutron probe calibration curve at two sites for 5-cm-diameter aluminum access
tubes set in tight-fitting boreholes (503DR hydroprobe, CPN Co., Martinez, CA).
(From McCord, 1985. With permission.)

rate or count ratio is recorded and compared against laboratory analyses of samples
collected from the tank. The soil in the tank must be removed, mixed with additional
water, and logged again. At a minimum, two points — one in the wet end and one
in the dry end — are required to prepare a calibration curve. The laboratory calibration can be inadequate if the compacted soil is unrepresentative of the field condition.

For most problems, the laboratory or field calibration curve is linear, but some
nonlinear behavior may be observed at very high or very low water content. However, the range of natural field moisture contents is typically limited and may not be
sufficiently wide to develop a good regression equation. The disadvantages of the
laboratory calibration are the extraordinary effort required to pack and wet the soil,
and the difficulty in reproducing the fine-scale field heterogeneity. Because of these
difficulties, the factory calibration curve for the probe may be adequate for some
purposes where high accuracy is not required.

There are important sources of error inherent with the neutron probe operation
(e.g., Greacen, 1981). First, it is important first to recognize the probabilistic nature
of the neutron thermalization process. The fast neutrons emanate from the source in
a random fashion and travel along different paths, colliding with a variety of atoms
along the way. So for a fixed probe depth, the number of collisions of the fast neutron
with hydrogen atoms and the resulting production of slow neutrons will also be a
random process. Consequently, two measurements with the probe at the same depth

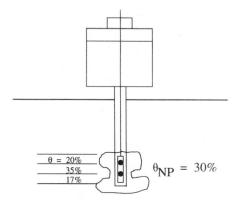

$\theta_{NP} = 30\%$

Figure 12   Schematic illustration of the influence of stratification and variable water content on the region of influence and water content measured by neutron logging.

will yield slightly different counts, with somewhat poorer precision at the wet end. To minimize this source of uncertainty, the counting period should be increased. Assuming that there are no errors in the calibration, the error in measuring water content attributable to counting statistics is less than about 0.7%, with 95% confidence, and the error in measuring a change in water content is less than about 1.0%, with 95% confidence (Gardner, 1986).

Soil heterogeneity also plays an important roll in errors in determining water content with the neutron probe. For example, when the profile consists of alternating layers of sand and clay, and the radius of influence of the neutron source is greater than the thickness of the layers, the probe measurement will be some average of the water content of the sand and clay (Figure 12). This measurement will not represent the water content of either layer, and it would be virtually impossible to calculate the size of the sphere of influence for the measurement. Moreover, the zone of influence would not be a sphere; instead, it would be irregular in shape, with greater penetration into the sand (lower water content) and smaller penetration into the clay (higher water content). Consequently, field calibration in such a layered media can lead to significant errors. This is especially important when there are differences in the depth to which the probe is held in the casing from one logging period to another.

Soil mineralogy also may have an influence the accuracy of the neutron probe measurement. This is because some elements are effective absorbers of the slowed neutrons, including boron, lithium, cadmium, iron, and chlorine. There will also be hydrogen in the organic matter of soil, which will attenuate fast neutrons. Soil contaminated by petroleum hydrocarbons would also tend to attenuate the fast neutrons and lead to a low density of slow neutrons around the detector; however, there has been no published literature on this subject. Clay minerals typically contain water bound within the lattice structure that is not driven off by oven-drying at 105°C for gravimetric analysis. Consequently, one might expect that the neutron probe method would tend to overestimate the field water content in clayey soils. The fact that the neutron probe measurements based on calibration curves for clay have produced very good results is probably explained by the coincidental balance between the increased attenuation of fast neutrons which occurs due to the clay-bound water and the absorption of some of the slowed neutrons due to the chemicals (e.g., lithium, boron, iron) commonly associated with the clay minerals (Gardner, 1986).

There has been a great deal of interest recently in the effect of the borehole characteristics on the neutron probe accuracy. These characteristics include casing material, borehole diameter, position of probe within the casing, and borehole annulus backfill material. The preferred borehole construction for shallow vadose zone investigations calls for 5-cm-diameter (2-in-diameter), thin-walled aluminum tubing within which the slightly smaller diameter neutron probe can be easily raised and lowered. The tubing fits tightly in a smooth-walled borehole to minimize air pockets outside the casing. Other casing materials, such as polyvinyl chloride and steel, tend to absorb the slow neutrons, but it is possible to prepare a separate calibration curve for each different casing material, with some loss of sensitivity. It is also important to be aware that casing couplings and variability in composition of the casing may also influence the neutron probe measurement. Muehlberger et al. (1992) reported success in detecting the true water table during drilling even when drilling mud was in the hole. Several researchers have demonstrated that calibration curves can be developed for boreholes having diameters up to 35.5 cm (14 in) and casing diameters up to 25.4 cm (10 in), even with a 2.6 cm (3 in) grout thickness in the annular space in one case (e.g., Hammermeister et al., 1985; Tyler, 1985; Amoozegar et al., 1989; Kramer et al., 1991). However, some sensitivity is lost and the sphere of influence is diminished relative to measurements in an access tubing of the preferred construction described above. Laboratory experiments logging inside a 10.1-cm-diameter (4-in-diameter) PVC casing set in a 25-cm-diameter (10-in-diameter) boring indicate that it is possible to detect a 9% change of water content in formations where the initial water content is less than 23% (Kramer et al., 1991). Where this level of sensitivity is not adequate for detection monitoring, then other construction methods should be considered to improve the sensitivity of the neutron probe measurements. When logging in large-diameter boreholes, to reduce measurement bias, it is preferred to use a centralizer or a device to keep the probe against one side of the casing. Klenke et al. (1987) also report improved results with neutron probe measurements in large-diameter casings by using a collimator to direct the neutron flux. Although in most applications the casing is vertical, angled installations have been completed adjacent to waste disposal sites (e.g., Reaber and Stein, 1990) and horizontal casings have been placed beneath landfills (e.g., Brose and Shatz, 1987).

Other sources of error in neutron probe logging occur when the probe is located near the land surface. To avoid erroneously large measured water contents, the sphere of influence should not extend across the land surface. For most cases, measurements within 30 cm of land surface should not be considered representative of the formation water content. Condensation inside the casing can also occur and affect the measured water content. Between monitoring events, condensation can be minimized by tightly capping the bottom and top of the casing and suspending a desiccant inside. Otherwise, condensation should be swabbed from the casing walls before logging. Anomalously high water content measurements at early time are also expected if grout that has not been completely cured is used to seal the annulus. Depending upon the cement-water mix, this could take days or months. Care should be taken to prevent cement bridging in the casing annulus or voids within the sphere of influence.

In some hydrogeologic settings, porous backfill around the neutron probe may offer some advantage in detecting the first arrival of seepage. For example, where seepage percolates preferentially along a local pathway, such as a single fracture, it may be difficult to detect because the amount of water stored in the fracture is small relative to the neutron probe sphere of influence. To improve detectability in this situation, a relatively fine textured, dry porous material such as silica flour could be placed in the neutron probe annulus (Daniel B. Stephens and Associates, Inc. 1989).

Much of the preceding discussion has been directed primarily at neutron logging using relatively low radiation sources (10 to 50 mCi) in tools most commonly used for shallow soils at depths to about 30 m. Considerably greater depths are commonly achieved in the petroleum and mining exploration industries using truck-mounted logging tools having source strengths of 2 to 5 Ci. The two types of tools have important differences and the results of the logging are also interpreted differently. With the shallow, low-strength tool, the source and detector are in close proximity to each other; in fact, in one design the source is contained in an annular ring around the detector. This allows for more accurate determination of the depth of the measurement. With the deep, high-strength neutron tool, the source and detector are separated on the tool by a distance of more than a meter. In this arrangement, the detector senses the number of fast, rather than slow, neutrons. Consequently, the counts received are inversely proportional to the formation water content. Because of the greater source-detector spacing, the high radiation source tool monitors a larger volume of material, and the depth to the center of the region of influence is more uncertain, in comparison to the shallow tool. An advantage of the deep logging tool is that its position is mechanically controlled using a winch to move the probe continuously. The tool position and output are synchronized electronically, so that spatially continuous results from even thousands of feet can be obtained rapidly in graphical stripchart records. On the other hand, the standard, commercially available shallow depth tools require manual positioning, with a pause at each discrete depth to take a reading for 0.25 to 4 min; however, mechanical modifications are feasible to position the shallow tool. For instance, Troxler Electronics (Research Triangle Park, NC) offers a logging tool (Model 4350, includes neutron- and gamma-density logs) that can be pulled through the horizontal casing by a winch; this system is tailored to applications of leak detection beneath landfills (Figure 13). Another useful aspect of deep geophysical logging is that the neutron log is often part of a suite of other geophysical logs that can be used to aid in interpretation of the effects of lithology and borehole construction on formation water status. For monitoring below depths of 30.5 m (100 ft), the truck-mounted continuous logging with the deep, high source strength tool seems to be the most practical approach. Unfortunately, the sensitivity of the method for detecting changes in water content has not been evaluated yet for typical borehole constructions used in environmental monitoring.

3. Gamma Ray Attenuation

It is well known that gamma ray radiation is attenuated by the mass of the material surrounding the source. Consequently, the gamma ray attenuation method

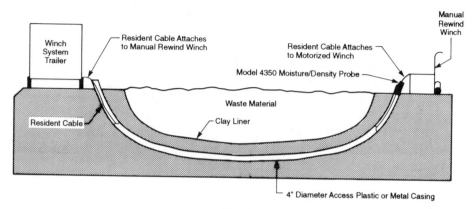

Figure 13   Application of a neutron probe to monitor water content beneath a landfill. (Courtesy of Troxler, 1993.)

is widely used in borehole geophysics and civil engineering to determine the *in situ* bulk density. Materials that can attenuate gamma rays include not only the soil solids, but also casing, water, and, to a negligible extent, air.

In applications to monitor soil water, the gamma ray attenuation method involves simply directing a collimated beam of gamma rays horizontally into the soil from one borehole and monitoring the amount of radiation that reaches a detector at the same depth in a parallel borehole located about 30 cm away. The radioactive source is usually cesium-137, and the detector is usually a sodium iodide crystal or scintillation counter (Hillel, 1980). When the gamma ray strikes the detector, the detector produces a photon that is counted in an analyzer on the land surface; this analyzer effectively filters out secondary sources of gamma ray energy. For fixed borehole conditions and source strength, the counts received are then dependent only upon the water content and the distance between the source and detector. To measure variations in water content, the mass of soil solid and casing must be constant from one monitoring event to the next.

The gamma ray attenuation technique is not commonly used in the field because it requires drilling and casing two holes, instead of one, and these boreholes must be constructed so that their alignment produces constant distances between both of the holes at all depths. This tends to limit the utility for deep logging. On the other hand, the gamma ray method measures water content over about 1 cm depth intervals and offers the highest resolution of the field methods to nondestructively monitor water content. Because of this capability, the gamma ray technique is more likely to be used in laboratory experiments to map out wetting fronts in thin soil columns or sand tanks, for example. Radiation protection is an additional concern with gamma ray equipment.

4. Frequency Domain Reflectometry (FDR)

Like time domain reflectometry, frequency domain reflectometry (FDR) is a nonnuclear method to nondestructively determine *in situ* soil water content from the dielectric properties. The technique is one of the newest of the commercially available field methods (Model Sentry 200-AP, Troxler Electronics Inc., Research Triangle

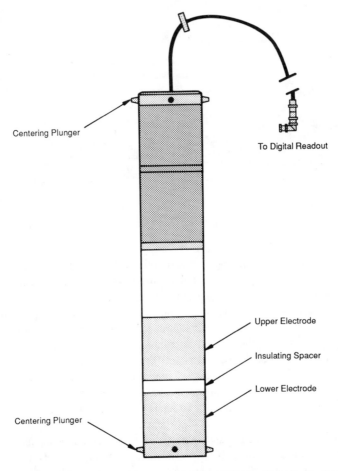

Figure 14   Schematic of the FDR (dielectric) probe (model Sentry 200-AP, courtesy of Troxler Electronics Laboratories, Inc., Research Triangle Park, NC).

Park, NC; Figure 14), but was developed years ago, apparently by Thomas (1966). As we use the term here, FDR is a borehole logging technique in which the sensor, comprised of two conductive electrodes separated by an insulating spacer, is lowered in a boring lined with nonconductive plastic (e.g., thin-walled polyvinyl chloride) casing. The sensor establishes a resonating electromagnetic field, the frequency of which depends upon the dielectric constant of the soil. The water content is determined through calibration with the apparent dielectric constant (e.g., Bell et al., 1987). The logging depth is limited to about 20 ft or so, because of concern that the PVC casing will deform at greater depths and restrict free movement of the tightly fitting tool. The FDR probe (Model Sentry 200, Troxler Electronic Laboratories, Research Triangle Park, NC) also can be buried in soil and engineered fill for permanent monitoring at any prescribed location or depth.

There has been little field experimentation with the FDR probe to delineate any serious limitations; however, the method seems to have distinct advantages over the neutron probe in logging within cased boreholes. The reported precision in measured water content is within 0.2% by volume, a clear improvement. Because the sphere of influence would seem to be only weakly dependent on water content, the method may give more accurate vertical resolution than a neutron probe in layered soils or where there are sharp wetting fronts. The nonnuclear aspect and the fact that measurements are essentially instantaneous are yet other advantages over neutron logging. Technical literature from the manufacturer suggests to avoid using the FDR probe in saline environments, where soil-water salinities exceed 10 g/L. Another weakness of the FDR probe is that it strongly weights water content near the access tube. Thus, irregular voids in the annular spaces near the tube may lead to large errors in apparent water content. Manufacturers' literature recommends careful calibration in the field for individual soils and specific borehole constructions. Because the calibration curve is nonlinear, sensitivity is lost at high water contents.

5. Other Subsurface Geophysical Methods

The position of the perched or regional water-table is a key requirement for many vadose-zone site characterization and monitoring programs. While methods, such as the neutron log, previously described may fill this data need, other borehole geophysical methods should be considered as well. In both clay-rich and sandy material, Keys et al. (1993) reported that the acoustic velocity logs gave the best indication of the water-table. Keys and MacCary (1971) described a field study where both neutron logging and natural gamma logs were run in a steel-cased borehole to locate a perched zone and perching clay bed within a thick basalt.

C. SURFACE GEOPHYSICAL METHODS

Surface geophysical methods are attractive as a means to obtain *in situ* water-content data without intruding into the subsurface. Although there are various surface geophysical methods in wide use for petroleum and mining exploration, such as seismic, gravity, and magnetic methods, most of these have found very limited use in vadose zone applications. Recently, however, many innovative remote-sensing techniques have been developed to detect buried tanks and conductive pipes as part of field investigations at sites of contaminated soil and groundwater. Our emphasis here is primarily on the use of geophysical methods that are appropriate in the evaluation of subsurface moisture. For this purpose, we highlight electrical methods and the ground-penetrating radar method and include a discussion of their relevance to detecting contamination.

1. Electrical Methods

The electrical conductivity of a soil, or its reciprocal, electrical resistivity, is influenced by the electrical conductivity of the soil particles as well as by the fluid characteristics. In the vadose zone, the pores are usually only partly filled with water and air. The ability of the soil to pass current is therefore dependent upon the amount of water, its conductivity, and the distribution of the water in the pores affecting the tortuosity of the path for the current. Where the salinity of the water is constant, then electrical methods are potentially useful to delineate zones of increased electrical conductivity attributable to water content. Within the root zone of an irrigated field,

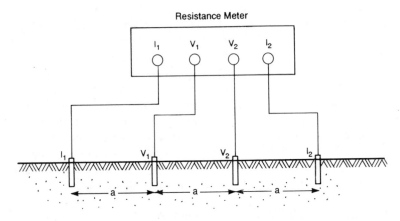

Figure 15   Direct resistivity survey showing the Wenner array with current electrodes (I) and potential electrodes (V).

the assumption of constant salinity is probably not reasonable; in this instance; however, electrical methods can be used to map zones of increased salinity beneath a field when the water content is nearly constant, such as after a few days following irrigation (Rhoades, 1978). Organic compounds tend to slightly decrease electrical conductivity, but the method is reported to be able to delineate phase-separated organic liquids when conductive inorganic liquids are absent.

There are two general types of electrical methods, direct current (DC) electrical resistivity, and electromagnetic induction. The former method uses electrodes in contact with the earth, while the latter uses coils of wire in above-ground equipment to induce a current in the earth. The electrical methods are usually more cost-effective for depth profiling, whereas the induction methods are often preferred for areal mapping (Olhoeft, 1986). By either resistivity or electromagnetic induction, the depth of the investigation is dependent upon the spacing between the electrodes or coils, as well as on the conductivity of the soil. Furthermore, the resolution of the measured apparent resistivity decreases with increasing depth (Olhoeft, 1986).

a. DC Resistivity and Induced Polarization

With the DC resistivity methods, four electrodes are set a fixed distance apart along a line (e.g., Edelfsen and Anderson, 1941; Zohdy et al., 1974). The outer two electrodes apply a constant current, while the inner two electrodes measure the potential difference created within the electromagnetic field (Figure 15). The apparent resistivity of the soil, ρ_{ar}, is determined from the ratio of voltage drop, ΔV, to applied current, I, according to

$$\rho_{ar} = K_e \frac{\Delta V}{I} \tag{5}$$

where K_e is a geometric factor that depends on the electrode configuration. If the soil is homogeneous and semi-infinite, the measured resistivity is called the true resistivity; otherwise, it is called the apparent resistivity. When the electrode spacing, a_e, between all four electrodes is constant, the arrangement is referred to as the Wenner array, and the resistivity is computed with the following equation:

$$\rho_{ar} = 2\pi a_c \frac{\Delta V}{I} \tag{6}$$

The method is used for sounding over an area to a fixed target depth to identify lateral variations in apparent resistivity and for depth profiling at a single location, by changing the distance between electrodes. For example, profiling could be applied to locate the depth to a perched aquifer, and sounding could be used to map its areal extent. The depth of the resistivity measurement, particularly for heterogeneous media, is best determined by computer-assisted mathematical analysis and type curves. After the resistivity profile is obtained, the soil water content could be determined using a calibration curve for the particular soil layer investigated. In most applications, the resistivity data are plotted directly, without conversion to water content, as reconnaissance tools to infer the spatial or temporal variability of water content at a large scale. Kean et al. (1987) used electrical resistivity to successfully monitor shallow soil moisture changes due to rainfall infiltration at four sites, but they could not detect moisture changes below a clay/silt layer that maintained a high antecedent water content.

When the applied current is shut off in the DC resistivity method, there is an initial rapid voltage drop followed by a gradual decrease in the measured voltage (Zohdy et al., 1974). This transient aspect of the measurement is referred to as induced polarization or complex resistivity. Olhoeft (1986) noted that the frequency dependence of the electrical properties, in terms of magnitude and phase, is attributable to chemical activity. Consequently, the induced polarization method may have broader applications in geochemistry, water quality, and contamination investigations. For example, complex resistivity has been effective in mapping organic contaminants, and it may also be a valuable tool to locate breaches in clay liners at landfills and hazardous waste sites where the clay physical and chemical properties have been altered by the organic liquids (Olhoeft, 1986).

b. Electromagnetic Induction

Electromagnetic induction surveys with portable equipment can be easily conducted by one or two technicians, depending upon the type of apparatus selected. The popular, commercially available equipment includes models with fixed intercoil spacing of 3.7 m (Model EM31) and a variable intercoil spacing of up to 40 m (Model EM34-3); both can be obtained from Geonics Limited (Mississauga, Ontario, Canada; Figure 16). The principles of operation are identical for both models. One coil transmits an alternating current at an audio frequency, which generates primary, H_p, and secondary, H_s, magnetic fields, which are sensed by a receiver coil. This ratio, along with the operating frequency of the alternating current, f, and the intercoil spacing, s, are used to compute the apparent electrical conductivity, σ_a:

$$\sigma_a = \frac{4\left(H_s / H_p\right)}{2\pi\mu_0 f s_c^2} \tag{7}$$

A

B

Figure 16 Apparent resistivity measurements in the field using (A) EM-31 and (B) EM-34-3 portable equipment. (Courtesy of Geonics, Ltd., Mississauga, Ontario, Canada.)

where μ_0 is the electrical permeability of free space (McNeill, 1980). Note the convention for induction methods to use units of conductivity (mhos per meter or siemens) instead of resistance (ohm-meters) as in the electrical resistivity methods. The effective depth of exploration with the EM31 is reported to be about 3 m if the coils are horizontal and 6 m in the vertical orientation, but with the EM34-3, at the 40 m intercoil spacing, the depths of exploration increase to 30 and 60 m in the respective orientations (McNeill, 1980). The electromagnetic induction method has been used by hydrologists mostly for locating buried conductors or for delineating contaminant plumes. But in uniform, undestructed soils, the method is applicable to identifying zones of increased soil moisture where recharge is most likely to occur.

Another type of resistivity survey based on electromagnetic induction, electromagnetic offset logging (EOL), places the source coil on the surface and the receiver coil is run down a monitor well to depths up to 100 m. The source coil is moved to distances as great as 100 m from the monitor well. Computer-processed data from this survey produce two-dimensional contours or three-dimensional isometric projections. Waste Microbes Inc. (Houston, TX) asserts that EOL is well suited to located floating and sinking immiscible liquids, as well as dissolved hydrocarbons, even where there is considerable cultural interference on the land surface. This method is still very new in application to environmental problems and is not widely tested; nevertheless, it appears to hold considerable promise.

2. Ground-Penetrating Radar (GPR)

This technique is based on the propagation of an electromagnetic energy pulse into the soil from a transmitter operating at radio frequencies between 10 and 1000 MHz (Redman et al., 1991). A receiver on the surface detects reflections of the energy pulse that occur where the radiowave velocity changes. Such velocity changes occur where the electrical properties of the soil change. The depth to the reflector is approximated from the travel time for the wave to propagate vertically downward and back to the surface. Olhoeft (1986) indicated that the GPR depths of penetration are as much as 30 m in clay-free coarse sand, with resolution to 10 cm, depending upon the frequency of measurement. Although in theory the resolution does not decrease with depth, the depth of penetration is strongly dependent upon soil texture. Clay severely restricts the effective depth of GPR. In fact, Olhoeft (1986) cited an example at an arid zone site near Henderson, NV, where GPR could not penetrate even 1 m, due to the presence of high soil electrical conductivity and high clay content. Inasmuch as the electromagnetic propagation is sensitive to dielectric permittivity, as well as the electrical resistivity, GPR can be used to detect organic liquids, especially above the water table (Olhoeft, 1986). Additionally, GPR does not suffer significantly from interference from buried pipelines and overhead wires that often cause electrical resistivity and electromagnetic induction methods to fail. As with the other surface geophysical methods, GPR is most useful as a reconnaissance tool for subsurface mapping of soil moisture zones, rather than as a direct method to quantify moisture content at a particular location. Portable field equipment and software for data processing are commercially available (e.g., Compu-Radar 2000, Century Geophysical Corp., Tulsa, OK; Figure 17).

D. OTHER METHODS

There are several other techniques to determine soil water content that have been developed. One of these includes standard airborne geophysical methods such as infrared imaging, which fall beyond the scope of this book. A very promising suite of new borehole techniques fall into the category of tomography and use an energy

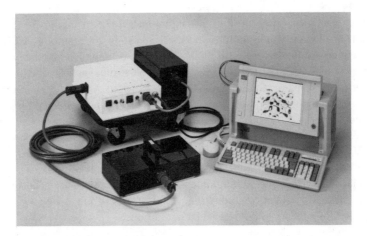

Figure 17 Ground-penetrating radar (GPR) portable field equipment and software for data processing using Compu-Radar 2000. (Courtesy of Century Geophysical Corp., Tulsa, OK.)

source in one borehole and multiple detectors in another borehole to produce three-dimensional images of the water content with the aid of sophisticated computer software. For example, electrical resistance tomography has been developed by Daily et al. (1992); X-ray computed tomography is described by Hainsworth and Aylmore (1983); and Tollner et al. (1987) discussed a special application of nuclear magnetic resonance (NMR) called magnetic resonance imaging (MRI). At this time the tomographic methods are prohibitively expensive for many routine commercial applications and essentially are still in the testing stages. Fiber-optic sensing (Prunty and Alessi, 1987) and microwave techniques (Rasmussen and Campbell, 1987) have also been proposed but are not yet available for field use.

III. SOIL GAS SAMPLING AND MONITORING

During the 1980s, the environmental industry has witnessed phenomenal growth, which has led to the development of innovative techniques to detect subsurface contamination. Soil gas monitoring is one of these techniques that is applicable to highly volatile organic chemicals that partition into the air-filled soil voids. There are both passive and active types of soil gas monitoring techniques. The passive devices are tubes containing absorbent material, usually activated carbon, which are placed in the soil for a period of time and later retrieved for chemical analysis (e.g., Bisque, 1984). A very recent passive soil gas detector is a porous GORE-TEX®* expanded polytetrafluoroethylene material filled with activated carbon, which is buried in the soil (Stutman, 1993). Like the other passive systems, the absorbing material picks up volatile and semivolatile chemicals as they diffuse from the soil and are adsorbed. The passive absorbers must be buried for approximately several days to 2 weeks before retrieval. Another very recent development is the EMFLUX trace-gas detection system (Quadrel Services Inc., Ijamsville, MD). This system uses different

* Registered Trademark of W. L. Gore & Associates, Inc., Elkton, Maryland

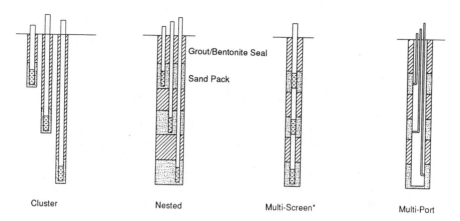

Cluster   Nested   Multi-Screen*   Multi-Port

Figure 18 Comparison of multidepth soil gas monitor well designs. (From Forbes et al., 1993. With permission.)

material to adsorb halogenated and fuel hydrocarbons, followed by thermal desorption and analysis by gas chromatography/mass spectroscopy (GC/MS). The EMFLUX canisters are placed on a cleared area of the land surface for about 3 d. The results, reported as a flux, incorporate the effects of earth tides, which can increase soil gas velocity during the day.

The active type of soil gas sampler is still by far the most popular. It uses vacuum to pump soil vapor from an *in situ* sampling probe directly into a chemical analyzer for a more rapid turnaround compared to the passive systems. The following discussion is directed toward the active soil gas sampling approaches.

A. EQUIPMENT AND INSTALLATIONS

There are several different installations for collecting soil gas in active sampling systems, including permanent soil gas monitor wells, a semipermanent monitoring system, soil gas surveys, and soil gas profiles.

Permanent soil gas sampling probes usually consist of cased boreholes constructed in the dry soil with screens open at discrete depths (Figure 18). The construction is essentially the same as a conventional groundwater monitor well or piezometer. Sampling over the entire depth of the vadose zone would usually require constructing separate soil gas monitor wells screened over different depths. To limit the expense of these installations, Forbes et al. (1993) developed a construction technique to obtain soil gas samples from multiple ports through a single well casing. A unique feature of this permanent, nested sampler is that this same casing allows sampling of groundwater from the base of the well if necessary (Figure 19).

Discrete depth sampling over the entire vadose-zone thickness is also achieved by a semipermanent vadose zone sampling system, called SEAMIST™*, which has recently been developed by Keller and Lowry (1990). The SEAMIST™ system consists of an impermeable nylon-coated liner that is pneumatically inflated in an open borehole (Figure 20). The liner can have soil gas sampling ports attached at

* Registered Trademark of Eastman Cherrington Environmental, Santa Fe, New Mexico.

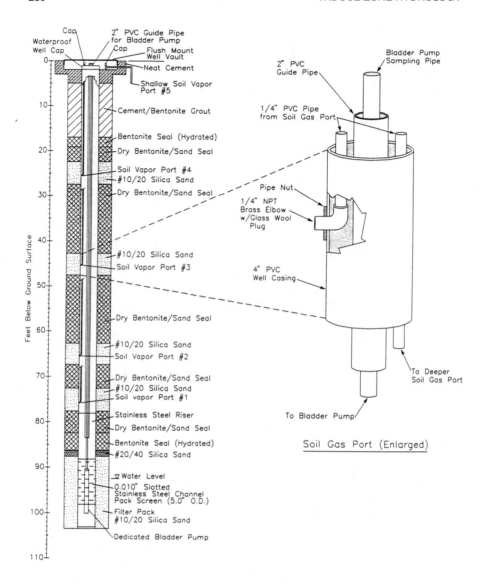

Figure 19   Details of multiport gas/groundwater well construction. (From Forbes et al., 1993. With permission.)

predetermined depths, which are pressed tightly against the wall of the borehole by the inflation pressure inside the liner. Flexible tubes leading from the sampling ports to the surface connect to vacuum pumps to produce gas samples for chemical analysis. Air permeability tests can be conducted by using the ports for pressure injection and pressure monitoring (Lowry and Narbutovskih, 1991). Mallon et al. (1992) reported very good results in using the SEAMIST™ system to monitor tritium in the vadose zone at Livermore, CA, and there is additional field testing ongoing now at several other sites in both horizontal and vertical borings. Two other advantages

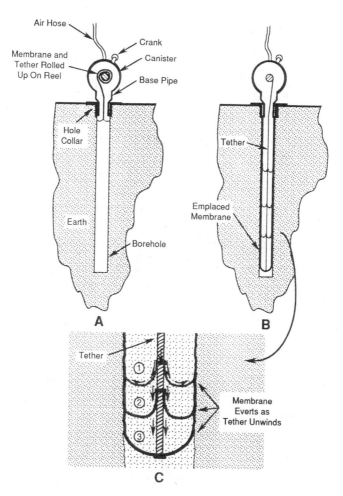

Figure 20 SEAMIST™ emplacement system. A, membrane begins to advance from cannister into bonehole; B, tether allows retrieval of membrane; and C, detail of tether and membrane emplacement. (From Keller and Lowry, 1990. With permission.)

of this system are that neutron logging can be conducted inside the liner, and absorbent pads attached to the outside of the liner can collect pore liquid samples by unrolling the liner. The disadvantage of the SEAMIST™ system is that gas pressure, for example, from bottled nitrogen or an air compressor, must be continuously supplied to maintain borehole integrity. However, for a more permanent installation, the liner can be filled in with dry sand that later could be removed by air jetting or vacuum. For installations at some sites the cost of the SEAMIST™ system will be another disadvantage, compared to the permanent installation. Additionally, side-wall leakage between soil gas parts may occur if the membrane does not tightly seal against the formation, such as in gravelly or cobbly soils or wherever borehole walls are very rough.

Keller (1992) designed a landfill monitoring system based on a gridwork of horizontal SEAMIST™ installations that is currently under testing (Figure 21). The

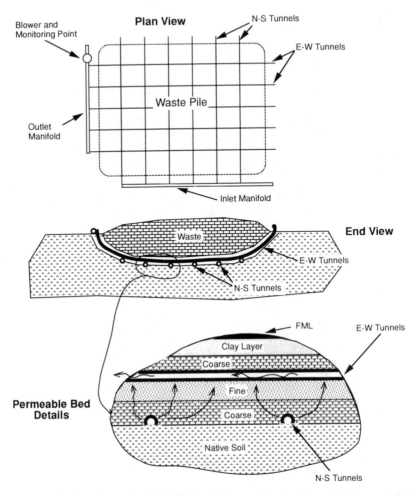

Figure 21   Conceptual design of a landfill monitoring system: plan view, end view, and permeable bed details. Tunnels allow airflow control or access by SEAMIST™ system. (Courtesy of Eastman-Cherrington.)

tunnels are constructed from sliced plastic or clay pipe, which provides support for the overburden and allows contact with the underlying soil. In this design, the tunnels are an integral part of a vapor monitoring system. Additionally, if a leak is detected in the vapor stream, the SEAMIST™ system could be set inside the tunnels to pinpoint the location of the leak. And if remediation of volatile organic chemicals (VOCs) is required, the tunnels may be connected to a vacuum pump to serve as horizontal vapor extraction wells.

Soil gas surveys are a means of rapidly collecting soil gas samples, usually at depths of less than few feet, over an area of a site. In soil gas surveys, small-diameter hollow pipe serves as a temporary casing, which is driven into the soil with hand tools, a pneumatic hammer (Kerfoot, 1989), or hydraulic jacks mounted to the rear of a van. A vacuum pump is used to exhaust about three casing volumes of gas before a sample is collected. For rapid field screening the sample is drawn from the exhaust stream and injected into the analyzer. For soil gas surveys the chemical analyzer can be a portable gas chromatograph with a photoionization detector for reconnaissance or field screening (e.g., Kerfoot, 1990), or it can be a truck-mounted laboratory-grade gas chromatograph with a mass spectrometer for high-resolution analyses of specific chemicals (e.g., Tillman et al., 1989; Tolman and Thompson, 1989). However, samples can also be collected into Tedlar™* bags or SUMMA™** canisters and transported to a laboratory for later analysis (e.g., Forbes et al., 1993).

Depth profiling for soil gas is accomplished either by the permanent or semipermanent installations described earlier or by driving probes and sequentially sampling as the drive point is advanced to greater depths. For most unconsolidated soils a cone penetrometer rig can collect soil gas samples as the cone is advanced by the hydraulic jack system. An advantage of the cone penetrometer method is that other information on soil texture can be collected continuously with depth at the same time and location as the soil gas sample is collected. This approach can lead to improved interpretations and more rapid decisions in the field program that may follow. In our own investigations, we have conducted depth profiling in moderately cemented soils to depths of about 30 m using a hollow-stem flight auger rig to create the boring and driving a conventional soil-gas sampling tube a few feet ahead of the auger. The latter profiling process is tedious because the soil gas equipment must be withdrawn each time the auger is advanced deeper.

B. INTERPRETATIONS

Interpretations of soil gas data require knowledge of the chemical characteristics of the gas, the hydraulic properties of the soil, hydrogeological conditions, and the nature of the chemical release, as well as soil air and barometric pressures in some instances. Most applications of soil gas surveys are intended for the purpose of (1) identifying areas where chemical releases to soil may have occurred, (2) determining the types of volatile chemicals present, (3) delineating groundwater contamination beneath a site, or (4) identifying locations for soil sampling borings or monitor wells.

In general, when the field data are plotted on a base map, the contoured results of shallow soil gas surveys may be expected to reveal high concentrations where volatile chemical releases have occurred. Results of field screening methods for gas chemical analysis, such as with flame ionization detectors, produce maps of total volatile gases that are not sufficient to unambiguously distinguish these groups of chemicals. For instance, Tillman et al. (1989) pointed out that naturally occurring methane and turpentine could complicate interpretations based on total volatile organic gas concentrations. When concentrations are obtained by a GC/MS, individual chemicals can be plotted to improve interpretations. This may be especially useful to identify responsible parties, and where both petroleum hydrocarbons and organic solvents are present, to guide the choice of chemical analyses for soil samples.

Shallow soil gas surveys also have been successful in defining the areal extent of groundwater contamination. As an example, Marrin and Thompson (1987) pushed

* Registered Trademark of E. I. du Pont de Nemours and Company, Inc., Wilmington, DE.
** Registered Trademark of Molectrex Corporation.

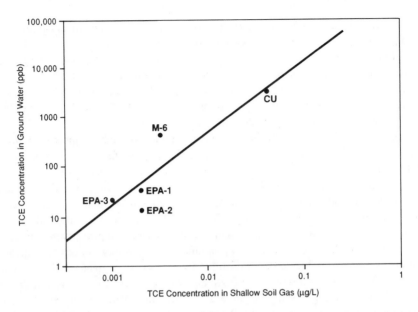

Figure 22   Regression line showing the relationship between trichloroethylene (TCE) concentration in groundwater and in shallow soil gas (<2 m deep) at the Tucson Airport site. *Note:* The water table depth at the Tucson Airport site is about 30 m below the land surface. (From Marrin and Thompson, 1987. With permission.)

soil gas sampling probes to depths of less than 2 m into the soil at a site in Tucson, AZ, and delineated trichloroethylene (TCE) in groundwater more than 30 m below land surface (Figure 22). The use of soil gas surveys is nearly standard in underground storage tank investigations, Phase II property assessments, and other site characterization programs where volatile chemicals are suspected. One of the advantages of the approach is that about 15 to more than 50 points can be sampled in a single day, depending on the suite of chemicals, the analyzer, and site conditions. From the maps that outline the extent of the soil gas contamination, monitor wells can be placed with increased confidence in the appropriate locations. The expense of conducting soil gas surveys is often easily justified by the savings in having to construct fewer monitor wells and collect fewer soil samples to define the nature and extent of contamination.

In interpreting soil gas survey data, as well as in designing a survey, one needs to bear in mind a number of factors that control the distribution of contaminants in the subsurface. The first of these is the degree of volatility, as determined by the vapor pressure and boiling point characteristics. Tolman and Thompson (1989) reported that compounds with boiling points greater than 150°C and vapor pressures less than 10 mm mercury (Hg) at 20°C usually will not be present in the vapor phase in sufficient quantities to be detected in most applications, except perhaps where there is free-phase residual hydrocarbon in the soil close to the sampling point. Compounds with high aqueous solubilities also are poor candidates for analysis by soil gas sampling, because the chemical has a greater affinity for the water phase in the soil pores. The degree of partitioning of a chemical between the gas and the dissolved aqueous phase is represented by Henry's constant, an empirical constant calculated from the ratio of the concentration of the component in the gas and the concentration in the water (Chapter 2). After the chemical partitions into the gas-phase, it may migrate due to diffusive, advective, or buoyant (gravity) forces.

Usually it is by diffusion that gases migrate through unsaturated soils. If the soil gas concentration increases with depth, the source of the detected volatile compounds is most often in the aquifer. For such a situation, Marrin and Thompson (1987) showed that the increase of TCE in soil gas was nearly linear with increasing depth below land surface. However, when the aquifer is the source of soil gas, Forbes et al. (1993) suggested that the effect of gas density on the equilibrium gas concentration may create a profile in which concentration increases exponentially with depth. Gases also diffuse radially from sources within the vadose zone, so they can be readily deleted at significant distances from a release. When the source is near the surface and the soil is very dry, the liquid-phase concentrations can be significantly depleted by diffusion into the atmosphere. On the other hand, after infiltration of precipitation, the upward gas diffusion from a near-surface VOC source can be slowed if the near-surface soil-water content increases and reduces the diffusion coefficient, as may be expected in particular in the finer-textured soils. When interpreting soil gas concentrations, especially after a recent release, it may be important to consider the transient nature of the diffusive process in order to assess, for example, the likelihood that a particular gas should be present at a particular location.

There is recent evidence to suggest that advection can play an important role in interpreting soil gas data. Advection is the process in which gas will migrate due to pressure gradients, in the same manner in which Darcy's equation is applied to water. When barometric pressure changes, there can be a gradual propagation of the pressure through the soil's air-filled voids. These oscillations in soil air pressure cause the gas to move in the direction of lower pressure, in a process called barometric pumping. Earth tides, caused by the attraction on the earth by the moon and, to a lesser extent, the sun also produce gas advection due to the semidiurnal dilation and contraction of the porous matric. Depending on the time of day, advection may produce significant variations in soil gas concentration. The rate of gas movement by advection is controlled by the permeability of the soil to the gas, a parameter that integrates the effects of pore tortuosity, connectivity, solid surface characteristics, and water content. Thus, the air pressure propagates most rapidly along highly permeable preferential paths such as fractures, macropores, and un-sealed casings. Clay soils typically have low air permeability due to their high water-holding capacity and small pores, whereas free-draining sandy soils would be expected to have high air permeability. Unusual distributions of soil gas concentrations often can be attributed to the influence of soil texture spatial heterogeneity on water retention, and hence gas migration, in the pores.

It is fairly common that deep test wells or groundwater production wells, some screened at least partially in the vadose zone, can transmit significant quantities of air. Undoubtedly, some of these wells could aid in the migration of volatile contaminants in the soil as well. At Los Alamos, NM, Purtymun et al. (1974) documented significant air transfers due to barometric pressure changes in cased wells completed in tuff and conglomerate to 433 m, in which the depth to the water-table was 332 m, and 4 m of the slotted casing extended above the water-table. Field investigations on the crest of Yucca Mountain, NV, have shown significant discharge of air from a well in a thick vadose zone comprised of fractured and bedded tuff that has been attributed to wind forces acting on the west (upward) face of the virtually barren mountain slope (Weeks, 1993). Recent work by Massmann and Farrier (1992) concluded that large changes in barometric pressure associated with weather fronts can cause "fresh" air to migrate several meters into the subsurface. They also concluded that horizontal gas transport can occur near soil gas monitor wells during atmospheric pressure cycles. A case study of a deep soil gas profiling investigation in Tucson, AZ, presented in the next chapter, showed evidence that barometric pumping had a significant influence on the vertical distribution of gas phase solvents within a 30-m-thick vadose zone (Forbes et al., 1993).

Whether gas migration occurs due to the diffusive, buoyant, or advective forces just described, the soil gas concentration can be influenced by interactions with the pore liquids and the solid surfaces, as discussed in Chapter 1. As the volatile chemical migrates into unaffected soil, some portion will partition back into the liquid, in accordance with Henry's law. Additionally, some of the gas-phase compounds can be adsorbed onto the solid dry soil (Shoemaker et al., 1990). Biological processes also can degrade volatile petroleum hydrocarbon compounds, especially near the soil surface where oxygen concentrations are high. Over time, especially in oxygen-deficient systems, halogenated organic compounds (e.g., organic molecules such as solvents with chloride, bromide atoms, etc.) can degrade into other chemicals with progressively fewer halogen atoms by processes such as hydrolysis, reductive dehydrohalogenation, and reduction (e.g., Olsen and Davis, 1990). Consequently, the interpretation of soil gas information can become exceeding complex, especially after significant time has elapsed since the release.

IV. PORE LIQUID SAMPLING

Up to this point we have discussed a variety of indirect methods that could be used to monitor leaks or seepage that potentially contain contaminants. However, neither changes in moisture content, pressure head, nor soil gas concentrations actually reveal whether liquid-phase contamination is present, or if it is, whether concentrations are sufficient to trigger some additional monitoring or remedial action. Therefore, pore liquid sampling is required to confirm the presence of liquid-phase contamination. Pore liquid sampling includes extracting liquids from the matrix of soil cores and *in situ* samplers to collect liquids from the vadose zone. In this section, we describe very briefly some of the issues related to soil core sampling and focus more on liquid collection devices, as well as on some of the geophysical indicators of soil liquid chemistry.

A. SOIL MATRIX ANALYSIS

Samples of soil can be collected from the soil by hand tools or mechanical drilling equipment. Hand augers and coring tools are usually viable only within about the upper 7 m of the soil, provided that the soil is unconsolidated and free of cobbles. Mechanical drilling and sampling equipment is the most common method, however, even for depths up to about 50 m. Conventional drilling equipment such as the hollow-stem flight auger with core sampling generally produces excellent results, except in very loose dry soils, and cemented or stony soils. Because this drilling and sampling technology is in everyday use by engineers and hydrogeologists and is thoroughly described in available literature (e.g., U.S. Bureau of Reclamation, 1974; U.S. Environmental Protection Agency, 1986; ASTM Guide D-4700), the mechanical details of how to collect soil samples are not discussed further.

In most solid or hazardous waste site sampling projects, soil cores are sent to a laboratory for chemical analysis, under strict preservation and quality assurance/quality control procedures such as those specified by EPA guidance manuals for RCRA and CERCLA. In the analysis, for instance by the Soxhlet extraction process, a toluene/methanol or acetone/hexane solvent is passed through the soil sample (U.S. Environmental Protection Agency, 1986). Essentially, the chemical analyses of the extract represents virtually all the chemicals extractable by the solubility mechanism, including chemicals sorbed onto the solid, chemicals in a free phase, and chemicals dissolved in the ambient pore water. This approach is essentially the same as that used in agricultural studies to analyze soluble salts, in which a known amount of distilled water is added to the soil to prepare a slurry called a saturated extract; then the solution to be analyzed is drawn out of the slurry through a filter by a vacuum (U.S. Department of Agriculture, 1954).

Pore liquids can also be extracted from soil cores directly by placing the sample either on a pressure plate and forcing the water out by a positive pressure or on Büchner funnel and drawing the water from the soil under a vacuum. Both these methods require that the sample pass through a water saturated ceramic plate into the collection vessel. Consequently, there is opportunity for mixing and dilution of the sample extracted. If a soil slurry or saturated extract is prepared, the liquid also can be separated from the solid fraction by centrifuge. While these methods may be adequate for extracting soluble inorganic compounds, they are inappropriate for organic compounds because of their low water solubility and high volatility.

The main point to recognize in considering soil cores for pore liquid sampling is that this is a destructive technique. For time-series monitoring, repeated coring to collect samples from the same general location potentially introduces significant bias, which may preclude distinguishing chemical changes attributable to contaminant migration from changes due to spatial variability in fluid chemistry. As illustrated in Chapter 3, there are a number of mechanisms governing fluid and contaminant migration that can lead to complex, three-dimensional subsurface pathways along or between which concentrations could be quite heterogeneous. This may explain, at least partly, a common pattern of detect/no-detect in vertical sampling sequences immediately below an organic contaminant release, especially when the analyte concentration is low.

A recent investigation (Hahne and Thomsen, 1991) illustrates this unusual pattern at the site of a release of the fumigant ethylene dibromide (EDB) and carbon tetrachloride due to a grain elevator explosion. Briefly, the authors of the study used a hollow-stem auger to collect soil samples with a split spoon sampler and to collect soil gas samples with a soil gas probe driven ahead of the auger, following protocol

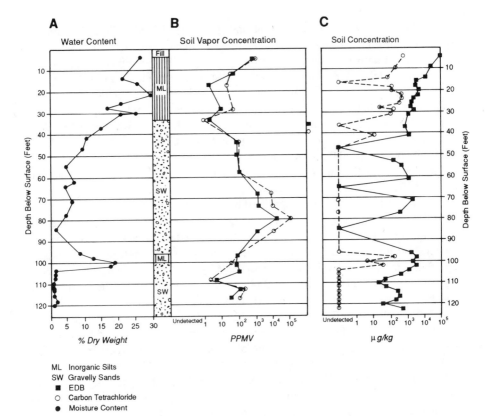

Figure 23   Comparison of soil concentration profiles for two chemicals, EDB and carbon tetrachloride, released simultaneously. A, water content and soil texture; B, concentration from soil gas analysis; and C, concentration from soil matrix analysis. (From Hahne and Thomsen, 1991. With permission.)

approved by the EPA for this Superfund site. The geologic description, moisture content, and analytical results are summarized in Figure 23. The upper 10 m (33 ft) is comprised of wind-blown silt and the lower part of the section is comprised of downward coarsening alluvial sand and gravel. The depth to the water table is about 38 m. The EDB and carbon tetrachloride concentrations in soil are highly variable, and there are several sample depth intervals where no contaminants were detected. That carbon tetrachloride is not detected in the soil where water content is relatively low and the soil is coarse suggests a dependence of porewater concentration on water content or soil texture. The same is true for EDB to a much lesser extent. Note, however, that soil gas concentration for carbon tetrachloride reaches a maximum at about 26 m (85 ft) in the interval where carbon tetrachloride was not detected in the soil matrix. Figures 23B and C indicate that in soil vapor carbon tetrachloride concentrations generally were greater than those of EDB, but in the soil matrix the opposite was true. The authors attributed the greater affinity for the gas phase of carbon tetrachloride relative to EDB to carbon tetrachloride's higher Henry's constant (0.016 atm-m^3/mol compared to 0.011 atm-m^3/mol), higher vapor pressure (113 mm Hg compared to 14 mm Hg), and lower solubility (800 mg/L compared to 4310 mg/L). This study and many similar experiences indicate that chemical analyses of soil matrix samples may not completely reveal contaminant pathways. Soil vapor analysis, on the other hand, may be more appropriate to delineate contaminant pathways where air permeability and air-filled pore space are high.

As this example showed, like many others (see Case Study 3 at the end of this chapter), the spatial distribution of a chemical's vapor is much more areally extensive than the distribution found from soil matrix sampling, owing to the tendency for gaseous diffusion. This is especially true for the nonspreading dense nonaqueous-phase liquids, which tend to migrate through texturally uniform unsaturated media as narrow fingers with minimal spreading. Abriola (1989) found that even for spills of limited volume, the vapors could diffuse rather quickly through the vadose zone and contaminate groundwater, without any liquid-phase contaminants reaching the water table. Consequently, there is potential that pore liquid sampling by matrix analysis alone may in fact fail to detect potentially hazardous releases of nonaqueous phase liquids (e.g., chlorinated solvents). It is also important to recognize the possibility that monitoring by soil matrix sampling can lead to false positive results due to cross-contamination during the drilling and sampling process (Hackett, 1987). As a consequence, in some instances, contaminants may be incorrectly interpreted to have contaminated deeper soil horizons or groundwater, leading to inappropriate decisions on selecting remedial action alternatives and perhaps to inappropriate allocation of responsibility. Fortunately, careful drilling and sampling procedures can alleviate most of this concern. However, one should avoid drilling through pools of dense nonaqueous-phase liquids that may be perched on clay layers in the vadose zone.

B. POROUS CUP SAMPLERS

Porous cup samplers, also referred to as suction lysimeters or porous suction samplers, consist of a porous tip attached with an airtight seal to the lower end of a thin plastic casing. The upper end is also tightly capped, so that a vacuum can be applied and maintained to the casing through connections in the upper cap. The sampler cup is designed with pores sufficiently fine that they remain saturated with water while the vacuum is applied to the casing section. The principle of operation is to place the sampler in a boring so that there is good communication with the formation. Usually a slurry of soil cuttings or 200-mesh silica flour is placed around the sampler, and bentonite seals are set above this interval to prevent channeling of runoff or perched fluids. A vacuum is applied to the sampler with the intent that it will establish a hydraulic gradient between the formation and the sampler. This system may fail for a variety of reasons, for example, if the applied vacuum exceeds the air-entry value for the porous cup or if the soil tension exceeds the air-entry pressure of the cup. In all these instances, consequently, the porous cup will dewater partially, the air in the sampler casing will be in communication with the soil air, and vacuum on the liquid phase will be lost. In the failure mode, there is no hydraulic gradient between the soil and sampler. Additionally, the hydraulic conductivity of the dewatered cup to liquid becomes so low that formation water does not enter the cup. Obviously, the porous cup samplers will also fail if there are leaks in the fittings.

Table 2   Pore Size and Air-Entry Pressures for Selected Porous
           Materials

Material	Nominal pore size (μm)	Air-entry pressure (bars)
Ceramic[a]	5.8	0.5
	2.9	1.0
Sintered stainless steel[b]	4.5	0.05
	2.0	0.20
	0.5	0.30
Cellulose acetate[c]	5.0	0.41
	1.2	0.62
	0.8	0.83
Nylon[c]	10.0	0.28
	5.0	0.41
	1.2	0.62
	0.8	0.97
Polycarbonate[d]	5.0	0.21
	3.0	0.48
	2.0	0.62
	1.0	0.96
Teflon (PTFE)[c]	1.0	0.21
	0.45	0.48
	0.22	0.90
Teflon (PTFE)[e]	0.2	0.86
	0.45	1.46
	1.0	0.56

[a] Soilmoisture Equipment Corp., PO Box 30025, Santa Barbara, CA 93105, (805) 964-3525.

[b] Mott Metalurgical Corp., Farmington Industrial Park, Farmington, CT 06032, (203) 677-7311; Soil Measurement Systems, 7266 N. Oracle Road, Tucson, AZ 85704, (602) 742-4471.

[c] Micron Separations, Inc., 135 Flanders Road, Westborough, MA 01581, (800) 444-8212.

[d] Nucleopore Corp., 7035 Commerce Circle, Pleasanton, CA 94588, (800) 882-7711.

[e] Gelman Sciences, 600 S. Wagner Road, Ann Arbor, MI 48103, (800) 521-1520.

Porous cup samplers differ in two general ways, in the material used to construct the cup and in the design of the component. Most of the cups consist of porous ceramic or Teflon™, although Alundum™ and nylon have been used (Creasey and Dreiss, 1985). Everett et al. (1984) reported that porous polyethylene also has been used, and Soil Measurement Systems Inc. (Tucson, AZ) distributes samplers made from sintered stainless steel. Table 2 summarizes the materials available and the pore size range available for each material. The ceramic cups are commonly manufactured with pore sizes in the 2 to 3 μm range, which, based on capillary theory in Equation 3 of Chapter 1, makes them potentially useful to withdraw soil water up to about 1 bar of tension (U.S. Environmental Protection Agency, 1986). Although the nominal size of pores in the Teflon™ material is similar to ceramic (Table 2) some of the pores in the Teflon™ are larger than the nominal size specified, in the 14 to 300 μm range. The Teflon™ samplers therefore would not function if suctions on the sampler or potentials in the soil water exceed about 0.2 bars. Clearly, the air-entry pressure of the sampler is critical to its performance and range of utility in the field. However, with respect to the tendency of the pores of the cup to fill and remain completely full of water, it is also important to recognize that the ceramic is hydrophilic, whereas the Teflon™ is hydrophobic. On the other hand, the wider pore size distribution of the Teflon™ leads to higher hydraulic conductivity of the cup and more rapid sample collection. However, for most applications, the ceramic cup will function satisfactorily to lower soil-water potentials than the Teflon™ cup.

Porous cups can also be used to collect nonaqueous-phase liquids, under some conditions. The type of fluid to be sampled should be identified first, so that the porous cup can be thoroughly wetted with the particular nonaqueous-phase liquid. If this is not done and water occupies the cup pores upon installation, it is unlikely that the nonaqueous-phase liquid will displace the water from the cup, owing to the small diameter of the pores and the greater interfacial tension between water and the pore surfaces. Recall that in most multiphase systems, that decreasing order of preferential wetting to the solid phase is water, nonaqueous-phase liquid (e.g., petroleum hydrocarbon or solvent), and air. Ballestero et al. (1990) found that when a petroleum hydrocarbon presoaked a porous ceramic cup sampler set in a water and hydrocarbon mix, the amount of hydrocarbon sampled gradually reduced because the water gradually displaced the hydrocarbon from the cup and reduced the number of continuous paths for hydrocarbon to enter the sampler. Although porous cup samplers and tensiometers may be used in the laboratory (e.g., Wilson et al., 1990) for nonaqueous-phase liquids, they do not seem to have found applications to the field where multiple liquid phases are expected.

In the simplest design of the porous cup sampler, the cup is located at the bottom of the sampler tube. A two-hole rubber stopper caps the upper end of the casing so that a short tube inserted through the stopper can be used to apply a vacuum to collect the liquid in the sampler. Then a vacuum is applied to the long discharge tube, which is set to the base of the sampler to bring the liquid to the surface (Figure 24). This design, referred to as the vacuum sampler, is usually used at depths of less than 2 m but potentially could be installed at depths less than about 8 m, the maximum practical suction lift.

For deeper sampling, a pressure-vacuum sampler (Parizek and Lane, 1970) is required. In this design, a short tube through the rubber stopper cap is used both to apply a vacuum to the casing to draw in the liquid, and to apply a positive pressure to force the liquid to the surface through the long discharge tube (Figure 25). One of the problems with this design was that the positive pressure could force some of the collected liquid back through the cup and into the formation, especially below depths of about 17 m (Wilson, 1990). And, in some instances, the high pressure could damage the cup. Another problem develops with both these shallow and deep designs if the long discharge tube does not extend completely to the base of the sampler; this leaves a dead space within the liquid collection reservoir where the newly sampled liquid will mix with the liquid left in the dead space from previous samplings (U.S. Environmental Protection Agency, 1986).

To overcome the problem of sampling at depth, the sampler tube body is compartmentalized with check valves. In one design, a plug is placed between the cup and chamber above it, with a short tube and check valve connecting the cup and upper chamber. When the vacuum is applied, the check valve opens and sample is drawn

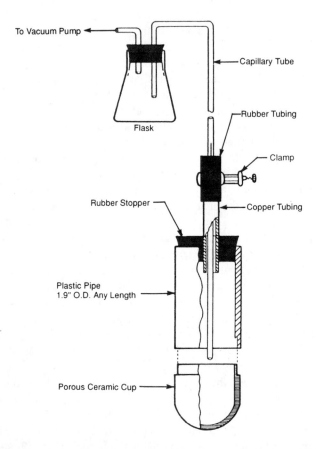

To Vacuum Pump

Capillary Tube

Rubber Tubing

Flask

Clamp

Rubber Stopper

Copper Tubing

Plastic Pipe
1.9" O.D. Any Length

Porous Ceramic Cup

Figure 24   Pore liquid sampling with porous cup vacuum sampler (lysimeter) *Note:* Capillary
tube is only inserted to recover the water sample. (From Parizek and Lane, 1970.
With permission.)

from the cup to the upper chamber, and when the positive pressure is applied the
check valve prevents liquid from the upper chamber from falling back into the cup
(Figure 26). The samplers with check valves are sometimes referred to as the high-
pressure/vacuum-suction cup samplers, because high pressures are required to lift the
water to the land surface in installations up to about 100 m.

1. Region of Influence and Formation Variability

The area of influence of the sampler is dependent upon the diameter of the boring,
length of backfill interval, applied vacuum, duration of vacuum, and soil properties.
Most sample cups are nominally 5 cm in diameter. For this size cup, Morrison and
Szecsody (1983) indicated that the radius of influence of the sample is about 10 cm
in coarse-textured soil and 65 cm in fine soil. Warrick and Amoozegar-Fard (1977)
developed a generalized steady-state analytical solution to compute the radius of the
measurement that is a function of the sampling rate, slope of the relative hydraulic
conductivity-pressure head relationship (α), hydraulic conductivity at zero pressure,

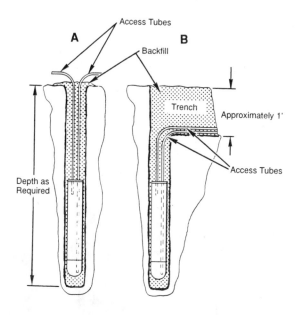

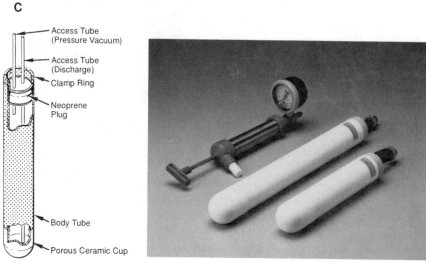

Figure 25   Pore liquid sampling with porous cup, pressure-vacuum samplers (lysimeters) showing: (A) installation in a backfilled borehole, (B) installation in a backfilled trench, and (C) sampler construction, plus (D) a photo. (Courtesy of Soilmoisture Equipment Corp.)

and initial pressure head. Figure 27 illustrates the influence of a solute sampler on the soil-water flow field. In highly heterogeneous and fractured systems, or where preferential flow is likely, the backfill interval should be sufficiently long to intercept such pathways and draw the formation liquid to the sampler. As the vacuum is applied, the radius of influence gradually expands either until the vacuum source is removed or until the rate of flow into the sampler is equal to the flow of downward percolation that is intercepted within the area of influence of the sampler. Numerical

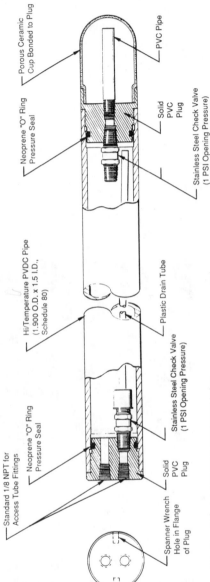

Porous Ceramic Cup Bonded to Plug

PVC Pipe

Neoprene "O" Ring Pressure Seal

Solid PVC Plug

Stainless Steel Check Valve (1 PSI Opening Pressure)

Hi/Temperature PVDC Pipe (1.900 O.D. x 1.5 I.D., Schedule 80)

Plastic Drain Tube

Standard 1/8 NPT for Access Tube Fittings

Neoprene "O" Ring Pressure Seal

Stainless Steel Check Valve (1 PSI Opening Pressure)

Spanner Wrench Hole in Flange of Plug

Solid PVC Plug

Note: Overall dimensions: 1.9 inches (O.D.) x 24 inches

Figure 26   Model 1940 high-pressure-vacuum soil water sampler. (Courtesy of Soilmoisture Equipment Corp.)

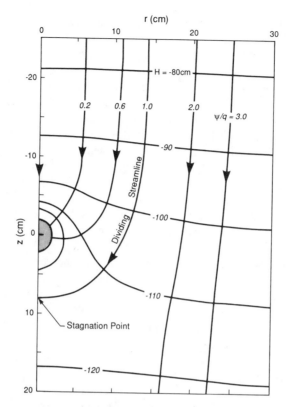

Figure 27   Flow field surrounding a porous cup pore liquid vacuum sampler (lysimeter). (From Warrick and Amoozegar-Fard, 1977. With permission.)

model simulations of the transient process of collecting a sample revealed that the volume of water sample and the region of influence are both dependent upon the volume of the sampler (Narasimhan and Dreiss, 1986).

This transient aspect of the sample space may be important to consider in designing monitor networks and interpreting results. Depending upon the sampler hydraulic characteristics and soil properties, several hours to days or weeks may be required to collect a sample of a few hundred milliliters of formation liquid.

In a contaminated vadose zone, typically there is chemical variability among the pore size classes, owing to the spatial variability in water content distribution, hydraulic conductivity, and variations of contaminant transport velocity. These factors, plus molecular diffusion, contribute to hydrodynamic dispersion (Chapter 1). The largest water-filled pores are capable of transporting contaminants faster than the small pores, but contaminants in these larger pores are also more readily flushed by infiltrating precipitation, for example. Likewise, the smaller pores are slower to receive contaminants from the source by advection. Water in the small pores also becomes contaminated by diffusion from chemicals in the larger pores. And, after the larger pores are drained free of chemicals, then chemicals from the small pores may diffuse slowly back into the large pores.

This media-induced spatial and temporal variability must be kept in mind, in order to differentiate it from sampler-induced variability. For instance, at high applied suctions, the porous cup sampler will drain both the large and small soil pores, whereas at low suction only the larger pores will be drained. Therefore, during long periods of sampling when the vacuum is falling and the sample intake rate is diminishing, chemical variability in a series of samples can be introduced because the samples are drawn from an increasing proportion of the large pores (Hansen and Harris, 1975). Hansen and Harris recommended keeping sample collection periods to less than a few hours if a falling vacuum is applied. Even if the applied suction in the sampler is constant so that pore scale variability is not a consideration, the composition of the extracted liquid can vary with time. This is because as the volume of influence surrounding the sampler increases with time (e.g., Severson and Grigal, 1976), the flow field encounters large-scale geologic heterogeneity and chemical variability. Just the opposite can occur if the permeability of the ceramic cup decreases over time due to plugging of the pores in the cup. Also, during long sample collection periods, temporal variations in formation liquid concentration can occur that are averaged into the measurement. These temporal variations can be natural seasonal fluctuations in pore liquid chemistry or contamination releases. Consequently, characterizing the background constituent concentrations and their pattern of variation is essential for reliable interpretations of pore liquid samples, particularly for the inorganic compounds.

2. Pore Liquid-Porous Cup Interaction

Almost since pore liquid samplers were first developed, there has been considerable interest in identifying chemical interactions between the pore liquid and the sampler. This concern led to standard procedures to flush the samplers with hydrochloric acid followed by a distilled water rinse before installation. For example, Zimmerman et al. (1978) tested ceramic and Teflon™ cups in known chemical solutions of ammonium, phosphate, nitrate, nitrite, and silica. They found that ceramic absorbed or excluded ammonium and phosphate, whereas Teflon™ did not. Because of concern for chemical interference with the ceramic cup, as well as with the PVC casing, Teflon™ cups and samplers were also developed into sleeve lysimeters (Fishbaugh, 1984), However, chemical interaction experiments by Creasey and Dreiss (1985) concluded that ceramic was superior to other materials, including Teflon™, but they also found that for any of the materials (also including Alundum™) the degree of interference in their tests was small relative to potential errors in laboratory analytical procedure. Consequently, they recommended any of the commercially available porous cups for use in monitoring at hazardous or solid waste sites. From the mix of laboratory and field results reported to date, there is still no clear picture on the nature of interaction. Additional testing for specific interactions may be warranted before application to a particular site.

Once the liquid is inside the sampler, other chemical processes may affect the pore liquid concentration. Obviously, dissolved volatile concentrations may be markedly reduced, due to the air that is inside the sampler at a vacuum above the liquid reservoir. Oxygen concentration in the air inside the sampler is likely to be much greater than in the soil, and as a result there may be concentration changes due to redox reactions. For example, Anderson (1986) speculated that chromium precipitation inside the sampler caused the concentration to decrease between 4 and 40 times the initial sample concentration. To limit oxide precipitation, Anderson recommended purging the sampler of oxygen.

A type of porous cup sampler called the BAT system (Torstensson, 1984) is a unique design to allow sampling of volatile chemicals and constituents involved in redox reactions (Figure 28). In this system, a sampling probe with porous sampling tip is pushed into the soil or is set in a borehole and backfilled. The porous section of the sampling probe is composed of a sintered ceramic with an average pore diameter of 1.8 μm and an average air-entry value of about 1.3 bars. The upper part of the buried probe is sealed by a flexible disk, like a septum rubber stopper. To collect the water sample, an evacuated gas cylinder (vial) with a flexible disk cap is lowered by cable into the casing along with a double-ended disposable hypodermic needle below it. When the sampling vial and needle contact the flexible disk on the top of the sampling probe, the hypodermic needle penetrates both flexible disks, thereby allowing water in the probe to be drawn into the cylinder. The gas under a vacuum in the cylinder could be nitrogen, which, for many applications, is essentially inert. As formation liquid is drawn into the porous cup, it flows into the gas cylinder without contacting the atmosphere, thereby preserving the integrity of the sample. The gas cylinder reportedly can be retrieved after as little as 5 min in permeable soil. Minor presampling purging is required because of the small volume of the porous cup.

3. Other Potential Problems

A number of problems with porous cup samplers have already been described, but there remain a few others that must be mentioned. Although there appear to be many more successful applications, plugging of porous cups has been reported in some cases, especially for the high-flow ceramic cups (Hansen and Harris, 1975). Plugging may occur if very fine clay-size particles are filtered during sampling or if chemicals precipitate in the cup. Periodic checking of the volume or rate of liquid yielded under a particular vacuum will reveal evidence of plugging. If bacteria are of interest, then porous cups are not recommended because bacteria are likely to be filtered out. However, virus will not be filtered by the pores of the cup (U.S. Environmental Protection Agency, 1986). Freezing conditions present some logistical difficulties for porous cup samplers. The soil may freeze, making sampling impossible, or the sampling line may have residual or condensed water that has frozen.

Perhaps the greatest problem with porous cup samplers is that they simply do not function if the soil water potential is less than about 1 bar. In humid climates this may not be a problem, except at shallow depths after prolonged dry conditions. But in semi-arid and mediterranean climates, many fine-textured soils are too dry for these samplers to function. Sandy soils in dry climates, on the other hand, may have sufficiently high natural moisture for samplers to function. For example, Stephens and Knowlton (1986) measured with tensiometers in situ soil water potentials of about –0.06 bar to depths of a few meters in an unconsolidated medium sand in New

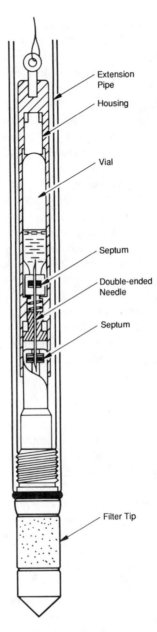

Figure 28 BAT system. (From Torstensson, 1984. With permission.)

Mexico where the mean annual precipitation is about 20 cm and potential evapotranspiration is about 178 cm. Such conditions may be atypical of many undisturbed field sites, so it is essential to characterize *in situ* soil water potential before designing the monitoring system. There is a common misconception that the soil-water potential at which porous cups fail to collect liquid corresponds to the soil-water potential at which the unsaturated hydraulic conductivity becomes zero. The fact that no pore liquid is collected by the porous cup sampler should not be interpreted as confirmatory evidence that no seepage is occurring; one should also consider that either the sampling system has failed or that the seepage is, in fact, occurring but at a soil-water potential which is less than about −1 bar.

If the porous cup and backfill dewater due to soil drying, the sampler eventually should function again (once the soil rewets) and detect seepage, provided that the contact between the cup and the soil has not been lost due to permanent shrinkage and that the soil-water potential increases sufficiently to allow liquid to saturate the cup pores. Owing to the low cost of a porous cup sampler, there may be merit to installing these samplers as part of a monitoring system even in soils with background soil-water potentials less than about −1 bar, in the event seepage occurs to wet the soil to the point where samples can be obtained. However, to better interpret the performance of a porous cup sampler and differentiate failure mechanisms, tensiometers or other soil-water status indicators are recommended as part of the overall monitoring system.

A related problem develops in coarse, uniform-textured soils that are sufficiently wet for the sampler to function, but when vacuum is applied there is virtually no liquid produced, even at high applied suctions. This problem is caused by removal of water from the large soil pores surrounding the cup, such that the unsaturated hydraulic conductivity is too low for significant transfer of liquid. If the volume of collected liquid is low, there may be a tendency for the operator to further increase the suction. Application of high suction in coarse, uniform-textured soils only exacerbates the problem, because the conductivity decreases very rapidly with decreasing water content for soils. This again illustrates the necessity for thorough site characterization, not only for soil-water potential but also for soil hydraulic properties, before implementing a monitoring program.

C. MONITORING PORE LIQUIDS IN SATURATED AND NEAR-SATURATED REGIONS

In Chapter 1, we indicated that the vadose zone can contain saturated regions where the pore liquids are under positive pressure. Examples of regions of positive pressure are inverted water-tables beneath free water bodies, perched groundwater, and, in some cases, macropores or fractures. Although porous cup samplers could easily obtain liquid under these conditions, there are other options such as vadose zone monitor wells, pan or barrel lysimeters, and wick samplers.

A vadose zone monitor well is a slotted plastic or steel casing completed in the vadose zone in the exact same manner as a regional groundwater monitor well. Extensive descriptions on monitor well design have already been published (e.g., U.S. Environmental Protection Agency, 1986) and are widely used by the groundwater industry. Vadose zone monitor wells are commonly used in hazardous and solid waste monitoring programs because the technology is simple and widely available. However, it is often forgotten that a vadose zone monitor well can only be used to collect pore liquid when the liquid is under a <u>positive</u> pressure, as illustrated by the following hyperbole. If a monitor well were situated within a waterfall cascading vertically downward by the force of gravity, the well would be dry because the water would only be at atmospheric (zero gage) pressure. A vadose zone monitor well open to the atmosphere would be dry if it were located within a flow field where the pore liquids are under a tension. If neutron logging, for instance, detected a saturated interval within the vadose zone during monitoring, a sample could not be collected from this zone by a vadose zone monitor well unless the liquid in this interval were under a positive pressure. Although vadose zone monitor wells may be a convenient approach to monitoring, they will not be effective in detecting seepage through unsaturated porous or fractured media.

Perched groundwater is easily monitored and sampled by monitor wells using the same technology that is applied to conventional, regional groundwater monitoring. Horizontal tile drains commonly used to alleviate water logging in irrigated field can also be effective in collecting samples of shallow perched water if the drain is located just above the perching layer. Because there is extensive literature already on collecting water from saturated zones, not much attention is devoted here to these methods. The difficult part of perched zone monitoring is first to be able to recognize truly perched conditions. All too often, the shallowest free water is called perched, usually because it was encountered at a shallower depth than expected or because there is cascading water in a well bore. Truly perched groundwater occurs above and within a relatively low-permeability layer that usually has limited areal extent, and it is underlain by unsaturated media. Frequently, perched conditions are misidentified in the uppermost, low permeability, saturated layer in heterogeneous aquifer systems. Neither geologic continuity of permeable beds nor well yield is a criterion to determine whether an aquifer is perched. Another common characteristic of perched aquifers is that some are ephemeral. Consequently, for continuous monitoring, perched zone monitor wells will need to be complemented by other devices to collect pore liquids when the soil is unsaturated.

Where soil water is under positive pressure, or wet but still unsaturated, downward percolating or draining water can be intercepted by pan lysimeters, barrel lysimeters, and wick samplers. This concept of intercepting the percolating liquid is along the same line as using plastic sheeting and sand-backfilled drain pipe for a leachate collection system beneath a landfill. However, when the soil is unsaturated, the percolating water will accumulate above the impermeable surface until the liquid pressure is positive before liquid will enter these types of lysimeters (or leachate collection systems), unless tension is applied with the lysimeter to extract the accumulated seepage.

The pan lysimeter, first described by Jordan (1968) and subsequently by Parizek and Lane (1970), is simply a metal pan or trough installed in soil with a nearly horizontal orientation. Usually the pan lysimeter is at some incline to allow the accumulated liquids to be collected in a sampling vial or other container located in the wall of an access trench. The trench facilitates equipment installation, protects the sampling system, and allows convenient access to collect the samples. Kmet and Lindorff (1983) used a variation of the pan lysimeter called a collection lysimeter to collect leachate beneath solid waste landfills in Wisconsin (Figure 29).

A hollow glass block also has been used as a lysimeter to collect pore liquids by installing the apparatus through the side of a trench in much the same way as the pan lysimeters (Barbee, 1983); however, this method does not seem to find wide appli-

A   **B**

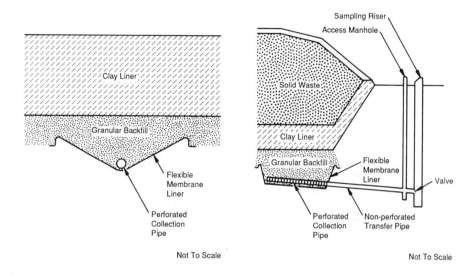

Section through a typical collection lysimeter   Collection lysimeter showing method of sample location

Figure 29   Typical designs for collection lysimeter at solid waste landfills. (From Kmet and Lindorff, 1983. With permission.)

cation in practice (Figure 30). A commercially available hollow glass block, about 30 cm × 30 cm × 10 cm deep, is drilled with small holes on the top side of the block to allow water to enter the chamber. A fiberglass top plate prevents soil from entering the holes and contaminating the sample. To keep percolating fluid from simply flowing off the top surface, there is a thin lip protruding above the edges of the upper face of the block that is set firmly in contact with the soil above the sampler. The collected fluid is removed under a vacuum by nylon tubing leading from the land surface, through a hole in the wall of the glass block, to a coil of the tubing on the bottom of the block. Installation details on this technique and pan lysimeters have been presented by the U.S. Environmental Protection Agency (1986).

One advantage of the pan lysimeter concept is that the volume of soil sampled is usually much larger than that of porous cup samplers. This large size minimizes the possibility that formation heterogeneity and macropores would cause the seepage to go undetected. However, a disadvantage is that percolating liquid will not enter the lysimeter until the pores above it become nearly saturated. To preserve the advantage of the concept, but to coax liquid into the sampler at pressures less than atmospheric, tension can be applied to improve the lysimeter design. For example, Hornby, et al. (1986) described a wick sampler (Figure 31) that consisted of a Plexiglas pan, about 30 cm × 30 cm in cross section, upon which there is a network of Herculon™ fibers that lead to a collection reservoir beneath the plate. The plate is covered by a porous

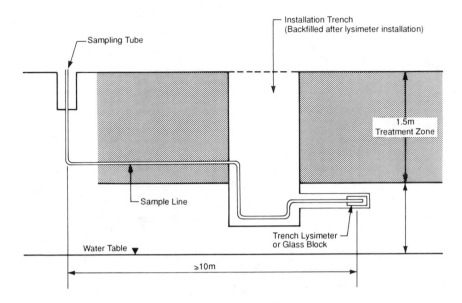

Figure 30   Glass-block lysimeter installation in a trench. (From U.S Environmental Protection Agency, 1986. With permission.)

geotextile. Moisture that contacts the Herculon™ fibers migrates to the receptacle through the fiber by capillary action, just as in a wick in a lantern, except the flow is downward. Once the fluid has collected in the reservoir, it is brought to the land surface with vacuum or pressure, similar to the porous cup vacuum lysimeter concept described previously. Undoubtedly due to considerable logistical difficulties in installation, there appear to be few examples of field applications.

More recently, a technique was developed to collect pore liquid samples at considerable depths in a vertical, uncased borehole using porous filters pressed against the wall of the boring and retrieved with the SEAMIST™ system described earlier in this chapter (Figure 16). The absorbers are attached to the outer side of the nylon casing at predetermined depths. When the flexible SEAMIST™ casing is rolled back up to the land surface, the absorbers are rewound from the casing and placed in containers for laboratory analysis. A similar concept, called a membrane filter sampler, was developed earlier by Stevenson (1978), in which glass fibers were set in contact with the soil at the base of borehole. Because vacuum is used to bring pore liquids to the surface, the membrane filter system is limited to shallow depths.

Two other concepts have been advanced that actively apply a tension to the sampler. The first is a design developed by Hoffman et al. (1978) where filter candles are emplaced in a horizontal access hole and pressed against the soil at the top of the hole by an inflatable pillow. A vacuum is applied to the filter candle and the pore liquids are collected in a flask located within the access trench (Figure 32). This system was successfully applied in a study of the effects of citrus irrigation on groundwater and surface-water salinity in western Arizona. The second design, called the barrel lysimeter, collects pore liquids from the base of a large column of

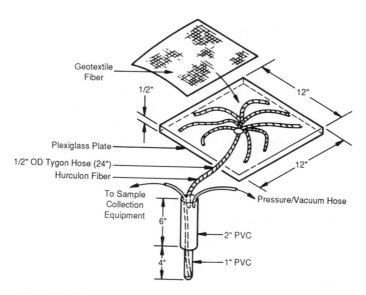

Figure 31 Wick sampler. (From Hornby et al., 1986. With permission.)

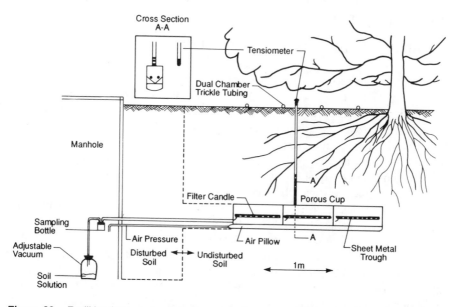

Figure 32 Facilities for sampling irrigation return flow via filter candles for research project at Tacna, AZ. (Modified from Hoffman et al., 1978. With permission.)

undisturbed soil designed in the same general manner as a weighing lysimeter, for example (Hornby et al., 1987). Porous ceramic cup samplers, located just above the sealed base of the barrel lysimeter, produce accumulated pore liquids.

There are a number of important factors to consider in the application of pan- and wick-type lysimeters. For designs in which a trench must be excavated to create an access for installation and/or sample collection, the first factor must be a concern for safety in entering any open excavation and the potential for side-wall collapse or the accumulation of dangerous vapors. There are federal requirements described by the Occupational Safety and Health Administration (OSHA) (29 CFR 1910) pertaining to worker safety in excavations that must be strictly followed. The second factor to consider is that the lysimeter must be located a distance sufficiently great from the access trench such that the sample is not biased by the influence of the excavation on the flow field. The collector should be located beneath an area that is representative of the conditions at the site to be monitored. Where lysimeters are proposed for regulatory compliance, guidance documents, such as the U.S. Environmental Protection Agency (1986) publication on land treatment facilities, should be consulted for specific applications on the location, size and number of samplers as well as sampler frequency. The third consideration is to locate the pan lysimeter where saturated conditions are likely to develop, such as above a low-permeability layer or where macropores and fractures are likely connected to potential sources of ponded water, for example. Pan lysimeters may not produce samples of downward percolating water until considerable time passes and the fluid pressures exceed atmospheric. Prior to this time, the percolating water will simply be held in the soil above or diverted around the sampler. Consequently, these types of lysimeters are most appropriate in relatively moist, shallow soils subject to periodic flooding and where the macropores are present.

D. INDIRECT METHODS OF PORE LIQUID SAMPLING

The previous sections have presented methods for collecting pore liquids from soil so that the liquid can be chemically analyzed. The costs of chemical analysis can be considerable after years of monitoring a system comprised of multiple discrete samplers. Moreover, collection of the samples must result in some invasion of the subsurface either to collect a soil core or to install an access boring for a monitoring device. In some applications, geophysical methods may be appropriate tools to indirectly monitor the quality of vadose zone pore liquids. In general, these methods are based on the electrical properties of the soil, such as resistance or dielectric permittivity, which change with chemical characteristics of the pore liquid. We discussed each of these methods in some detail in the previous section on water-content monitoring. However, here we want to highlight those methods that also have been shown to be useful for indirectly monitoring pore liquids as well.

Direct current electrical resistance methods to monitor changes in soil salinity have been reviewed by Rhoades (1978). He described soil salinity sensors, the four-electrode method, and probe versions of the latter technique. The soil salinity sensor, a small thumb-tip-size device buried in the soil, consists of electrodes embedded in a porous ceramic that respond to changes in salinity as pore liquids migrate into the ceramic cup. In one design a spring pushes the sensor in contact with the soil. Rhoades reported that the commercially available sensors have an accuracy of about ±0.5 mmho/cm with a good reliability record even after 5 years of field use. Because the ceramic cup must remain saturated, the range of accurate application depends upon the pore size of the cup; however, specially designed units are reported to have operated to about −20 bars. Where the soil remains at nearly a constant water content, the four-electrode method can detect changes in electrical conductivity that are attributed to salinity, as described by Rhoades (1978). A regression method is used to establish a relationship between the measured soil electrical conductivity and the actual conductivity of the pore liquid. For evaluating the electrical conductivity in a small volume, Rhoades (1978) described a portable salinity probe that is pushed into shallow soils and also a buried probe for semipermanent installations. Both probes contain four electrodes that are annular rings molded in a plastic probe.

Yong and Hoppe (1989) noted that although DC resistivity methods can provide an inexpensive method to locate subsurface contamination in simple geologic systems, in complex geologic settings it is difficult to separate the resistivity of different soils from the spatial variations in resistivity due to pore water chemistry. However, by comparing successive DC resistivity surveys, one potentially could delineate areas of resistivity change even with complex geology. Yong and Hoppe (1989) also indicated another complication, that a variety of chemically different materials exhibit essentially similar resistivities. Additionally, most organic liquids reduce electrical conductivity by a small amount so that contamination due to inorganic chemicals in pore liquids can mask the presence of organic compounds (Olhoeft, 1986). Consequently, monitoring over a very wide range of electrical frequencies, such as with complex resistivity, can overcome some of his lack of selectivity (Yong and Hoppe, 1989).

Olhoeft (1986) presented a good review of the complex resistivity method as well as the ground-penetrating radar technique, with examples of their use for both inorganic and organic pore liquid monitoring. He concluded that ground-penetrating radar works well in the vadose zone where the electrical resistivity exceeds 30 ohm-m, and that this technique offers a means to detect petroleum hydrocarbons floating on the water table. Furthermore, complex resistivity is able to map some organic contaminants where they have altered the physicochemical properties of clay layers. TDR, based on the dielectric properties of the soil, also is sensitive to pore liquid chemistry. While the travel time of the electrical pulse determines the water content, the amplitude of the signal is used to quantify salinity (e.g., Dalton, 1987). Redman et al. (1991) successfully tested TDR to detect perchloroethylene (PCE) injected into a 3 m × 3 m × 4 m deep test cell filled with silty beach sand and a basal clay. The organic liquid content was obtained by calibrating their unique TDR probe to organic liquids exhibiting a range of dielectric permittivities. Figure 33 shows the decreases in dielectric permittivity with time and depth. PCE with a density of 1620 kg/m^3 was injected for 27.4 h into the center of the test cell 65 cm below the ground surface and was first detected at the TDR probe shown after about 14 h (log time = 1.1). At 40 h (log time = 1.6), the PCE zone extended to depths between 0.7 and 1.6 m, and at 500 h (log time = 2.7), the liquid reached the clay aquitard at the 3.5 m depth. Ground penetrating radar was also used in this same experiment (Figure 34). Comparing the reflection before injection (Figure 34B) with that after 59 h (Figure 34C) reveals reflectors attributable to the PCE, particularly at the 0.7 and 1.3 m depths, which are probably the top and bottom of the PCE pool (Redman et al., 1991). They concluded that the GPR method was successful in monitoring the migration of the PCE over time.

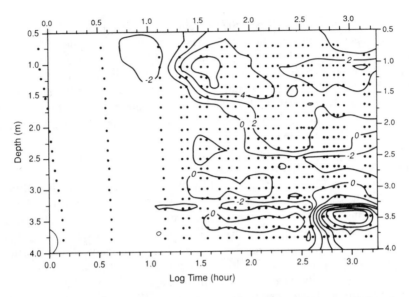

Figure 33   The changes in dielectric permittivity with respect to a background measurement taken prior to the perchloroethylene (PCE) injection. Measurements were taken at times and depths indicated by dots. The injection ended at 27.4 h. This TDR probe consisted of 20 measurement intervals, each a pair of 15-cm-long parallel brass conductors attached to a polyvinyl chloride pipe, over the depth of 0.74 to 3.79 m, with the lower 0.3 m in a clay aquitard. (From Redman et al., 1991. With permission.)

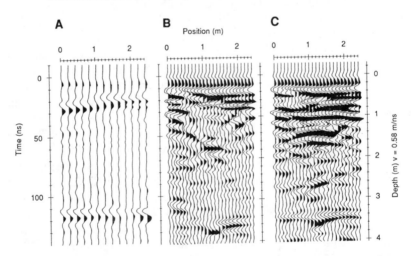

Figure 34   Ground-penetrating radar (GPR): (A) 100-MHz data collected prior to construction of the cell walls, (B) 200-MHz data after wall construction and prior to PCE injection, and (C) 200-MHz data 59 h after start of injection. (From Redman et al., 1991. With permission.)

Tomographic methods that have a source and detector in adjacent boreholes also appear very promising for detecting subsurface changes in pore fluid chemistry. An electromagnetic tomographic imaging system was used to interpret depths of soil contamination between four 33-m-deep boreholes that straddled a chromic acid disposal pit at Sandia National Laboratories (Borns et al., 1993). At another site, electrical resistance tomography successfully mapped spatial and transient distributions of steam injected into a borehole during a pilot test to remove volatile organic compounds from the vadose zone (Ramirez et al., 1993). The electromagnetic induction methods (Chapter 5, Section II.C.1.b) are also valuable reconnaissance tools for monitoring soil salinity.

CHAPTER 7

Vadose Zone Monitoring Strategy and Case Studies

Vadose zone monitoring is rapidly gaining acceptance as a means to protect human health and the environment. Two reasons for this seem to be advances in research to demonstrate the feasibility of vadose zone monitoring, and arguments for environmental protection and economic benefits, which have been recently presented in the technical literature (Kirschner and Bloomsburg, 1988; Cullen et al., 1992; Durant et al., 1993). Perhaps the more important motivation is that groundwater regulations are changing, and operating and closure permits are being written to require vadose zone monitoring as part of the overall monitoring plan. For example, Chapter 6.7 of the California Health and Safety Code (Bonkowski and Rinehart, 1986) on underground storage of hazardous substances gives owners of underground storage tanks (USTs) eight options for monitoring their installations, three of which include vadose zone monitoring (Bonkowski and Rinehart, 1986).

At the federal level, operators of land treatment facilities are required to conduct vadose zone monitoring under RCRA 40 CFR 264 Subpart M (U.S. Environmental Protection Agency, 1982). A land treatment facility, or what is sometimes referred to as a land farm, is a treatment technology to degrade, transform, or immobilize hazardous constituents within less than 1.8 m of the land surface and where the depth to the seasonal high water table exceeds 1 m. EPA currently is developing guidance documents for vadose zone monitoring at these and other RCRA facilities, such as surface impoundments and landfills (Durant et al., 1993).

I. MONITORING SYSTEM DESIGN

There are several steps in the process of designing a vadose zone monitoring system:

- Develop a release scenario
- Select instrumentation
- Determine location for instrumentation

As described in Chapter 5, Section I, for projects where implementing vadose zone monitoring is a goal, site characterization activities should be tailored to provide the information required to design the appropriate system. The following discussion is predicated upon the assumption that site characterization objectives are clearly tied to the data needs for each of the following steps in the monitoring system design process.

A. SCENARIO DEVELOPMENT

In order to design a monitoring system, we must first anticipate what type of release is going to be monitored. The potential source may be a point source, such as a tank leak, line leak, or liner puncture; it may be a finite area source such as an impoundment, pit, or landfill; or it could be a diffuse source such as irrigated fields or natural landscapes. The next aspect is to estimate the fluid characteristics and the magnitude and duration of the potential release. With this information, we can begin to develop a site characterization program that will lead to a conceptual model of the system, including potential fluid migration pathways.

Site characterization involves a comprehensive description of the subsurface materials such as geologic evaluations of lithology, stratification, and structure, including faults, joints, fractures, and macropores. Site characterization also includes quantification of the hydrologic properties of the representative geologic units, as well as the engineered materials in a landfill cover liner, for instance. These hydrologic properties, which are perhaps the most important aspect of site characterization, typically include hydraulic conductivity, soil-water characteristic curves, initial water content, and initial pressure head.

The fluid and media properties, along with a description of the probable release, allow one to delineate the likely migration pathways and to evaluate the potential for fracture flow, unstable flow, and colloidal transport, among other processes. Along these likely pathways, at some sites one may intuitively identify vulnerable areas, such as perched zones, springs, or canyon walls, which may require more intense monitoring efforts to prevent excursions to the accessible environment. At more complex sites, models may be useful to predict the likely pathways and how the media will respond to the release. For instance, analytical or numerical models can predict the likely increase in moisture content or concentration; they can assess where the seepage may migrate and the area potentially affected; and models can be used to predict when the release may migrate to a particular depth or location. To illustrate, at an RCRA waste site where the depth to groundwater was roughly 300 m, regulators questioned whether seepage from the facility would migrate horizontally to a vertical fracture and enter the aquifer downgradient of the proposed location of the groundwater monitor wells. We applied a numerical model of variably saturated flow, VS2D (Lapalla et al., 1987), to demonstrate, if seepage did in fact occur, that the likely pathways would be expected to reach the aquifer upgradient of the groundwater monitor wells. Ironically, the vadose zone modeling in this case was a tool to support the viability of the site groundwater monitoring program that was already in place.

Field experiments can also indicate probable vadose zone pathways. For example, Unruh et al. (1990) described a constant-head borehole infiltration test our firm conducted in a thick clay sequence at a hazardous waste disposal facility in Westmoreland, CA. The water and dye patterns observed in an excavation through the borehole infiltration test plot revealed preferential horizontal flow in a thin sandy layer, which later became a focus of the vadose zone monitoring program by neutron logging around the perimeter of the facility. The results of the field test were interpreted to show that if a significant leak occurred in the landfill clay liner, and if this leak fed directly into a macropore, then this sand layer would retain seepage in the shallow vadose zone, thereby minimizing the potential endangerment to the groundwater below.

B. INSTRUMENTATION

The second step in designing a vadose zone monitoring system is to select the appropriate instrumentation. What is to be detected: changes in moisture content or pressure head, or changes in chemical concentration of pore liquid or vapor? Usually more than one type of device is employed to provide an increased degree of assurance that the system will function properly. Next, there are two important aspects to be considered in selecting the appropriate instrumentation. One is to assess the magnitude and duration of the expected response to the release; the other is the sensitivity of the sensor. The nature of the response can be predicted by models, when system geometry, hydraulic properties, and initial and boundary conditions are known. In selecting the monitoring sensor, consider that a slow release from the source may induce a change in the soil parameter that is too small to be detected by some instruments. The primary purpose in most vadose zone monitoring efforts is to detect soil contamination and to prevent a significant mass flux from contaminating the aquifer. Therefore, we should be more interested in whether the monitored parameter reveals a significant change in the rate of deep percolation than in the magnitude of the water content change, for instance.

To illustrate how hydraulic properties affect sensitivity of a water-content monitoring system to changes in seepage rate, we can approximate the unsaturated seepage rate, assuming a unit hydraulic gradient, as equal to the unsaturated hydraulic conductivity at the *in situ* water content, that is, $q = -K(\theta)$. Therefore, sensitivity of the water-content monitoring system, the change in water content for a unit change in seepage rate, can be approximated as

$$\frac{d\theta}{dq} \approx -\frac{d\theta}{dK(\theta)} \tag{1}$$

Thus, sensitivity is defined here as the inverse of the slope of the unsaturated hydraulic conductivity and water content curve. The inverse of the K-θ slope generally will be smaller for the coarser than for the finer textures. Therefore, water content in sandy soils may be expected to be less sensitive to changes in the seepage rate than clay soils typified in Chapter 1, Figure 4, A through C. In considering the effect of hydraulic properties in the choice of a monitoring parameter, recall that water actually moves by differences in soil-water potential, rather than water content. Under very dry conditions, where the soil-water characteristic curve becomes asymptotic, there can be rather significant changes in potential that accompany very small changes in water content (Chapter 1, Figure 4A). If water content is to be monitored, rather than potential, the selected sensor must be able to detect a significant change in a water content that correlates to a significant change in seepage rate. It may not be sufficient to set arbitrary standards, such as that a 5% change in water content will trigger a regulatory response, unless quantitative analyses using the site-specific soil hydraulic properties indicate that such a change in water content will not adversely impact underlying aquifers.

The magnitude and duration of the release are also important considerations that influence instrument choice. Large releases may induce a response that overcomes any concerns for instrument sensitivity. Releases of short duration may require more frequent monitoring events than if the release is continuous over a long period of time. In some cases automation may be required to collect crucial transient information, such as the response to infiltration following a thunderstorm.

When the hydraulic properties and release are known or can be estimated, analytical or numerical models are useful system performance assessment tools. In the previous section on scenario development, we indicated that models can usually aid in defining probable fluid pathways. They can also be used to select the appropriate instrument by evaluating the likely changes in soil-water physical or chemical status. Thus, a performance assessment becomes an integral part of the vadose zone monitoring system design. This same model may be used later to demonstrate the likely impacts to groundwater if a release actually occurs, and it can also be an essential aspect of remedial design. All too often, in practice, quantitative modeling is only initiated after a release has occurred. To get more value from the investment in modeling, a cost-effective approach would be to build a predictive model early in the site investigation that can be a tool throughout the project, including monitoring system design, remedial design, and site closure.

The foregoing discussion is clarified by the very recent work of Yeh et al. (1994) who simulated a leak (5.79×10^{-3} m³/h-m²) from a synthetic liner into a three-layered sequence of engineered soils set beneath the liner. The soil layers and model grid are shown in Figure 1 and the hydraulic properties are shown in Figure 2. The pressure-head distribution in Figure 3 clearly illustrates seepage progressively migrating through each of the three layers, with the maximum lateral spreading in the clay layer at early time. However, note how the lower coarse sand layer caused significant pressure build up (from about $\psi = -6.1$ m to -0.3 m) and lateral flow in the medium sand above it after about 168 h (Figure 3D,E). At these same times, there is a somewhat different water-content distribution predicted by the model in Figure 4D and E. The increasing volumetric water content in the middle fine sand is very slight, only about 3%, owing to the nature of the soil-water characteristic curve (Figure 2A).

If the regulatory criterion for leak detection is a measured increase in volumetric water content of at least 5%, monitoring water content within the fine sand material would not trigger a response in this example. Although this numerical simulation is for a soil system that is only about 90 cm thick, the results are highly relevant to thicker natural or engineered soils. If water-content monitoring is to be conducted in a setting like the one described here, the preferred location would be just above the interface of the coarse layer. Near the interface, the unsaturated hydraulic conductivity

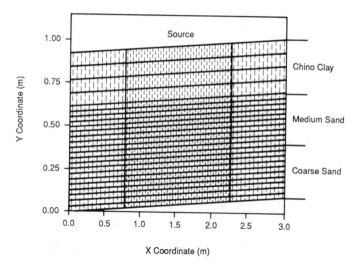

Figure 1 Finite element discretization of the three-layer liner cross section. (From Yeh et al., 1994. With permission.)

of the coarse sand is initially much lower than that of the overlying medium sand layer (Figure 3A); consequently, water movement into the large pores is impeded until the pressure head builds up in the medium sand to near the water-entry value of the coarse-sand (Figure 3A). The most significant implication of this example with respect to vadose zone monitoring is that water content trigger levels must be based on the specific hydraulic properties of the soils. Additionally, based on these results, leaks may be more readily detected by monitoring pressure head than water content. The selection of the vadose zone monitoring instrumentation is also dependent upon the subsurface area likely to become contaminated relative to the area of the source. For diffuse sources where contamination may be widespread over large areas, such as pesticides beneath an irrigated field, point samplers (e.g., tensiometers, soil cores) may be appropriate. If the soil is very uniform in texture and infiltration is uniform across the site, the location of the sampler within the field may not be critical to the determination of whether irrigation water or pesticides migrate below the root zone. On the other hand, if flow beneath the field moves via local macropores or if there is a potential point source release somewhere beneath a landfill, then spatially integrating samplers (e.g., neutron probe, pan lysimeters) or more intense point instrumentation would be required to improve the probability of detecting a release.

Under RCRA, the guidance on the minimum number of groundwater monitor wells is quite clear: three downgradient and one upgradient. In the vadose zone, there is no specific minimum requirement, however. The number of instruments depends upon a number of factors, the most important of which is perhaps the release scenario. Obviously, to monitor beneath a sump of known location requires fewer instruments than to monitor beneath a large, lined surface impoundment where the location of the leak is unknown. Formation spatial variability in hydraulic properties, which is almost always imperfectly characterized, induces uncertainty in the flow path that leads to uncertainty in whether the release can be detected at a discrete location.

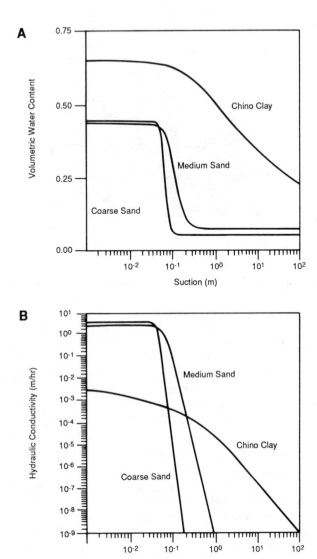

Figure 2   (A) Hydraulic conductivity as a function of suction for the three-layer system, and (B) moisture release curves for the three-layer system. (From Yeh et al., 1994. With permission.)

The nature of spatial variability in chemical transport in the vadose zone has been extensively investigated over the last two decades. Most of the research reveals complex pathways and subsurface distributions of solutes and contaminants. If one were to obtain representative local scale samples beneath an areally extensive site of a contaminant release, one would likely find at a particular depth that the fluid pore liquid velocity and spatial distribution of contaminants would be lognormally distributed (Figure 5). For example, Biggar and Nielsen (1976) intensively sampled 20

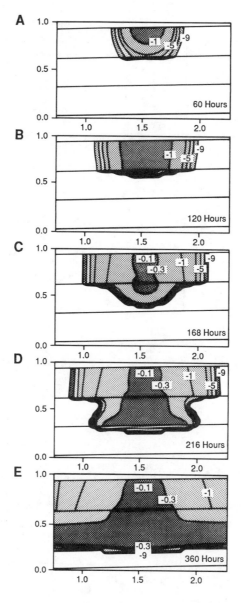

Figure 3 Matric potential distribution from constant flux application at 5.79 m/h for times (A) 60, (B) 120, (C) 168, (D) 216, and (E) 360 h. (From Yeh et al., 1994. With permission.)

6.5-m² instrumented plots beneath a 150-hectare irrigated field. Chloride and nitrate tracers were applied to each test plot under ponded conditions after the flow field reached steady state. On the basis of field data, they concluded that to determine the true mean tracer velocity within ±50% at the 95% confidence level, one would need to obtain 100 measurements (Figure 6). Lognormal spatial distributions of velocity suggest that high transport rates occur over only a small area, even though the source

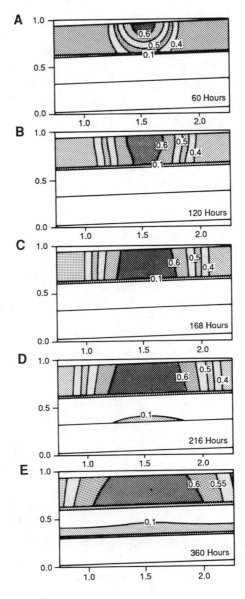

Figure 4   Water-content distribution from constant flux application at 5.79 m/h for times (A) 60, (B) 120, (C) 168, (D) 216, and (E) 360 h. (From Yeh et al., 1994. With permission.)

is areally extensive. Therefore, it is important to have a sufficiently large number of samplers or a sampler that integrates a rather large area, in order to adequately meet the protection objective.

The following example is offered to show how site characterization data, when coupled to a simple vadose zone model, can assist in the determining the number of monitoring stations. Warrick et al. (1977) used the statistical properties (mean,

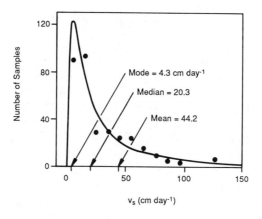

Figure 5 Log normal distribution of fluid pore velocity. (From Biggar and Nielsen, 1976. With permission.)

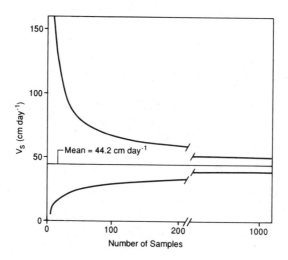

Figure 6 Mean tracer velocity and ±95% confidence limits as a function of number of samples taken through the entire field. (From Biggar and Nielsen, 1976. With permission.)

variance, and probability density function) of field-measured hydraulic properties to calculate the number of samples necessary to compute the mean specific discharge (Darcy velocity) at different times of drainage following a thorough saturation of a field soil. Their analysis is based on the following form of unsaturated hydraulic conductivity:

$$K = K_0 \ \exp\left[\beta(\theta - \theta_0)\right] \tag{2}$$

where β is the slope of the ln K-θ curve, and K_0 and θ_0 are the hydraulic conductivity and water content at maximum field saturation. They used the following equation to compute the Darcy velocity for a draining profile:

$$q = K_0 \left[1 + \left(\beta K_0 t / L\right)\right]^{-1} \qquad (3)$$

where t is the time after infiltration stopped, and L is the depth below land surface where q is calculated. The hydraulic property data were obtained from instantaneous profile tests at the same experiment site described previously by Biggar and Nielsen (1976). Statistical tests showed that β and K_0 were lognormally distributed, whereas θ_0 was normally distributed. Based on these distributions and the mean and variance of the properties, Monte Carlo simulations were run with Equation 3. The model results indicate, for example, that at time zero one would need to make 100 measurements of Darcy velocity to obtain the true mean within ±51%, at the 95% confidence limit. After 1 d of drainage, one would need only 50 measurements to characterize the true mean within ±20%. As time progresses and water content decreases, fewer samples are needed to achieve a given level of accuracy in this example, apparently due to the greater spatial variability at saturation due to macropore flow. These results are similar to the field tracer velocity data of Bigger and Nielsen (1976). One weakness in this Monte Carlo approach is the assumption that the hydraulic properties are independent of one another when actually it is more likely that they are correlated. Nevertheless, this is a good example to illustrate how to predict the gains in accuracy that accompany increased monitoring density.

There are a few other practical considerations in deciding on the number of vadose zone monitoring devices. First, what are the consequences if the proposed number of devices fails to perform? Is there a backup device within the vadose zone monitoring system? Is there a groundwater monitor well system? And finally, and often most important, what is the available budget? Owing to the nature of spatial variability, a prohibitively large number of vadose zone monitoring devices or samples may be needed to estimate site statistics, such as may be desired by regulators for formal risk assessments. However, for detection monitoring, accurate pathway delineation, rather than sample statistics, is probably the key issue; consequently, the number of instruments or sampling locations is commonly guided by professional judgement, knowledge of vadose zone processes, and predictive modeling.

C. INSTRUMENT PLACEMENT AND MONITORING FREQUENCY

The location of the vadose zone instrumentation depends on some of the same factors discussed earlier, including the location of the likely release areas, the vadose zone pathways, and the environmentally sensitive areas at the site. Wherever the location, the procedures to install the instrumentation should avoid drilling through contaminated media and should avoid the potential for contaminants to migrate along the monitoring installation to the sensor. In some cases, angle hole or horizontal drilling is a viable means to overcome this potential problem. The disadvantage of angle hole drilling is sometimes that additional time is required to complete the installation or that there is less confidence in the quality of instrument installation of the sensor, due to concerns for the contact with the formation, backfill location, etc. For new facilities it is most practical to embed the vadose zone monitoring system

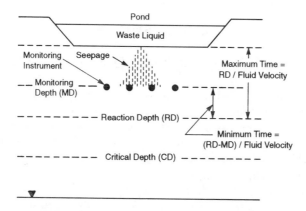

Figure 7   Monitoring depth and time between monitoring criteria. (From Robbins and Gemmell, 1985. With permission.)

within the design of the engineered structure. For example, neutron probe access tubes can be installed in horizontal trenches excavated in the foundation before a landfill liner is emplaced.

For successful detection monitoring beneath hazardous waste facilities, the horizontal location and vertical emplacement depths of the instrumentation are crucial. At some sites, particularly where the soils are highly stratified and dry, and where the leak rate is small, considerable horizontal liquid movement can occur (McCord et al., 1991). If the sensors are too deep, lateral migration could go undetected. For most applications, the primary advantage of vadose zone monitoring is to provide an early warning of seepage, before the aquifer is impacted. Therefore, the instrumentation must be placed above the depth where a suitable response such as remediation is feasible (Figure 7). If contamination is detected, sufficient time must be allowed to plan and implement a remedial response before the contamination migrates below a critical depth. This critical depth could be the maximum depth that excavation is feasible, for example. Site characterization, release scenario development, and performance assessment modeling are essential tools to predict the potential rate of migration between the sensor depth and the critical depth. Depending upon the hydraulic properties and degree of saturation, the preferred sensor location may often be in the fine-textured layers.

The maximum and minimum frequency for monitoring can be determined from site characteristics and release scenarios (Robbins and Gemmell, 1985). In Figure 7, the maximum time between measurements should be less than the time for seepage from the source to migrate to the reaction depth. If seepage should reach just beyond the reaction depth, there would be insufficient time to implement vadose-zone remediation that would prevent the seepage from reaching the critical depth and eventually migrating to the aquifer. The minimum time between monitoring events is based on the time for seepage to travel from the monitoring depth to the reaction depth. More frequent monitoring would allow one to implement a remedial action before seepage reaches the reaction depth. This may be unnecessary and costly, but more frequent monitoring, or continuous monitoring, potentially could lead to lower remediation costs.

II. CASE STUDIES

A. CASE STUDY 1: FLOW AND TRANSPORT TEST FACILITY*

This case study (modified from Stephens et al., 1988) describes a controlled field experiment in which water and tracer were introduced over a test plot at a constant rate. The water application rate was fixed so the soil beneath the wetted plot would remain unsaturated during infiltration. The primary objective of the experiment was to evaluate the significance of geologic heterogeneity on three-dimensional soil-water and contaminant movement.

The field site is located in the semi-arid southwestern United States at Socorro, NM, on the west side of the New Mexico Tech campus golf course. The vadose zone at the site is approximately 24 m thick, and the native, sandy soil had never been irrigated prior to the experiment.

1. Instrumentation

Water was applied over a 10.5 m × 10.5 m area using a drip irrigation system installed approximately 60 cm below the land surface in the center of the field site (Figure 8). The area for the water application system was excavated using a backhoe, and final leveling was accomplished by hand. A thin (2 cm) layer of uniform sand was spread over the leveled area, and 21 drip irrigation lines were placed on the flat surface to create a grid of emitters spaced at 50-cm intervals. To prevent evaporation, plastic was placed over the irrigation lines. Next, the irrigation system was covered with hay for insulation, and earthen fill was added to a grade slightly above land surface. A second layer of plastic covered the fill to prevent infiltration of precipitation.

Municipal water used for the infiltration experiment was piped to a small water supply reservoir at the site. Flow to the test plot was controlled by a positive displacement pump, which delivered a prescribed volume of water to the drip lines from the water supply reservoir. The timer-controlled pump fed water to the drip lines for approximately 1 min each hour. Water flow and distribution through the drip system was monitored by eight totalizing flow meters. A data logger in the field office recorded precipitation; wind speed; soil, air, and water temperature; pumping duration; and in-line water pressure. A flux rate of approximately 1×10^{-5} cm/s (0.86 cm/d) was chosen for the infiltration rate. This rate was based on saturated hydraulic conductivity from laboratory analyses of soil core samples. With the exception of minor clay lenses, the lowest saturated hydraulic conductivity found in the soil profile was roughly 100-fold more than this flux rate. Therefore, we anticipated by design that the soil would be unsaturated.

Moisture movement through the soil was monitored by a neutron probe (Model 503DR, CPN Co., Martinez, CA) and tensiometers (Figure 8). The tensiometers were constructed using 1.9-cm polyvinyl chloride (PVC) pipe, 1 bar standard ceramic cups, and a septum rubber stopper. Pressure head was recorded by a pressure transducer connected to a hypodermic needle inserted through the rubber stopper.

---

* Modified from Stephens et al., 1988.

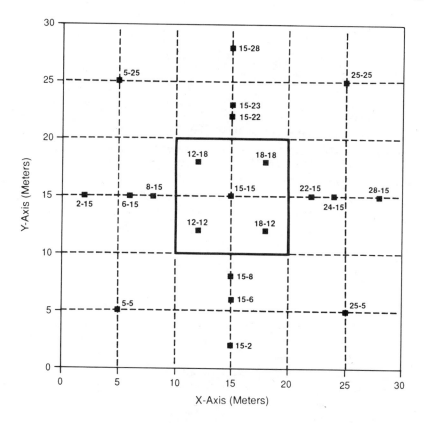

Figure 8 Plan view of field instrumentation and infiltration source area at an experimental site on the campus of New Mexico Tech, Socorro, NM. Source area is outlined by the heavy line. Each square represents a monitoring station consisting of one or two tensiometer nests and one access tube. (From Stark, 1992. With permission.)

Twenty-one water content and pressure head monitoring stations were located in the field site (Figure 8). Every station contained a 5-cm-diameter (2-in-diameter) by 9.1-m-long (30-ft-long) aluminum neutron access tube and tensiometers installed in duplicate nests at approximately 50-cm intervals to 4 m depth below land surface. An auger drill rig was used to install the neutron access tubes and tensiometers in 20.3-cm-diameter (8-in-diameter) boreholes. Native soil was compacted around the instrumentation in the borehole annulus. Layers of bentonite were placed between tensiometer intervals and at the land surface to deter channeling of infiltrated water.

Prior to the experiment start on January 29, 1987, the instrumentation was monitored weekly to provide background information. After water began infiltrating into the soil, data were collected daily, biweekly, and then weekly as the experiment progressed.

2. Site Geology

Geologic cross sections of the east-west and north-south transects were determined by correlating visual characteristics of soil samples collected during drilling.

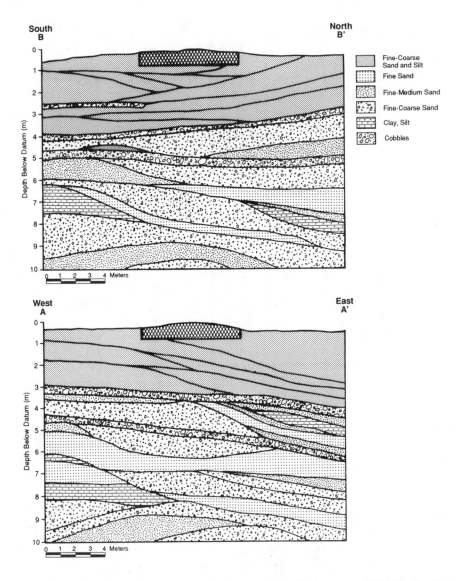

Figure 9 Geologic cross sections through the New Mexico Tech campus field infiltration test plot. (From Stephens et al., 1988. With permission.)

No samples could be obtained of the cobble zones, but the locations of cobble layers were determined from drilling characteristics. Figure 9 shows the soil profiles exaggerated twice on the vertical scale to illustrate textural contrasts in more detail. The profile is stratified, consisting of two general soil zones: an upper zone and a lower zone. The upper zone is comprised of red-brown silty sands and pebbles interbedded with cobbles to a depth of about 4 to 5 m. The lower zone is comprised of clean, tan, fine sand and fine to coarse sand and pebbles. Clay lenses of undetermined extent are present at all depths. Both major soil zones represent facies of the Sierra Ladrones formation (Machette, 1978). The red-brown silty sands and gravel are piedmont slope facies (mudflow/debris flows) derived from the Socorro Range to the west, and the underlying tan sands represent ancient Rio Grande fluvial deposits.

Several exposures that have similar geologic characteristics to the study area profile are found in an arroyo a few hundred meters west of the field site. Individual layers exposed in these outcrops cannot be directly correlated to the field-site soil profile, but these exposures give an indication of the sedimentary structures likely to be found beneath the site. One of the outcrops in the piedmont facies exhibits a sequence of horizontally continuous fine and coarse sandy layers ranging from 30 to 80 m thick. Major textural features are laterally continuous for meters across the outcrops. Within each dominant horizon there are thin layers of fining upward sequences consisting of gravel to thin sand and silt.

3. Hydrologic Characteristics

Split-spoon and core samples were analyzed in the laboratory to determine hydrologic and physical characteristics. The split-spoon (disturbed) samples were analyzed for gravimetric water content by oven-drying and for particle size by mechanical sieving and the hydrometer method. The core samples (less disturbed) were analyzed to determine a number of parameters, including volumetric moisture content, bulk density, porosity, saturated hydraulic conductivity by constant- and falling-head permeameters, soil moisture characteristic curve by hanging column method, 15-bar moisture content by pressure-plate method, and particle size distribution from sieve and hydrometer analysis. Unsaturated hydraulic conductivities were calculated from moisture retention data (imbibition cycle) using Mualem's theory (van Genuchten, 1980). Table 1 summarizes the hydraulic properties of the piedmont and fluvial facies.

4. Water Content and Pressure Head-Monitoring

Infiltration began January 29, 1987 and has continued for more than 33 months. Infiltration rate was approximately 1.0×10^{-5} cm/s for the first 29 months and increased 10-fold thereafter. Evaluation of data from the flow meters that monitor segments of the flow system indicate that the distribution of water was reasonably uniform over the 10.5 m $\times$ 10.5 m area. However, there was a leak in one of the emitter line connections to the water supply manifolds located in the northeast part of the plot. This leak went undetected from d 26 to 39 after infiltration began. The leak may have reduced the infiltration to the plot by about one-third during this time.

Water-content and pressure-head measurements immediately beneath the infiltrated area suggest that most of the water moved vertically downward. The location of the wetting front was determined from the depth of greatest moisture content gradient in water-content profiles such as those in Figure 10. The rate of advance of the wetting front for station 15-15 in the center of the 10.5 m $\times$ 10.5 m wetted area, as determined from neutron probe logging, is shown in Figure 11. The rate of wetting front advance in the first 3.5 m ranged from about 9 cm/d at location 12-12 to 23 cm/d at location 18-18, and the mean of the five stations was about 17 cm/d.

Table 1 Summary of Mean Hydraulic Properties[a] at the New Mexico Tech Golf Course Site, Socorro, NM

Facies	Saturated hydraulic conductivity in K_s (cm/s)	Porosity[b] n (cm³/cm³)	Initial water content[c] θ_i (cm³/cm³)	Water content[d] θ_{15} (cm³/cm³)	Fitting parameters to retention curves α (cm⁻¹)	N	10% Passing particle size d_{10} (mm)
Piedmont slope	−5.48 <4.08> [67]	0.434 <0.387> [69]	0.134 <0.268> [112]	0.124 <0.220> [36]	0.123 <0.099> [63]	1.719 <0.412> [63]	0.067 <0.008> [60]
Fluvial[e]	−4.35 <1.70> [15]	0.360 <0.206> [21]	0.062 <0.005> [72]	0.052 <0.049> [9]	0.087 <0.008> [15]	2.704 <1.906> [15]	0.163 <0.002> [21]
Clay[f]	−10.45 <—> [1]	0.479 <—> [1]	— <—> [—]	−0.28 <—> [1]	0.085 <—> [1]	1.118 <—> [1]	4.3×10^{-4} <2.2×10^{-7}> [11]

[a] Numbers in angled brackets are variances, and those in square brackets show number of samples.

[b] Calculated from bulk density.

[c] From neutron probe data.

[d] At 15 bars.

[e] All but eight fluvial sand samples were repacked to in situ bulk density.

[f] Clay occurs occasionally in both the piedmont slope and fluvial facies. It is shown separately and has not been included in statistics of these major facies.

Source: From Stephens et al., 1988. With permission.

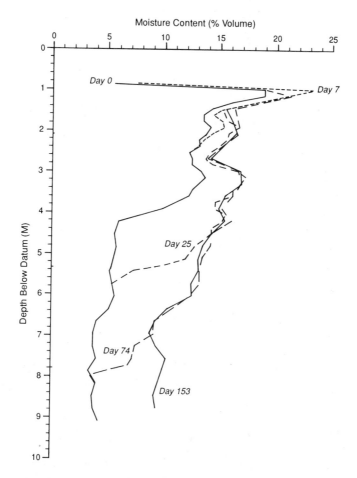

Figure 10 Moisture content profiles for selected days after infiltration began at Station 15-15. (From Parsons, 1988. With permission.)

Because the rate of wetting front advance seemed to have considerable variability, we wanted to evaluate whether at least the mean velocity was reasonable. Warrick et al. (1971) indicated that for a simple one-dimensional flow field, the increase in moisture content behind the wetting front should be approximately equal to the ratio of steady applied flux to rate of wetting front advance (Chapter 2, Equation 5). Using this simple calculation (0.86 cm/d ÷ 17 cm/d), we should expect that the average moisture content would increase by about 5% beneath the plot. The observed increase in moisture content at the five stations is quite variable (Figure 12), but does average about 5%. Consequently, there is some consistency between the mean wetting front velocity and observed increase in water content in response to the applied infiltration.

Pressure-head measurements with tensiometers beneath the infiltration area indicate that the mean hydraulic gradient is near unity, as one would expect for a vertical flow system (Figure 13). Although the one-dimensional model is quite appropriate

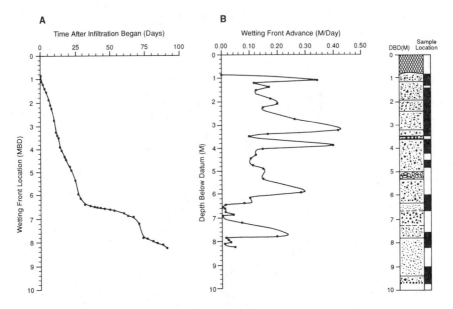

Figure 11   Wetting front location, rate of wetting front advance, and the geologic log for Station 15-15. (From Parsons, 1988. With permission.)

for predicting mean behavior, some lateral flow does occur. Neutron logging outside the perimeter of the 10.5 m × 10.5 m plot shows the multidimensional nature of the flow field (Figure 14). After 80 d, the wetting front progressed approximately 7 m vertically and 2 to 3 m laterally from the edge of the infiltration plot. This multidimensional behavior was also illustrated by the hydraulic head data mapped at the end of infiltration (Figure 15).

In plan view (Figure 16), the distribution of moisture content appears to be quite symmetric about the plot in the upper unit, the piedmont slope facies. In contrast, in the fluvial facies, the distribution is axisymmetric, with the wetting front extending more than several meters from the edge of the wetted plot. It is interesting to note in Figure 16 that the moisture content at a particular depth appears to be less variable at locations near the source where the soil is wetter than in the drier region. This type of variability may be expected based on stochastic theory (e.g., Yeh et al., 1985).

5. Pore Liquid Monitoring

A calcium bromide tracer was added to the injection system for 151 h between February 25 and March 2, 1988, without any appreciable change in the application rate during or following the tracer application. Flanigan (1989) described details of the instrumentation installations and the tracer experimental procedures. Fourteen pressure-vacuum porous cup samplers, approximately 5 cm in diameter, were installed in boreholes located beneath and surrounding the plot to depths of 6 m below the emitters (Figure 17). To collect a sample, the applied vacuum ranged from about 20 to 25 cbars for samplers inside the wetted zone to 60 to 70 cbars for the outside samplers situated in drier soil. Pressure was applied to force the sample to the surface

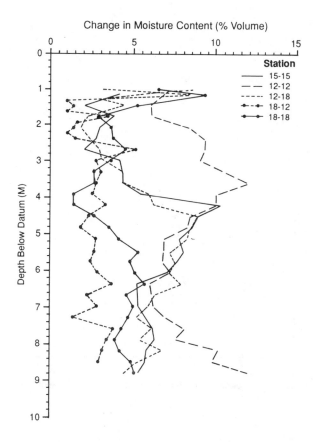

Figure 12   Change in moisture content at five locations beneath the wetted plot after about 150 d of infiltration. (From Stephens et al., 1988. With permission.)

through a production tube. Samples, which ranged in volume from about 20 to 200 ml, were analyzed by high-performance liquid chromatography. Laboratory experiments by Flanigan (1989) showed that the bromide exhibited slight anion exclusion characteristics.

The solute samplers successfully monitored the pore liquid and clearly indicated the migration of the bromide tracer (Figures 18, 19, and 20). Three samplers, G, H, I, at different depths near the center of the plot reveal breakthrough curves that are consistent with expected behavior (Figure 18). Note that the peak concentration diminishes and the curves widen with increasing depth as one would predict from the advection-dispersion equation. Figure 19, on the other hand, illustrates some unusual behavior in the early arrival of tracer at sampler B, which is located below samplers F and J. This may be attributed to heterogeneity in the media or to channelling along the sampler (Stark, 1992). The outer suite of samplers illustrates (Figure 20) extreme variability in arrival time and peak concentration in the dry range of water contents. This tendency for increasing variability in the flow paths with increasing dryness may be expected based on stochastic theories cited during the preceding discussion of plan-view moisture-content contours.

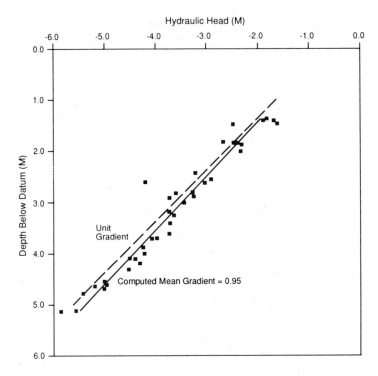

Figure 13 Hydraulic head beneath wetted area at the end of infiltration. (From Stark, 1992. With permission.)

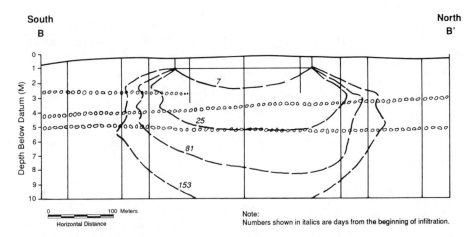

Figure 14 North-south cross section of wetting front positions. (From Stephens et al., 1988. With permission.)

In summary, this case study illustrates that under controlled experimental conditions vadose zone monitoring can delineate flow paths as well as detect seepage and tracers. The experiment also revealed that beneath the source the hydraulic gradient was near unity in the vertical direction; however, lateral flow components were also

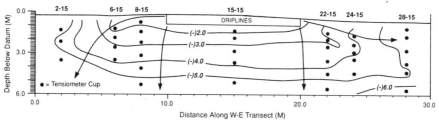

Contour Interval = (-) 1.0 Meter

Figure 15 Hydraulic head and flow lines after 2 years and 217 d of infiltration.

important. Furthermore, at increasing distance from the source, and as the water content decreased, water-content and tracer data suggest that the variability in flow paths and solute transport increases.

B. CASE STUDY 2: LANDFILL MONITORING*

Consultants to operators of a Class I hazardous waste landfill in Westmoreland, CA, undertook site characterization and vadose zone monitoring to support the operators' permit application. This section (modified from Reaber and Stein, 1990) describes the installation of the vadose monitoring networks surrounding one of the landfill cell areas, LC-3, and discusses some of the initial findings.

1. Hydrogeologic Setting

The facility is located in the Colorado Desert geomorphic province south of the Salton Sea as shown on Figure 21. The geologic units beneath the facility consist of well-stratified lacustrine deposits overlain by a thin veneer of surficial alluvium (Qa). In the vicinity of LC-3, six lacustrine units (Ql_1 through Ql_6) have been identified within the uppermost 33 m of soils below the site. These deposits are generally fine grained, ranging between plastic clays and silty fine sands. Groundwater is encountered at approximately 18.3 m below land surface. Operating permits require monitoring in both the unsaturated zone and the uppermost continuous saturated unit.

2. Vadose Zone Monitoring Program

The purpose of the LC-3 vadose zone monitoring program is to provide for early detection of waste constituent migration from the landfill toward underlying groundwater. The monitoring program consists of measuring *in situ* soil moisture with a CPN Co. model 503DR hydroprobe at four access tubes around the landfill (VZ-5 through VZ-8) as shown on Figure 22.

Access tubes were installed at 45 degrees with both hollow-stem auger and ODEX drilling. Pilot borings were drilled with a 0.12-m ($4^7/_8$-in) core barrel. Soil samples were collected at 0.76 m (2.5 ft) intervals by advancing a 0.76-m-diameter (2.5-in-diameter) modified California split-barrel sampler into the undisturbed soil

* Modified from Reaber and Stein, 1990.

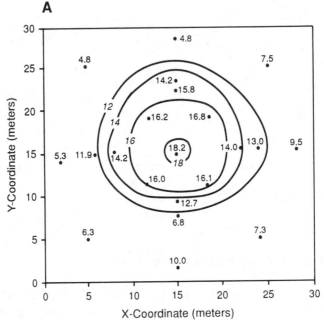

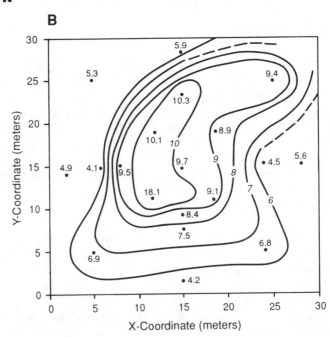

Figure 16   Water-content distribution at depth intervals of (A) 2.9 to 3.2 m and (B) 8.3 to 8.6 m below datum after 329 d of infiltration. (From Stephens et al., 1988. With permission.)

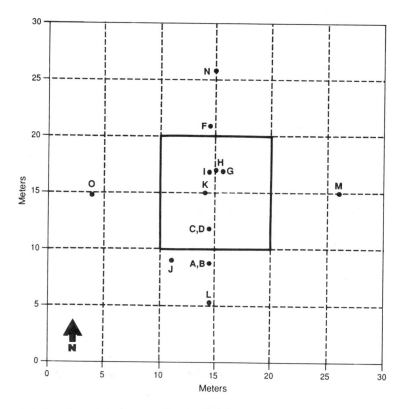

Figure 17   Location of porous cup soil-water samplers for bromide tracer experiment. (From Flanigan, 1989. With permission.)

immediately in front of the core barrel. The sampler was advanced using a 63.6-kg (140-lb) hammer and an 0.46-m (18-in) drop. The sampler was lined with two brass tubes and one stainless steel tube, with the brass tubes used for physical testing and soil classification and the stainless steel tube used for chemical testing of the soils. All soils were classified using the Unified Soil Classification System (USCS) and a Munsell soil color chart. The soil samples retained for physical and chemical testing were sealed with plastic end caps and taped to both ends with silicone tape. The pilot borings were extended to the approximate depth of the lowest clay liner.

The 0.14-m ($5^5/_8$-in) OD Schedule 40 steel access tubes were installed using the ODEX air percussion drilling method. The tubes are 21 to 23 m long and are oriented at a 45 degree angle so that they extend below the landfill into the basal sand of unit Ql_2. The ODEX system was equipped with a 0.13-m ($5^1/_4$-in) drill bit that attaches to the end of the steel casing. As the drill stem rotates, an eccentric cutting edge extends past the edge of the steel casing. The steel casing is advanced by percussion, resulting in a tight fit between the casing and the borehole wall. The steel casing was installed in 3.3-m (10-ft) lengths that were butt-welded together to minimize borehole disturbance. The access tubes were completed to depths between 1.6 and 3.3 m (5 and 10 ft) above the water-table and were fitted with Neoprene end plugs to minimize the possibility of soil moisture condensation within the access tubes. The

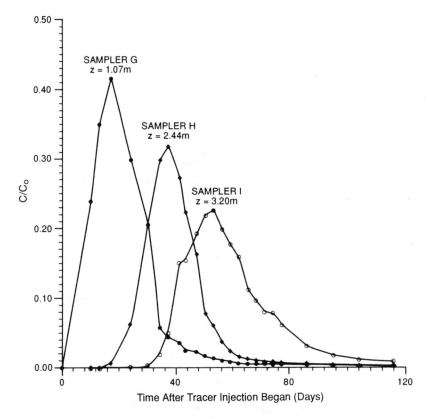

Figure 18   Bromide breakthrough curves for samplers G, H, and I, located at various depths
below the drip lines beneath the plot. *Note*: *z* is the depth below the lines of each
porous cup. (From Flanigan, 1989. With permission.)

access tubes were completed 0.3 m (1 ft) above ground surface and were protected
with an outer locking steel casing. A typical completion diagram is shown on
Figure 23.

A total of 55 samples was collected during drilling of the four neutron probe
access tubes. These samples were tested for moisture content (weight percent) and
dry density. Five soil samples from each vadose-zone access tube boring representing
each of the major geologic units at the site were submitted for chemical analysis. The
samples were analyzed for metals, as well as chloride, fluoride, sulfate, total organic
carbon, volatile organics, and pH.

3. Water-Content Monitoring

Water content was measured at 0.76-m (2.5-ft) increments by lowering the
neutron probe down the access tubes. The neutron probe count rate was recorded in
the field and was then compared to the laboratory results of the soil-water content in
order to develop separate calibration curves for the sands and clays. The calibration

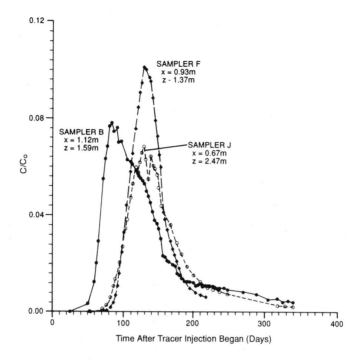

Figure 19   Bromide breakthrough curves for samplers B, F, and J, located at various depths below the drip lines on the perimeter of the plot. *Note*: x is the horizontal distance from the edge of the plot; z is the depth of the sampler cup below the plane of the drip lines. (From Flanigan, 1989. With permission.)

curve was developed by linear regression on the ratio of measured counts to the standard counts of the instrument versus the laboratory water contents converted to a volume basis (Figure 24).

Figure 25 shows a plot of laboratory-determined and field-determined soil moisture content for VZ-6, juxtaposed to a lithologic log for VZ-6. In general, the moisture content of soils in the unsaturated Ql_1 and Ql_2 units ranged between 5 and 40%. The sandier soil units at the top of Ql_1 typically contained 10 to 20% moisture by volume, while the clay-rich soils of the Ql_2 unit tended to range between 25 and 40% moisture by volume.

Soil moisture content monitoring was proposed on a quarterly basis. The quarterly monitoring results will be compared with the background data and previous results to assess whether a significant change in soil moisture content has occurred. An increase in moisture content of 5% will be considered to represent a significant change. A 5% increase of water content was chosen on the basis of previous studies (Daniel B. Stephens & Associates, Inc., 1988; Tyler, 1985) that have concluded that small percentage increases may be detected by neutron probes in large diameter steel casings.

If a significant change in the soil moisture content is detected, the appropriate agencies would be notified, and an assessment program would be implemented to evaluate whether the change may be related to leakage from the landfill.

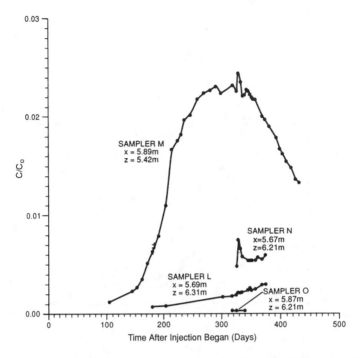

Figure 20   Bromide breakthrough curves for samplers L, M, N, and O, located about 6 m below
the plane of the drip lines, one on each side of the plot. *Note: x* is the horizontal
distance from the edge of the plot; *z* is the depth of the sampler cup below the plane
of the drip lines. (From Flanigan, 1989. With permission.)

Figure 21   Location map for Class I hazardous waste facility in Westmoreland, CA (From
Reaber and Stein, 1990. With permission.)

C. CASE STUDY 3: SOIL GAS INVESTIGATION*

Soil gas surveys have become indispensable in the delineation of subsurface
volatile organic compounds (VOCs). To date, the majority of soil gas surveys have
been aerial screening investigations of the shallow vadose zone, usually to depths of
less than 3.3 m (10 ft). There has recently been an increasing need for information

* Modified from Forbes et al., 1993.

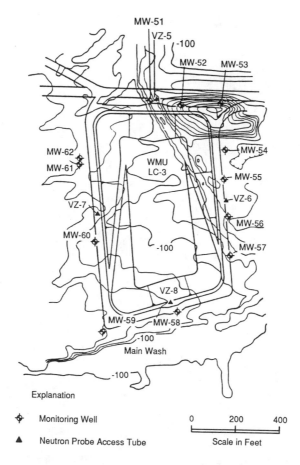

Figure 22 LC-3 monitoring network. (From Reaber and Stein, 1990. With permission.)

regarding the distribution and movement of VOCs at greater depths. This is particularly true in the western United States where depths to groundwater frequently exceed 33 m (100 ft).

This case study (modified from Forbes et al., 1993) presents the results of a deep soil gas investigation conducted at a semi-arid site in the Southwest as part of a Remedial Investigation/Feasibility Study (RI/FS).

The site is located in an area underlain by several thousand feet of basin fill alluvium. The site stratigraphy consists primarily of silty sand with some gravel and clay zones. The upper 10 m of the profile is calichified in a layer that is rather extensive throughout the site (Figure 26). The depth to groundwater is approximately 30 m (90 ft). The topography is nearly flat, and the climate is semi-arid, with an annual average precipitation of 0.3 m (11 in), over half of which occurs during summer cloudbursts during July and August. Past releases of chlorinated solvents and fuel hydrocarbons are believed to have occurred near the site, and previous shallow soil gas surveys (Figure 27) and deep soil borings have shown that trichloroethylene (TCE) is the predominant subsurface contaminant. The ongoing RI/FS is

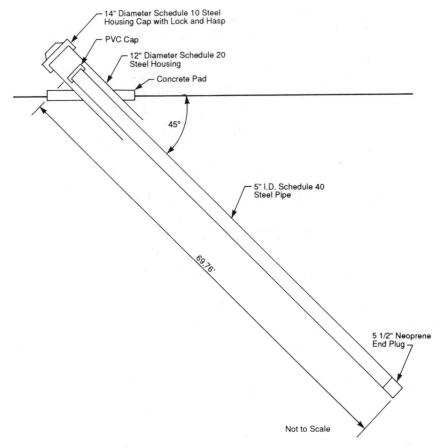

Figure 23   Typical vadose-zone access tube construction. (From Reaber and Stein, 1990. With permission.)

being conducted in order to determine the nature, extent, and magnitude of VOC contamination in the vadose zone and shallow groundwater.

The shallow soil gas survey was conducted to depths about 1 to 1.5 m by a commercial truck-mounted mobile laboratory to detect volatile organic compounds. The principal chemical of interest is the solvent trichloroethylene (TCE). The shallow soil gas survey revealed several areas of concern where there was potential prior solvent use (Figure 27). Because the primary purpose of the RI/FS is on soil remediation, probably by soil vapor extraction, the vertical distribution of soil contamination is important. As part of the next phase of investigation, a deep soil gas survey was designed and initiated in some of the areas delineated by the shallow soil gas survey.

1. Soil Gas Well Designs

As was discovered during the planning phase of this study, no standard design exists for the construction of soil gas monitor wells. With the exception of shallow

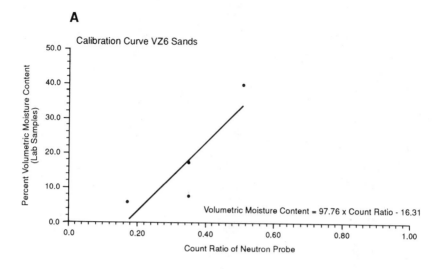

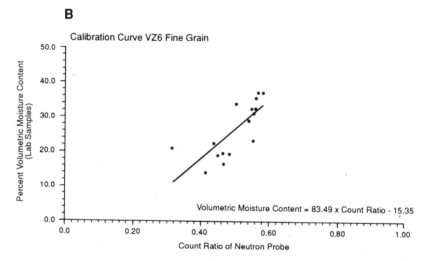

Figure 24 Calibration curves for VZ-6: (A) sands, and (B) fine grain. (From Reaber and Stein, 1990. With permission.)

methane monitoring at landfills, few permanent soil gas monitor wells have been installed in the past. A handful of recent investigations have employed permanent clustered or nested well constructions (Chapter 6, Figure 18) to monitor the vertical distribution of VOCs in the unsaturated zone (e.g., Rosenbloom et al., 1992). The main disadvantage of clustered wells is that separate boreholes must be drilled for installation of a number of adjacent monitor wells, and drilling costs can become prohibitively high if the depth to groundwater is great.

The deep soil gas monitoring program consisted of seven multiport type wells, three of which were completed in angle borings (A-series), and one, V-1, was a SEAMIST™ installation (Chapter 6, Figure 20). The other three were a unique

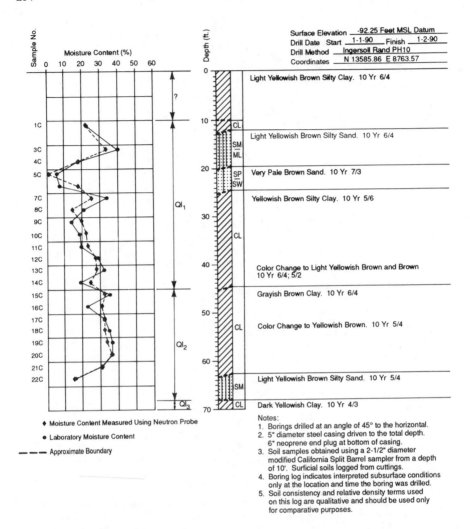

Figure 25   Moisture profile and lithologic log for VZ-6. (Modified from Reaber and Stein, 1990. With permission.)

design to monitor both soil gas as in the multiport design, but also groundwater (Chapter 6, Figure 19). The obvious advantage in this innovative design is considerable cost savings in avoiding drilling separate groundwater monitor wells. Forbes et al. (1993) describes details of the design.

2. Results and Discussion of Soil Gas Analyses

Three of the seven wells were found to contain very high TCE concentrations in soil gas. For these wells, TCE was the only VOC detected. In nearly every case, VOC concentrations were found to increase with depth down to a shallow water-bearing at approximately 30 m (90 ft), indicating that the shallow water-bearing zone contains

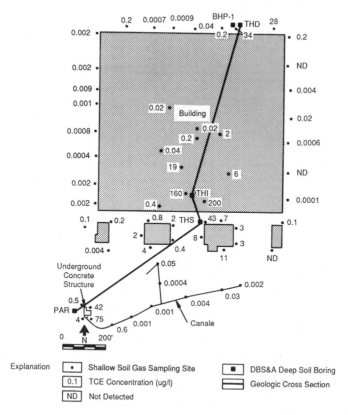

Figure 26 Trichloroethylene (TCE) concentrations in shallow soil gas surrounding existing structures.

the bulk of the VOCs present in the subsurface (Figure 28). This finding was significant because it had previously been hypothesized that the majority of the chlorinated solvents might be found at shallower depths as a result of past spills. It had also been proposed that VOCs volatilizing from subsurface sources might tend to accumulate beneath the 0.15-m-thick (6-in-thick) concrete slab that caps the ground surface throughout the study area. Although a few instances were observed that may suggest trapping of VOCs under the concrete, this was generally found not to be significant. In most cases, the lowest VOC concentrations were found at the shallowest depths sampled, implying that the concrete slab does not represent a significant barrier to transport of VOCs from the soil to the atmosphere.

The shapes of the TCE gas concentration profiles were similar for each of the three wells that contained high TCE concentrations (>1000 μg/L). Figure 28 shows the concentration of TCE versus depth for one of the wells (A-2) oriented at an angle to vertical. As shown in Figure 28, TCE concentrations increased exponentially with depth below the surface. This observation was surprising, because most previous studies have suggested that linear soil gas VOC concentration gradients are to be expected inasmuch as diffusion has been said to be the dominant transport mechanism (e.g., Devitt et al., 1987; Marrin and Kerfoot, 1988).

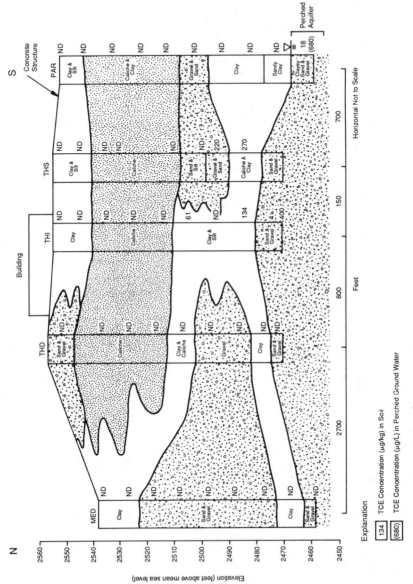

Figure 27 Soil profile at the site where chlorinated solvents were detected in soil gas.

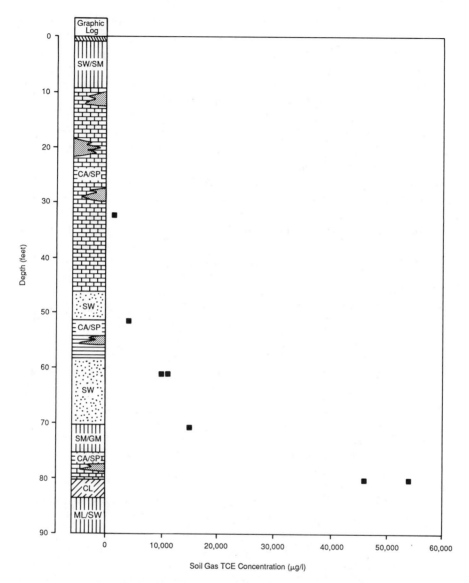

Figure 28 Trichloroethylene (TCE) vapor concentration profile for MPSG Well A-2 (November 1992).

Various explanations have been previously proposed to explain the observance of nonlinear concentration profiles. It has been suggested by several authors that rapid biodegradation at shallow depths may result in lower than expected VOC concentrations in the upper part of the vadose zone (Marrin and Kerfoot, 1988; Kerfoot, 1991). However, biodegradation is not expected to be as significant for halocarbon compounds, such as TCE, as compared with aromatic fuel hydrocarbons. Nonlinear VOC concentration profiles with depth have also been explained as a result of variable soil texture, and hence water content, that can lead to anomalously low VOC concentrations in the more fine-grained, wetter strata (Marrin and Thompson, 1987).

While the processes just described may certainly influence soil gas VOC concentrations, we believe that the TCE profiles observed in this study are attributable to a different, previously undiscussed phenomenon: soil vapor density stratification. We calculated that at the maximum measured soil gas TCE concentration of 54,000 μg/L from the deepest port in MPSG well A-2, the soil vapor from this well (RVD = 1.037) would be approximately 3.7% denser than moist soil vapor not containing TCE (RVD = 1). Thus, the observed TCE concentration profiles are consistent with a diffusional model, except that the density effect does not yield a linear concentration-depth function. Instead, dense soil vapor appears to be continuously produced at the bottom boundary of the unsaturated zone, as a result of volatilization of VOCs from the shallow water-bearing zone. Once produced, the dense vapor tends to remain at the bottom of the soil vapor column unless diffusion/advection rates are sufficient to overcome the density contrast.

As observed in some previous studies (e.g., Hahne and Thomsen, 1991; Kerfoot, 1991), poor correlation was evident between VOC concentrations in soil gas and soil matrix samples collected at approximately the same depths (e.g., Chapter 6, Figure 23). In our case study, TCE was detected in 36 of 39 soil gas samples (92%) from the seven soil gas monitor wells, but only 10 of 83 soil matrix samples (12%) collected during drilling of these wells contained TCE above the laboratory detection limit (0.05 mg/kg). Nearly all of the soil matrix samples in which TCE was detected were collected from the lower portion of the vadose zone, which is consistent with the hypothesized deep VOC source based on the distribution of TCE in soil gas (Figure 28).

The more frequent detection of TCE in soil gas as compared to soil is probably partially due to the much lower detection limits achieved for soil gas as compared to the soil matrix. In addition, the soil matrix is undoubtedly far more heterogeneous with respect to contaminant distribution than the soil gas. Nevertheless, for those soil matrix samples with detectable TCE, the concentrations reported by the laboratory were far below those that would be predicted based on equilibrium with the measured TCE concentrations in adjacent soil gas and making reasonable assumptions regarding soil moisture, bulk density, and organic carbon content. If vapor-phase equilibrium between soil gas and soil matrix is assumed, then the discrepancy most likely reflects the shortcomings of soil matrix analytical procedures, which often result in reported VOC concentrations that are lower than expected (Siegrist and Jenssen, 1990).

3. Soil Gas Pressure Monitoring

Soil gas pressures were measured in several of the wells using two different techniques: (1) instantaneous measurements using an inclined manometer and (2) continuous measurement using a multichannel data logger and pressure transducers. All pressure measurements were measured relative to atmospheric pressure, and barometric pressure was also monitored using an absolute pressure transducer in order to allow correction of soil gas gage pressures to absolute pressures.

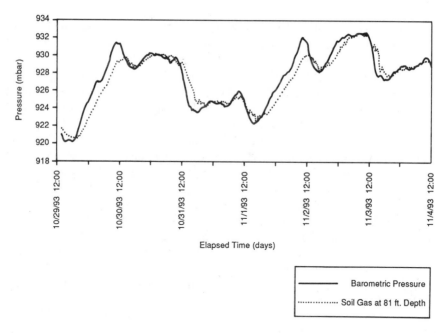

Figure 29   Soil gas pressure in vadose zone at 81-ft depth compared to surface barometric pressure at MPSG well A-2 during a 6-d period.

Figure 29 shows absolute soil gas pressures at the 24.6 m (81 ft) depth in MPSG well A-2 over a 6-d period, as measured with pressure transducers and a data logger. Absolute barometric pressure data for the same time period, as measured using the absolute pressure transducer, are also shown on the graph. The barometric pressure curve shows the normal twice-daily oscillation that occurs primarily due to atmospheric tides, with high pressures at approximately 10:00 a.m. and 10:00 p.m. and low pressures at 4:00 a.m. and 4:00 p.m. The soil gas pressure recorded by the transducer relative to barometric pressure (gage pressures) was corrected to absolute pressure by adding the absolute barometric pressure values to the gage (differential) pressures within the well.

A strong correlation between barometric pressure and soil gas pressure is evident (Figure 29). Pressure fluctuations at the 81 ft depth in the vadose zone closely track the semidiurnal atmospheric pressure wave (atmospheric tides), indicating the soil vapor at this depth is in good communication with the atmosphere. However, the fluctuations in soil gas pressure are slightly attenuated and out of phase with respect to surface pressure, with a lag of approximately 1 h. Soil gas pressure measurements at shallower depths were found to be less attenuated and more in phase with the atmosphere. Pressure data such as these provide a means of determining the depth and magnitude of pressure fluctuations in the vadose zone due to "barometric pumping." Where the soil gas and barometric pressure curves cross (Figure 29), a reversal of the pressure gradient has occurred. At these points, the direction of gas flow changes, from soil toward the atmosphere to atmosphere toward the soil and vice versa. Such information is important for shallow soil gas surveys because false negative results could be obtained if soil vapor samples are collected during periods when downward air pressure gradients exist and clean atmospheric air has penetrated into the soil.

Careful measurements of this type, corrected for transducer drift, might allow calculation of the average vertical air permeability of the unsaturated zone at ambient soil moisture conditions (Weeks, 1978), as well as the field-scale permeabilities of individual strata between the various ports.

A final observation is that the 0.15-m (6-in) concrete slab that caps the entire study area does not appear to provide significant impedance to the transmission of barometric pressure pulses into the subsurface. This suggests that the concrete slab possesses sufficient air permeability to allow rapid fluxes of air into and out of the soil through cracks and joints. This is potentially an important observation, because soil gas "breathing" through a permeable surface cover should result in greater flux of VOCs to the atmosphere. Contrasts in pneumatic diffusivity suggest that more breathing of VOCs may occur through a fractured, relatively impermeable surface than through a permeable surface. This is because slugs of fresh air may rapidly reach considerable depths in a fracture, whereas a uniformly permeable surface merely results in shallow piston flow.

References

Abriola, L. M., Modeling multiphase migration of organic chemicals in groundwater systems — a review and assessment, *Environ. Health Perspect.*, 83, 11, 1989.

Adrian, D. P., and Franzini, J. B., Impedance to infiltration by pressure build-up ahead of the wetting front, *J. Geophys. Res.*, 27, 5857–5862, 1966.

Aldrich, M. J., Bowers, K. D., Brorby, G. P., Campbell, K., Charles, R. W., Coons, L. M., Filemyr, R., Harrington, C. D., Johnson, P. S., McFadden, L., and Stephens, D. B., RFI Work Plan for Operable Unit 1071, Environmental Restoration Program, LANL, Los Alamos, NM, LA-UR-92-810, M. J. Aldrich, Compiler and Editor, May 1992.

Allison, G. B., and Hughes, M. W., The use of environmental chloride and tritium to estimate total local recharge to an unconfined aquifer, *Aust. J. Soil Res.*, 16, 181, 1978.

Allison, G. B., The use of natural isotopes for measurement of groundwater recharge — a review, in *Proc. AWRC Groundwater Recharge Conference*, Townsville, Australia, July 1980, AWRC Conf. Series No. 3, 1981.

Allison, G. B., Stone, S. J., and Hughes, M. W., Recharge in Karst and dune elements of a semi-arid landscape as indicated by natural isotopes and chloride, *J. Hydrol.*, 76, 1, 1985.

Allison, G. B., Barnes, C. J., Hughes, M. W., and Leaney, F. W. J., Effect of climate and vegetation on oxygen-18 and deuterium profiles in soils, *Proc. Int. Symp. on Isotope Hydrology in Water Resource Development*, IAEA-SM-270/20, Isotope Hydrology, 1983, International Atomic Energy Agency, Vienna, Austria, 1984, 105–124.

Allison, G. B., Gee, G. W., and Tyler, S. W., Vadose-zone techniques for estimating ground-water recharge in arid and semiarid regions, *SSSAJ*, 58(1), 6–14, 1994.

American Ground Water Consultants, 1980, p. 130.

American Standards for Testing and Materials, Various Standards.

D-3404	Measuring matric potential in the vadose zone using tensiometers
D-4700	Soil sampling from the vadose zone
D-4996	Pore-liquid sampling from the vadose zone
D-5126	Comparison of field methods for determining hydraulic conductivity in the vadose zone
D-1950	Measure surface tension with a du Nuoy precision tensiometer or du Nuoy interfacial tensiometer
D-2434-68 and D-5084-90	Method for permeability of granular soils; measurement of hydraulic conductivity of saturated porous materials using a flexible wall permeameter. Laboratory tests on granular and fine-textured soils
D-3385-88	Measurements in lower permeability materials and clay liners by the addition of a top cap over the water reservoir to prevent evaporation and disturbance to the water supply
D-5298-92	The filter paper method
D-5220-92	Method for water content of soil and rock in-place by the neutron depth probe method

Amoozegar, A., Martin, K. C., and Hoover, M. T., Effect of access hole properties on soil water content determine by neutron thermalization, *Soil Sci. Soc. Am. J.*, 53, 330, 1989.

Amoozegar-Fard, A., Comparison of the Glover solution with the simultaneous-equations approach for measuring hydraulic conductivity, *Soil Sci. Soc. Am. J.*, 53, 1362–1367, 1989.

Anderson, L. D., Problems interpreting samples taken with large-volume, falling-suction soil-water samplers, *Ground Water*, 24, 761, 1986.

Anderson, W. G., Wettability literature survey — part 1: rock/oil/brine interactions and the effects of core handling on wettability, *J. Petrol. Technol.*, October, 1125, 1986.

Anderson, W. G., Wettability literature survey — part 2: wettability measurement, *J. Petrol. Technol.*, October, 1246, 1986.

Ankeny, M. D., Kaspar, T. C., and Horton, R., Design for an automated tension infiltrometer, *Soil Sci. Soc. Am. J.*, 52, 893, 1988.

Ankeny, M. D., Ahmed, M., Kaspar, T. C., and Horton, R., Simple method for determining unsaturated hydraulic conductivity, *Soil Sci. Soc. Am. J.*, 55, 467–470, 1991.

Ankeny, M. D., Forbes, J. R., McCord, J., and Leenhouts, J., *Reverse Osmosis in Vadose Zone: Effects on Recharge Estimates Using the Chloride Mass Balance Method.* 1995, in preparation.

Armstrong, J. E., Frind, E. O., and McClellan, R. D., Nonequilibrium mass transfer between the vapor, aqueous, and solid phases in unsaturated soils during vapor extraction, *Water Resources Res.*, 30(2), 355–368, 1994.

Avon, L., and Durbin, T. J., Evaluation of the Maxey-Eakin method for estimating recharge to groundwater basins in Nevada, *Water Resources Bull.*, American Water Resources Association, 30(1), 99–111, February 1994.

Aylor, D. E., and Parlange, J. Y., Vertical infiltration into a layered soil, *Soil Sci. Soc. Am. J.*, 37, 673–676, 1973.

Baehr, A. L., and Hult, M. F., Evaluation of unsaturated zone air permeability through pneumatic tests, *Water Resources Res.*, 27 (10), 2605-2617, 1991.

Baker, R. S., and Hillel, D., Observations of fingering behavior during infiltration into layered soils, in *Proc. Natl. Symp. on Preferential Flow*, Gish, T. J., and Shirmohammadi, A., Eds., ASAE, St. Joseph, MI, 1991, 87.

Ballestero, T. P., McHugh, S. A., and Kinner, N. E., Monitoring of immiscible contaminants in the vadose zone, in *Ground Water and Vadose Zone Monitoring, ASTM STP 1053*, Nielsen, D. M., and Johnson, A. I., Eds., ASTM, Philadelphia, 1990, 25.

Barbee, G. C., A Comparison of Methods for Obtaining "Unsaturated Zone" Soil Solution Samples, M. S. thesis, Texas A & M University, College Station, TX, 1983.

Barnes, C. J., and Allison, G. B., The distribution of deuterium and ^{18}O in dry soils, 1. Theory, *J. Hydrol.*, 60, 141, 1983.

Bauer, H. H., and Vaccaro, J. J., Documentation of a deep percolation model for estimating groundwater recharge, U. S. Geological Survey, Open-File Report 86-536, Tacoma, WA, 1987.

Bauer, H. H. and Vaccaro, J. J., Estimates of groundwater recharge to the Columbia Plateau regional aquifer system, Washington, Oregon, and Idaho, for predevelopment and current land-use conditions, USGS Water-Resources Investigations Report 88-4108, Tacoma, WA, 1990.

Bear, J., *Dynamics of Fluids in Porous Media*, Environmental Science Series, American Elsevier, New York, 1972.

Bear, J., *Dynamics of Fluids in Porous Media*, Environmental Science Series, American Elsevier, New York, 1975.

Beckwith, G., Personal communication, Sergeant, Hawkins, and Beckwith, 1994.

Bell, J. P., Dean, T. J., and Baty, A. J. B., Soil moisture measurement by an improved capacitance technique, part II, field techniques, evaluation and calibration, *J. Hydrol.*, 93, 79, 1987.

Bell, R. W. and Schofield, N. J., Design and application of a constant head well permeameter for shallow high saturated hydraulic conductivity soils, *Hydrol. Proc. J.*, 4, 327, 1990.

Bentley, H. W., Phillips, F. M., Davis, S. N., Gifford, S., Elmore, D., Tubbs, L. E., and Gove, H. E., Thermonuclear Cl-36 pulse in natural water, *Nature* 30 (23/30), 737–740, 1982.

Bentley, H. W., Phillips, F. M., Davis, S. N., Habermehl, M. A., Airey, P. L., Calf, G. E., Elmore, D., Grove, H. E., and Torgersen, T., Chlorine 36 dating of very old groundwater 1. The great artesian basin, Australia, *Water Resources Res.,* 22(13), 1991–2001, 1986.

Bentley, H. W., Phillips, F. M., and Davis, S. N., Chlorine-36 in the terrestrial environment, in *Handbook of Environmental Isotope Geochemistry*, Vol. 2B, Fontes, J. C., and Fritz P., Eds., Elsevier, Amsterdam, 427–480, 1986.

Bergstrom, R. E. and Aten, R. E., Natural recharge and localization of fresh groundwater in Kuwait, *J. Hydrol.,* 2, 213, 1964.

Besbes, M., Delhomme, J. P., and DeMarsily, G., Estimating recharge from ephemeral streams in arid regions: a case study at Kairouan, Tunisia, *Water Resources Res.,* 14(2), 281–290, 1978.

Beven, K., Modeling preferential flow: an uncertain future?, in *Proc. Natl. Symp. on Preferential Flow*, Gish, T. J., and Shirmohammadi, A., Eds., ASAE, St. Joseph, MI, 1991, 1.

Biggar, J. W., and Nielsen, D. R., Spatial variability of the leaching characteristics of a field soil, *Water Resources Res.,* 12, 78, 1976.

Bisque, R. E., Migration rates of volatiles from buried hydrocarbon sources through soil media, in *Proc. Natl. Conf. on Petroleum Hydrocarbons and Organic Chemicals in Groundwater — Prevention, Detection, and Restoration*, NWWA, Worthington, OH, 1984, 267.

Blaney, H. F., and Criddle, W. D., Determining water requirements in irrigated areas from climatological and irrigation data, Tech. Paper No. 96, USDA Soil Conservation Service, Washington, DC, 1950.

Boast, C. W., and Robertson, T. M., A "micro-lysimeter" method for determining evaporation from bare soil: description and laboratory evaluation, *Soil Sci. Soc. Am. J.,* 46, 689, 1982.

Bohn, H., McNeal, B., and O,Conner, G., *Soil Chemistry,* John Wiley & Sons, New York, 1979.

Bond, R. D., The influence of the microflora on the physical properties of soils, II. Field studies on water repellent sands, *Aust. J. Soil Res.,* 2, 123, 1964.

Bonkowski, M. S., and Rinehart, R., Underground storage tank monitoring in California: An evaluation of recent legislation, in *Proc. Sixth Natl. Symp. and Expo. on Aquifer Restoration and Ground Water Monitoring*, NWWA, Dublin, OH, 1986, 185.

Borns, D. J., Newman, G., Stolarczyk, L., and Mondt, W., Cross borehole electromagnetic imaging of chemical and mixed waste landfills, in *Proc. Symp. on the Application of Geophysics to Engineering and Environmental Problems*, Vol. 1, Bell, R. S., and Lepper, C. M., Eds., Environmental and Engineering Geophysical Society (EEGS), San Diego, CA, 1993.

Bouma, J., Baker, F. G., and Veneman, P. L. M., Measurement of water movement in soil pedons above the water-table, Info. Circular No. 27, University of Wisconsin, Madison, 1974.

Boutwell, G. P. and Tsai, C. N., The two-stage field permeability test for clay liners, *Geotech. News*, 10(2), 32, 1992.

Bouwer, H., A double tube method for measuring hydraulic conductivity of soil in sites above a water table, *Soil Sci. Soc. Am. Proc.,* 25, 334, 1961.

Bouwer, H., Rapid field measurement of air-entry value and hydraulic conductivity of soil as significant parameters in flow system analysis, *Water Resources Res.,* 2, 729, 1966.

Bouwer, H., *Groundwater Hydrology*, 1st ed., McGraw-Hill, New York, 1978.

Bouwer, H., Estimating and enhancing groundwater recharge, in *Proc. Symp. on Groundwater Recharge*, Sharma, M.L., Ed., Balkmas, Rotterdam, 1989, 1–9.

Bouyoucos, G. J., and Mick, A. H., An electrical resistance method for continuous measurement of soil moisture under field conditions, Mich. Agric. Exp. Stn. Tech. Bull. No. 172, East Lansing, 1940.

Boyle, J. M., and Saleem, Z. A., Determination of recharge rates using temperature-depth profiles in wells, *Water Resources Res.*, 15(6), 1616–1622, December 1979.

Bredehoeft, J. D., and Papadopulos, I. S., Rates of vertical groundwater movement estimated from earth's thermal profile, *Water Resources Res.*, 1, 325–328, 1965.

Briscoe, R. D., Thermocouple psychrometers for water potential measurements, in *Advanced Agricultural Instrumentation: Design and Use*, NATO ASI Series, Gensler, W. G., Ed., Martinus & Nijhoff, Dordrecht, 1986, 3.

Brookings, D. G., and Thomson, B. M., Geochemical considerations for disposal facilities: west disposal in desert playas, in *Deserts as Dumps*, Reith, C. C. and Thomson, B. M., Eds., University of New Mexico Press, Albuquerque, 1992, 199.

Brooks, R. H., and Corey, A. T., Properties of porous media affecting fluid flow, *ASAE J. Irr. Drainage Div.*, IR2, 61, 1964.

Brose, R. J., and Shatz, R. W., Neutron monitoring in the unsaturated zone, in *Proc. First Natl. Outdoor Action Conf. on Aquifer Restoration, Ground Water Monitoring, and Geophysical Methods*, NWWA, Dublin, OH, 1987, 455.

Brown, D. A., and Scott, H. D., Dependence of crop growth and yield on root development and activity, in *Roots, Nutrient and Water Influx, and Plant Growth*, ASA Spec. Pub. 49, Barber, S. A., and Bouldin, D. R., Eds., ASA, CSSA, and SSSA, Madison, WI, 1984, chap. 6.

Brown, R. W., and Chambers, J. C., Measurements of *in situ* water potential with thermocouple psychrometers: A critical evaluation, in *Int. Conf. on Measurement of Soil and Plant Water Status*, Vol. 1, *Soils,* Utah State University, Logan, 1987, 125.

Bruce, R. R., and Klute, A., The measurement of soil-water diffusivity, *Soil Sci. Soc. Am. Proc.,* 20, 458, 1956.

Brunini, O., and Thurtell, G. W., An improved thermocouple hygrometer for *in situ* measurements of soil water potential, *Soil Sci. Soc. Am. J.*, 46, 900, 1982.

Brusseau, M. L., Jessup, R. E., and Rao, P. S. C., Modeling solute transport influenced by multiprocess nonequilibrium and transformation reactions, *Water Resources Res.,* 28(1), 175–182, 1992.

Brustkern, R. L., and Morel-Seytoux, H. J., Analytical treatment of two-phase infiltration, *J. Hydrol. Div.*, ASCE, 96, HY12, 2535–2548, 1970.

Brutsaert, W., A functional iteration technique for solving Richard's equation applied to two-dimensional infiltration problems, *Water Resources Res.* 6, 1971.

Brutsaert, W., Vertical infiltration in dry soil, *Water Resources Res.*, 13, 363, 1977.

Buckingham, E., Studies on the movement of soil moisture, U.S. Department of Agriculture Bureau Soils, 38, 1907.

Burdine, N. T., Relative permeability calculation from pore size distribution data, *AIMME Petroleum Trans.*, 198, 71, 1953.

Burkham, D. E., Depletion of streamflow by infiltration in the main channels of the Tucson Basin, southeastern Arizona, USGS Water-Supply Paper 1939-B, U.S. Government Printing Office, Washington, DC, 1970.

Busenberg, E., and Plummer, L.N., Chlorofluorocarbons (CCl_3F and CCl_2F_2): use as an age dating tool and hydrologic tracer in shallow groundwater systems, *U.S. Geol. Surv. Water Resources Invest.*, 91-4034, 542–547, 1991.

Busenberg, E., and Plummer, L. N., Use of chlorofluorocarbons (CCl_3F and CCl_2F_2) as hydrologic tracers and age-dating tools: the alluvium and terrace system of central Oklahoma, *Water Resources Res.*, 28(9), 2257–2283, September 1992.

Byers, E., and Stephens, D. B., Statistical and stochastic analysis of hydraulic conductivity and particle size in a fluvial sand, *Soil Sci. Soc. Am. J.*, 47, 1072, 1983.

Calf, G. E., The isotope hydrology of the Mereenie Sandstone aquifer, Alice Springs, Northern Territory, Australia, *J. Hydrol.*, 38, 343, 1978.

Campana, M., Application of carbon-14 groundwater ages in calibrating a flow model of the Tucson Basin Aquifer, Arizona, in *Hydrology and Water Resources in Arizona and the Southwest*, Proc. 1976 Meetings of the Arizona Section: American Water Resources Association and Hydrology Section, Arizona Academy of Science, Tucson, 6, 197–202, Arizona Section, American Water Resources Association, 1976.

Campana, M. E., and Simpson, E. S., Groundwater residence times and recharge rates using a discrete-state compartment model and ^{14}C data, *J. Hydrol.*, 72, 171–185, 1984.

Campana, M. E., and Mahin, D. A., Model-derived estimates of groundwater mean ages, recharge rates, effective porosities and storage in a limestone aquifer, *J. Hydrol.*, 76, 247–264, 1985.

Campbell, G. S., and Gee, G. W., Water potential: miscellaneous methods, in *Methods of Soil Analysis, Part 1*, 2nd ed., Klute, A., Ed., Agron. Monogr. 9, ASA, CSSA, and SSSA, Madison, WI, 1986, 619.

Campbell, E. C., Campbell, G. S., and Barlow, W. K., A dewpoint hygrometer for water potential measurement, *Agric. Meteorol.*, 12, 113, 1973.

Carlson, T. N., and El Salem, J., Measurement of soil moisture using gypsum blocks, in *Intl. Conf. on Measurement of Soil and Plant Water Status*, Vol. 1, *Soils*, Utah State University, Logan, 1987, 193.

Carlsson, L., Winberg, A., and Grundfelt, B., Model calculations of the groundwater flow at Finnsjön, Fjällveden, Gideå and Kamlunge, KBS, Technical Report 83-45, SKB, Stockholm, Sweden, 1983.

Carlsson, L., Bromley, J., and Smellie, J., Groundwater resource in a poorly transmissive sandstone in semiarid environment — 1. Estimation of recharge, in *State of the Art of Hydrology and Hydrogeology in the arid and semiarid areas of Africa*, Stout, G. E. and Demissie, Eds., Proc. Sahel Forum, Ouagadougou, Burkina Faso., Feb. 18–23, 1989, UNESCO, New York.

Cartwright, K., Groundwater discharge in the Illinois Basin as suggested by temperature anomalies, *Water Resources Res.*, 6, 912–918, 1970.

Cartwright, K., Measurement of fluid velocity using temperature profiles: experimental verification, *J. Hydrol.*, 43, 185–194, 1979.

Cary, J. W., Gee, G. W., and Simmons, C. S., Using an electro-optical switch to measure soil water suction, *Soil Sci. Soc. Am. J.*, 55, 1798, 1991.

Cassell, D. K., and Klute, A., Water potential: tensiometry, in *Methods of Soil Analysis, Part 1*, 2nd ed., Klute, A., Ed., Agron. Monogr. 9, ASA, CSSA, and SSSA, Madison, WI, 1986, 563.

Cecen, K., *The Investigation of the Coefficient of Permeability in Connection with Construction Engineering Soil Investigations*, Verlag Wasser and Boden, Hamburg, 1967.

Charbeneau, R. J. and Daniel, D. E., Contaminant transport in unsaturated flow, *Handbook of Hydrology*, Maidment, D. R., Ed., McGraw Hill, New York, 1993.

Chiew, F. H. S., McMahon, T. A., and O'Neill, I. C., Estimating groundwater recharge using an integrated surface and groundwater modelling approach, *J. Hydrol.*, 131, 151–186, 1992.

Childs, E. C., and Collis-George, N., The permeability of porous materials, *Proc. R. Soc. London*, 201A, 392, 1950.

Clapp, R. B., and Hornberger, G. M., Empirical properties for some soil hydraulic properties, *Water Resources Res.*, 14, 601, 1978.

Clark, I. D., Fritz, P., Quinn, O. P., Rippon, P. W., Nash, H., and Bin Ghaib Al Said, S. B., Modern and fossil groundwater in an arid environment: a look at the hydrogeology of southern Oman, in *Proc. Intl. Symp. on Isotope Techniques in Water Resources Development*, International Atomic Energy Agency, Vienna, Austria, 1987, IAEA-SM-299/15.

Cohen, R. M., and Mercer, J. W., *DNAPL Site Evaluation*, C. K. Smolley, Boca Raton, FL, 1993.

Conca, J. L., and Wright, J., Diffusion coefficients in gravel under unsaturated conditions, *Water Resources Res.*, 26, 1055, 1990.

Conca, J. L., and Wright, J., Diffusion and flow in gravel, soil, and whole rock, *Appl. Hydrogeol.*, 1, 5, 1992.

Constantz, J., Temperature dependence of unsaturated hydraulic conductivity of two soils, *Soil Sci. Soc. Am. J.*, 46, 466, 1982.

Constantz, J., and Herkelrath, W. N., Submersible pressure outflow cell for measurement of soil water retention and diffusivity from 5 to 95°C, *Soil Sci. Soc. Am. J.*, 48, 7, 1984.

Cook, P. G., Edmunds, W. M., and Gaye, C. B., Estimating paleorecharge and paleoclimate from unsaturated zone profiles, *Water Resources Res.*, 28(10), 2721–2731, October 1992.

Cook, P. G., and Kilty, S., A helicopter-borne electromagnetic survey to delineate groundwater recharge rates, *Water Resources Res.*, 28,(11), 2953–2961, November 1992.

Cook, P. G., Walker, G. R., Buselli, G., Potts, I., and Dodds, A. R., The application of electromagnetic techniques to groundwater recharge investigations, *J. Hydrol.*, 130, 201–229, 1992.

Cooper, J. B., and John, E. C., Geology and groundwater occurrence in Southeastern McKinley County, New Mexico, Technical Report 35, New Mexico State Engineer Office, Santa Fe, NM, 1968.

Corey, A. T., Measurement of water and air permeability in unsaturated soil, *Soil Sci. Soc. Am. Proc.*, 19, 7, 1959.

Corey, A. T., Mechanics of Heterogeneous Fluids in Porous Media, Water Resources Publications, Ft. Collins, CO, 1977.

Creasey, C. L., and Dreiss, S. J., Soil water samplers: do they significantly bias concentrations in water samples?, in *Proc. Conf. on Characterization and Monitoring of the Vadose (Unsaturated) Zone*, NWWA, Dublin, OH, 1985, 173.

Cullen, S. J., Kramer, J. H., Everett, L. G., and Eccles, L. A., Is our groundwater monitoring strategy illogical?, *Ground Water Monitoring Rev.*, 12, 103, 1992.

Czarnecki, J. B., Geohydrology and evapotranspiration at Franklin Lake Playa, Inyo County, California, Open-File Rept. 90-356, U.S. Geological Survey, Denver, CO, 1990.

Daily, W., Ramirez, A., LaBrecque, D., and Nitao, J., Electrical resistivity tomography of vadose water movement, *Water Resources Res.*, 28, 1429, 1992.

Dalton, F. N., Measurement of soil water content and electrical conductivity using time-domain reflectometry, in *Int. Conf. on Measurement of Soil and Plant Water Status*, Vol. 1, *Soils*, Utah State University, Logan, 1987, 95.

Daniel, D., Hamilton, J., and Olson, R., Suitability of thermocouple psychrometers for studying moisture movement in unsaturated soils, in *Permeability and Groundwater Contaminant Transport, ASTM STP 746*, Zimmie, T. F., and Riggs, C. O., Eds., ASTM, Philadelphia, 1981, 84.

Daniel, D. E., Predicting the hydraulic conductivity of compacted clay liners, *ASCE J. Geotech. Eng.*, 110, 285, 1984.

Daniel, D. E., and Trautwein, S. J., Field permeability test for earthen liners, in *Proc. In Situ Conf.*, Blacksburg, VA, 146, 1986.

Daniel, D. E., *In situ* hydraulic conductivity tests for compacted clay, *ASCE J. Geotech. Eng.*, 115, 1205, 1989.

Daniel, D. E., Burton, P. M., and Hwang, S.-D., Evaluation of four vadose zone probes used for leak detection and monitoring, in *Current Practices in Ground Water and Vadose Zone Investigations, ASTM STP 1118,* Nielsen, D. M., and Sara, M. N., Eds., American Society for Testing and Materials, Philadelphia, 1992.

Daniel, D. E., State-of-the-art: laboratory hydraulic conductivity tests for saturated soils, in *Hydraulic Conductivity and Waste Contaminant Transport in Soils, ASTM STP 1142,* Daniel, D. E., and Trautwein, S. J., Eds., ASTM, Philadelphia, 1993.

Daniel B. Stephens & Associates, Inc., Final report on vadose zone characterization at IT Corporation's Imperial Valley Facility, Imperial County, California, unpublished report to IT Corporation/Irvine, DBS&A, Albuquerque, NM, 1988.

Daniel B. Stephens & Associates, Inc., Casmalia Resources Hazardous Waste Management Facility Vadose Zone Investigation for Woodward-Clyde Consultants, Oakland, CA, DBS&A, Albuquerque, NM, 1989.

Daniel B. Stephens & Associates, Inc., Water balance and impact assessment, Nu-Mex landfill, Sunland Park, New Mexico, DBS&A, Albuquerque, NM, 1992.

Daniel B. Stephens & Associates, Inc., 1994.

Daniel B. Stephens & Associates, Inc., Unpublished Consultant's Report for Los Alamos National Laboratories, TA-73 Landfill, 1995.

Daniel, J. F., Estimating groundwater evapotranspiration from streamflow records, *Water Resources Res.*, 12(3), 360-364, 1976.

Darcy, H., Les fontaines Publiques de la Ville de Dijon, Victor Dalmont, Paris, 1856.

Darling, W. G., Edmunds, W. M., Kinniburgh, D. G., and Kotoub, S., Sources of recharge to the basal Nubian Sandstone aquifer, Butana Region, Sudan, *J. Hydrol.*, 131, 1, 1992.

Davidson, E. S., Geohydrology and water resources of the Tucson Basin, Arizona, USGS Water-Supply Paper 1939-E, U.S. Government Printing Office, Washington, DC, 1973.

De Smedt, F., and van Beker, A., The groundwater flow in the unsaturated zone, Department of Hydrology, Faculty of Applied Sciences, Voije Universitait, Pleinlaan, Brussels, Belgium, 1974.

Devitt, D. A., Evans, R. B., Jury, W. A., Sparks, T. H., Eklund, B., and Gnolson, A., *Soil Gas Sensing for Detecting and Mapping of Volatile Organics*, NWWA, Dublin, OH, 1987.

Dincer, T., Al-Mugrin, A., and Zimmerman, V., Study of the infiltration and recharge through the sand dunes in arid zones, with special reference to the stable isotopes and thermonuclear tritium. *J. Hydrol.* (Amsterdam) 23, 79–109, 1974.

Dirksen, C., Unsaturated hydraulic conductivity, in *Soil Analysis: Physical Methods*, Smith, K. A., and Mullins, C. E., Eds., Marcel Dekker, New York, 1991, 209.

Doering, E. J., Soil-water diffusivity by the one-step method, *Soil Sci.*, 99, 322, 1965.

Domenico, P. A., and Schwartz, F. W., *Physical and Chemical Hydrogeology*, Wiley & Sons, New York, 1990, 824.

Doorenbos, J., and Pruitt, W. O., Crop water requirements, in *Irrigation and Drainage Paper*, FAO, Rome, 1975.

Dreiss, S. J. and Anderson, L. D., Estimating vertical soil moisture flux at a land treatment site, *Ground Water*, 23(4), 503–511, July–August 1985.

Duffy, C. J., Gelhar, L. W., and Gross, G. W., Recharge and groundwater conditions in the western region of the Roswell Basin, Partial Technical Completion Report, Proj. No. A-055-NMEX, New Mexico Water Resources Research Institute, 1978.

Duffy, C. J., and Al-Hassan, S., Groundwater circulation in a closed desert basin: Topographic scaling and climatic forcing, *Water Resources Res.*, 24, 1675, 1988.

Dunkle, S. A., Plummer, L. N., Busenberg, E., Phillips, P. J., Denver, J. M., Hamilton, P. A., Michel, R. L., and Coplen, T. B., Chlorofluorocarbons (CCl_3F and CCl_2F_2) as dating tools and hydrologic tracers in shallow groundwater of the Delmarva Peninsula, Atlantic Coastal Plain, United States, *Water Resources Res.*, 29(12), 3837–3860, 1993.

Durant, N. D., Myers, V. B., and Eccles, L. A., EPA's approach to vadose zone monitoring and RCRA facilities, *Ground Water Monitoring & Remediation*, 13, 151, 1993.

Eagleson, P. S., Climate, soil, and vegetation, 7. A derived distribution of annual water yield, *Water Resources Res.*, 14(5), 765–776, October 1978.

Eagleson, P. S., The annual water balance, *J. Hydraul. Div., Proc. Am. Soc. Civ. Eng.*, 105(HY8) 923–941, 1979.

Eakin, T. E., Price, D., Harril, J. R., Summary appraisals of the nation's groundwater resources — Great Basin region, USGS Professional Paper 813-G, U.S. Government Printing Office, Washington, DC, 1976.

Edlefsen, N., and Anderson, A., The four-electrode resistance method for measuring soil moisture content under field conditions, *Soil Sci.*, 51, 367, 1941.

Ekwurzel, B., Schlosser, P., and Smethie Jr., W.M., Dating of shallow groundwater: comparison of the transient tracers $^3H/^3He$, chlorofluorocarbons, and ^{85}Kr, *Water Resources Res.*, 30(6), 1693–1708 (Figure 2, p. 1695), June 1994.

Elkins, J. W., Thompson, T. M., Swanson, T. H., Butler, J. H., Hall, B. D., Cummings, S. O., Fisher, D. A., and Raffo, A. G., Decrease in growth rates of atmospheric chlorofluorocarbons 11 and 12, *Nature*, 364, 780–783, 1993.

Elrick, D. E., Reynolds, W. D., and Tan, K. A., Hydraulic conductivity measurements in the unsaturated zone using improved well analysis, *Ground Water Monitoring Rev.*, IX(3), 184, 1989.

Elrick, D. E., and Reynolds, W. D., Infiltration from constant-head well permeameters and infiltrometers, in *Advances in Measurement of Soil Physical Properties: Bringing Theory into Practice*, SSSA Special Pub. No. 30, Topp, G. C., Reynolds, W. D., and Green, R. E., Eds., ASA, CSSA, and SSSA, Madison, WI, 1992, 1.

Enfield, C. G., Hsieh, J. J. C., and Warrick, A. W., Evaluation of water flux above a deep water-table using thermocouple psychrometers, *Soil Sci. Soc. Am. Proc.*, 37, 968, 1973.

Environmental Systems & Technologies, Inc., *MOTRANS: A finite Element Model for Multiphase Organic Chemical Flow and Multispecies Transport*, Version 1.1, Program Documentation, Environmental Systems & Technologies, Inc., Blacksburg, VA, 1990.

Evans, D. D., and Kirkham, D., Measurement of air permeability of soil *in situ*, *Soil Sci. Soc. Am. Proc.*, 14, 65, 1949.

Evans, D. D., Sammis, T. W., and Warrick, A. W. Transient movement of water and solutes in unsaturated soil systems, Tucson, Arizona, Phase II Project Completion Report, OWRT Proj. No. B-040-ARIZ, University of Arizona, Department of Hydrology and Water Resources, 40, 1970.

Evans, D. D. and Warrick, A. W., Time in transit of water moving vertically for ground-water recharge, Proc. of the SW Rocky Mountain Division of the American Association for the Advancement of Science, 1970.

Everett, L. G., Wilson, L. G., and Hoylman, E. W., *Vadose Zone Monitoring for Hazardous Waste Sites*, Pollution Technology Review No. 112, Noyes Data Corporation, Park Ridge, NJ, 1984.

Everett, L. G., Hoylman, E. W., Wilson, L. G., and McMillion, L. G., Constraints and categories of vadose zone monitoring devices, *Ground Water Monitoring Rev.*, 4(1), 26, 1984.

Falconer, K. L., Hull, L. C., and Mizell, S. A., Site criteria for shallow land burial of low-level radioactive waste, in *Waste Management*, Vol. 2, Post, R. G., Ed., University of Arizona, Tucson, 1982, 199.

Falta, R. W., Javandel, I., Pruess, K., and Witherspoon, P. A., Density-driven flow of gas in the unsaturated zone due to the evaporation of volatile organic compounds, *Water Resources Res.*, 25(10), 2159–2169, 1989.

Falta, R. W., and Preuss, K., STMVOC Users Guide, LBL-30758, UC-402, Lawrence Berkeley Laboratory, Berkeley, CA, 1991.

Fayer, M. J., and Jones, T. L., Unsaturated soil water and heat flow model version 2.0, Report PNL-6779/UC-702, Battelle, Pacific Northwest Laboratory, Richland, WA, 1990.

Fishbaugh, T., Monitoring in the vadose and saturated zones utilizing fluoroplastic, *Ground Water Monitoring Rev.*, 4, 183, 1984.

Flanigan, K. G., Field Simulation of waste impoundment seepage in the vadose zone: non-reactive solute transport through a stratified, unsaturated field soil, M.S. thesis, Department Geoscience, New Mexico Institute Mining and Technology, Socorro, 1989.

Flühler, H., Ardakani, M. S., and Stolzy, L. H., Error propagation in determining hydraulic conductivities from successive water content and pressure head profiles, *Soil Sci. Soc. Am. J.*, 40, 830, 1976.

Forbes, J., Havlena, J., Burkhard, M., and Myers, K., Monitoring of VOCs in the deep vadose zone using multi-port soil gas wells and combination multi-port soil gas/groundwater wells, in *Proc. Seventh Natl. Symp. and Expo. on Aquifer Restoration, Ground Water Monitoring, and Geophysical Methods*, NGWA, Dublin, OH, 1993, 557.

Frederick, R. B., A laboratory experiment of uniform infiltration into a sloping, stratified and uniform sandy soil, M.S. independent study, Department Geoscience, New Mexico Institute of Mining and Technology, Socorro, 1988.

Freeze, R. A., The mechanism of natural groundwater recharge and discharge, II. One-dimensional, vertical, unsteady, unsaturated flow above a recharging or discharging groundwater flow system, *Water Resources Res.*, 5, 153, 1969.

Gardner, R., A method of measuring the capillary tension of soil moisture over a wide moisture range, *Soil Sci.*, 43, 277–293, 1937.

Gardner, R., Relation of temperature to moisture tension of soil, *Soil Sci.*, 79, 257, 1955.

Gardner, W. H., Water content, in *Methods of Soil Analysis, Part 1*, 2nd ed., Klute, A., Ed., Agron. Monogr. 9, ASA, CSSA, and SSSA, Madison, WI, 1986, 493.

Gardner, W. R., Some steady-state solutions of the unsaturated moisture flow equation with application to evaporation from a water-table, *Soil Sci.*, 85, 228–232, 1958.

Gardner, W. R., Water uptake and salt distribution patterns in saline soils, in *Isotope and Radiation Techniques in Soil Physics and Irrigations Studies*, Proc. Symp. Istanbul, Vienna, IAEA, 1967, 335.

Gardner, W. R., Water content: An overview, in *Int. Conf. on Measurement of Soil and Plant Water Status*, Vol. 1, *Soils*, Utah State University, Logan, 1987, 7.

Gardner, W., Israelsen, O. W., Edlefsen, N. E., and Clyde, D., The capillary potential function and its relation to irrigation practice, *Phys. Rev.*, 20, 196, Abstract, 1922.

Gardner, W. R., Hillel, D., and Benyamini, Y., Post-irrigation movement of soil water, I. Redistribution, *Water Resources Res.*, 6, 851–861, 1970.

Gee, G. W., Stiver, J. F., and Borchert, H. R., Radiation hazard from americium-beryllium neutron moisture probes, *Soil Sci. Soc. Am. J.*, 40, 492, 1976.

Gee, G., and Hillel, D., Groundwater recharge in arid regions: review and critique of estimation methods, *Hydrol. Proc.*, 2, 255, 1988.

Gee, G. W., Rockhold, M. L., and Downs, J. L., Status of FY 88 soil-water balance studies on the Hanford Site, PNL-6750/UC-70, Battelle, Pacific Northwest Laboratory, Richland, WA, 1989.

Gee, G. W., Wierenga, P. J., Andraski, B. J., Young, M. H., Fayer, M. J., and Rockhold, M. L., Variations in water balance and recharge potential at three western desert sites, *Soil Sci. Soc. Am. J.,* 58(1), 63-72, 1994.

Gelhar, L. W., Mantoglou, A., Welty, C., and Rehfeldt, K. R, A review of field scale subsurface solute transport processes under saturated and unsaturated conditions, prepared for Electric Power Research Institute, Palo Alto, CA, Environmental Physics and Chemistry Division, RP-2485-05, 1984.

Gile, L. H., Hawley, J. W., and Grossman, R. B., Soils and geomorphology in the basin and range area of southern New Mexico — guidebook to the desert project, Memoir 39, New Mexico Bur. Mines and Mineral Resources, New Mexico Institute of Mining and Technology, Socorro, 1981.

Gish, T. J., and Shirmohammadi, A., *Proc. Natl. Symp. on Preferential Flow*, ASAE, St. Joseph, MI, 1991.

Glass, R. J., Laboratory research program to aid in developing and testing the validity of conceptual models for flow and transport through unsaturated porous media, SAND 89-2359C, Sandia National Laboratory, Albuquerque, NM, 1991.

Glover, R. E., Flow for a test hole located above groundwater level, in *Theory and Problems of Water Percolation*, Engineering Monograph 8, Zangar, C. N., Ed., U.S. Bureau of Reclamation, 1953, 69.

Glover, R. E., Mathematical derivations as pertain to groundwater recharge, Rept. CER60REG70, U.S. Department of Agriculture, Fort Collins, CO, 1960.

Goodspeed, M. J., Neutron moisture meter theory, in *Soil Water Assessment by the Neutron Method*, Greacen, E. L., Ed., CSIRO, East Melbourne, Australia, 1981, 17.

Graham, W. D., and Neff, C. R., Optimal estimation of spatially variable recharge and transmissivity fields under steady-state groundwater flow, part 2. Case study, *J. Hydrol.,* 157, 267–285, 1994.

Graham, W. D., and Tankersley, C. D., Optimal estimation of spatially variable recharge and transmissivity fields under steady-state groundwater flow, part 1. Theory, *J. Hydrol.,* 157, 247–266, 1994.

Greacen, E. L., *Soil Water Assessment by the Neutron Method*, CSIRO, East Melbourne, Australia, 1981.

Greacen, E. L., Walker, G. R., and Cook, P. G., Evaluation of the filter paper method for measuring soil water suction, in *Intl. Conf. on Measurement of Soil and Plant Water Status*, Vol. 1, *Soils,* Utah State University, Logan, 1987, 137.

Green, W. H., and Ampt, G.A., Studies in soil physics, I. The flow of air and water through soils, *J. Agric. Sci.,* 4, 1–24, 1911.

Grismer, M. E., McWhorter, D. B., and Klute, A., Monitoring water and salt movement in soils at low solution contents, *Soil Sci.,* 141, 163, 1986.

Groenewoud, H. V., Methods and apparatus for measuring air permeability of the soil, *Soil Sci.,* 106, 275–279, 1968.

Grover, B. L., Simplified air permeameters for soil in place, *Soil Sci. Soc. Am. Proc.,* 19, 414, 1955.

Gureghian, A. B., A two-dimensional finite-element model for the simultaneous transport of water and reacting solutes through saturated-unsaturated porous media, Rept. ANL/ES-114, Argonne National Laboratory, Argonne, IL, 1981.

Hackett, G., Drilling and constructing monitoring wells with hollow-stem augers, part 1: drilling considerations, *Ground Water Monitoring Rev.,* 7, 51, 1987.

Hahne, T. W., and Thomsen, K. O., VOC distribution in vadose zone soil gas resulting from a fumigant release, in *Proc. Fifth Natl. Outdoor Action Conf. on Aquifer Restoration, Ground Water Monitoring, and Geophysical Methods*, NGWA, Dublin, OH, 1991, 699.

Hainsworth, J. M., and Aylmore, L. A. G., The use of computer-assisted tomography to determine spatial distribution of soil water content, *Aust. J. Soil Res.*, 21, 435, 1983.

Hakonson, T. E., Lane, L. J., and Springer, E. P., Biotic and abiotic processes, in *Deserts as Dumps*, Reith, C. C., and Thomson, B.M., Eds., University of New Mexico Press, Albuquerque, 1992, 101.

Hamilton, J. M., Daniel, D. E., and Olson, R. E., Measurement of hydraulic conductivity of partially saturated soils, in *Permeability and Groundwater Contaminant Transport, ASTM STP 746*, Zimmie, T. F., and Riggs, C. O., Eds., ASTM, Philadelphia, 1981, 182.

Hammermeister, D. P., Kneiblher, C. R., and Klenke, J., Borehole-calibration methods used in cased and uncased test holes to determine moisture profiles in the unsaturated zone, Yucca Mountain, Nevada, in *Proc. Conf. on Characterization and Monitoring of the Vadose (Unsaturated) Zone*, NWWA, Dublin, OH, 1985, 542.

Hansen, C. V., Estimates of freshwater storage and potential natural recharge for principal aquifers in Kansas, USGS Water-Resources Investigations Report 87-4230, Lawrence, Kansas, 1991.

Hansen, E. A., and Harris, A., Validity of soil-water samples collected with porous ceramic cups, *Soil Sci. Soc. Am. Proc.*, 39, 528–536, 1975.

Havlena, J. A., and Stephens, D. B., Vadose zone characterization using field permeameters and instrumentation, in *Current Practice in Ground Water and Vadose Zone Investigations, ASTM STP 1118*, Nielsen, D. M., and Sara, M. N., Eds., ASTM, Philadelphia, 1991.

Heermann, S., A laboratory experiment of axisymmetric infiltration into a layered soil, M.S. independent study, Department of Geoscience, New Mexico Institute Mining and Technology, Socorro, 1986.

Hendrickx, J. M. H., and Dekker, L. W., Experimental evidence of unstable wetting fronts in homogenous non-layered soils, in *Proc. Natl. Symp. on Preferential Flow*, Gish, T. J., and Shirmohammadi, A., Eds., ASAE, St. Joseph, MI, 1991, 22.

Hendrickx, J. M. H., Dekker, L. W., and Boersma, O. H., Unstable wetting fronts in water-repellent field soils, *J. Environ. Qual.*, 22(1), 109, 1993.

Herst, W. E., The borehole infiltration method for determining saturated hydraulic conductivity in the vadose zone: a deep water-table case in low permeable soil, M.S. independent study, Department of Geoscience, New Mexico Institute of Mining and Technology, Socorro, 1986.

Hill, M. C., A computer program (MODFLOWP) for estimating parameters of a transient, three-dimensional, groundwater flow model using non-linear regression, USGS Open File Report 9-484, U.S. Government Printing Office, Washington, DC, 1992.

Hillel, D., Krentos, V. D., and Stylianou, Y., Procedure and test of an internal drainage method for measuring soil hydraulic characteristics *in situ*, *Soil Sci.*, 114, 395, 1972.

Hillel, D., *Fundamentals of Soil Physics*, Academic Press, New York, 1980.

Hillel, D., *Applications of Soil Physics*, Academic Press, New York, 1980.

Hillel, D., Unstable flow in layered soils: a review, *Hydrol. Proc.*, 1, 143, 1987.

Hoffman,G. J., Dirksen, C., Ingvalson, R. D., Maas, E. V., Oster, J. D., Rawlins, S. L., Rhoades, J. D., and van Schilfgaarde, J., Minimizing salt in drain water by irrigation management, in *Agricultural Water Management*, 1, 233–252, Elsevier, Amsterdam, 1978.

Hornby, W. J., Zabcik, J. D., and Crawley, W., Factors which affect soil-pore liquid: a comparison of currently available samplers with two new designs, *Ground Water Monitoring Rev.*, 6(2), 61, 1986.

Horton, J. H., and Hawkins, R. H., Flow path of rain from the soil surface to the water-table, *Soil Sci.*, 100, 377, 1965.

Horton, R. E., An approach toward a physical interpretation of infiltration capacity, *Soil Sci. Soc. Am. Proc.*, 5, 399–417, 1940.

Hussen, A. A., and Warrick, A. W., Algebraic models for disc tension permeameters, *Water Resources Res.*, 29, 2779, 1993.

Huyakorn, P. S., Kool, J. B., Robertson, J. B., VAM2D — *Variably Saturated Analysis Model in Two Dimensions*, Version 5.0 with hysteresis and chained decay transport, Documentation and User's Guide, NUREG/CR-5352 (HGL/89-01), Hydrogeologic, Inc., Herndon, VA, 1989.

Huyakorn, P. S., Wu, Y. S., Panday, S., Park, N. S., *MAGNAS3: Multiphase Analysis of Groundwater, Non-aqueous Phase Liquid and Soluble Component in 3 Dimensions*, Version 1.0, Documentation and User's Guide, HydroGeoLogic, Inc., Herndon, VA, 1993.

Iwata, Shingo, Tabuchi, T., and Warkentin, B. P., *Soil-Water Interactions Mechanisms and Applications*, Marcel Dekker, New York, 1988.

Jabro, J. D., and Fritton, D. G., Simulation of water flow from a percolation test hole in a layered soil, *Soil Sci. Soc. Am. J.*, 54, 1214, 1990.

Jacob, C. E., Correlation of groundwater levels and precipitation on Long Island, New York, *Trans. Am. Geophys. Union*, 24, 564–573, 1943.

Jackson, R. D., and Taylor, S. A., Thermal conductivity and diffusivity, in *Methods of Soil Analysis, Part 1, Physical and Mineralogical Methods,* 2nd ed., Klute, A., Ed., American Society of Agronomy, Soil Science Society of America, Madison, WI, 1986, 945–956.

Jarvis, N. J., Leeds-Harrison, P. B., and Dosser, J. M., The use of tension infiltrometers to assess routes and rates of infiltration in a clay soil, *J. Soil Sci.*, 38, 633–640, 1987.

Jaynes, D. B., Temperature variations effect on field-measured infiltration, *Soil Sci. Soc. Am. J.*, 54, 305, 1990.

Jensen, M. E., and Haise, R., Estimating evapotranspiration from solar radiation, *ASCE J. Irrig. Drainage Div.*, 89, 15, 1963.

Jensen, M. E., Robb, D. C. N., and Franzoy, C. E., Scheduling irrigation using climate-crop soil data, *ASCE J. Irrig. Drainage Div.*, 96, 25, 1970.

Jeppson, R. W., Rawls, W. J., Hamon, W. R., and Schreber, D. L., Use of axisymmetric infiltration model and field data to determine hydraulic properties of soils, *Water Resources Res.*, 11, 127, 1975.

Johnson, P. C., Stanley, C. C., Kemblowski, M. W., Byers, D. L., and Colthart, J. D., A practical approach to the design, operation, and monitoring of *in situ* soil-venting systems, *Groundwater Mon. Rev.*, X(2), 1990, 159.

Jordan, C. F., A simple, tension-free lysimeter, *Soil Sci.*, 105, 81, 1968.

Jury, W. A., Spencer, W. F., and Farmerm, W. J., Behavior assessment model for trace organics in soils. I. model description. *J. Environ. Qual.*, 12(4), 558–564, 1983.

Jury, W. A., and Miller, E. E., Measurement of the transport coefficients for coupled flow of heat and moisture in medium sand, *Soil Sci. Soc. Am. Proc.*, 38, 551–557, 1974.

Jury, W. A., and Ghodrati, M., Overview of organic chemical environmental fate and transport modeling approaches, *Reactions and Movement of Organic Chemicals in Soils*, SSSA Special Publication Number 22, Madison, WI, 271–304, 1989.

Kean, F. K., Waller, M. J., and Layson, H. R., Monitoring moisture migration in the vadose zone with resistivity, *Groundwater*, 25, 562, 1987.

Keller, C., Engstrom, D., and West, F., *In situ* permeability measurements for the event in U1-C, LA-5425-MS (UC-35), Informal Report, Los Alamos National Laboratory, Los Alamos, NM, 1973.

Keller, C., and Lowry, B., A new vadose zone fluid sampling system for uncased holes, in *Proc. Fourth Natl. Outdoor Action Conf. on Aquifer Restoration, Groundwater Monitoring, and Geophysical Methods*, May 14–17, 1990, Las Vegas, NV, NWWA, Well Water Publishing Co., 1990, 3.

Keller, C., A serious vadose defense of groundwater from landfill contamination — new concept, old principles, *Proc. Sixth National Outdoor Action Conf. on Aquifer Restoration, Groundwater Monitoring, and Geophysical Methods*, NGWA, Dublin, OH, 1992, 73.

Kelly, T. E., Link, R. L., and Schipper, M. R., Effects of uranium mining on groundwater in Ambrosia Lake area, New Mexico, in *Memoir 38: Geology and Mineral Technology of the Grants Uranium Region 1979*, Rautman, C. A., Compiler, New Mexico Bureau of Mines and Mineral Resources, Socorro, NM, 1980, 313.

Kerfoot, H. B., Field evaluation of portable gas chromatographs, in *Proc. Fourth Natl. Outdoor Action Conf. on Aquifer Restoration, Ground Water Monitoring, and Geophysical Methods*, NGWA, Dublin, OH, 1990, 247.

Kerfoot, H. B., Soil gas surveys in support of design of vapor extraction systems, in *Soil Vapor Extraction Technology*, Pollution Technology Review No. 204, Pedersen, T. A., and Curtis, J. T., Eds., Noyes Data Corporation, Park Ridge, NJ, 1991.

Kerfoot, W. B., Pneumatic hammer soil vapor probes and miniature piezometers in use at a gasoline spill site, in *Proc. Third Natl. Outdoor Action Conf. on Aquifer Restoration, Ground Water Monitoring, and Geophysical Methods*, NGWA, Dublin, OH, 1989, 15.

Keys, W. S., and MacCary, L.M., Application of borehole geophysics to water-resources investigations, *Techniques of Water-Resources Investigations of the United States Geological Survey*, 2, (E1), 1971.

Keys, W. S., Crowder, R. E., and Henrich, W. J., Selecting geophysical logs for environmental applications, Groundwater management, in *Proc. Seventh Natl. Outdoor Action Conf. and Exposition, Aquifer Restoration, Ground Water Monitoring, and Geophysical Methods,* Las Vegas, NV, 1993, 15, 267–283.

Kickham, B., A field study on the water use of a *Dalea scoparia* plant in the northern Chihuahuan desert, M.S. thesis, New Mexico Institute of Mining and Technology, Socorro, 1987.

Kirk, S. T., and Campana, M. E., A deuterium-calibrated groundwater flow model of a regional carbonate-alluvial system, *J. Hydrol.*, 119, 357–388, 1990.

Kirkham, D., Field methods for determination of air permeability of soil in its undisturbed state, *Soil Sci. Soc. Am. Proc.*, 11, 93, 1946.

Kirkham D., and Powers, W. L., *Advanced Soil Physics*, John Wiley and Sons, New York, 1972.

Kirkham, R. R., Gee, G. W., and Jones, T. L., Weighing lysimeters for long-term water balance investigations at remote sites, *Soil Sci. Soc. Am. J.*, 48(5), 1203–1205, 1984.

Kirschner, F. E., Jr., and Bloomsburg, G. L., Vadose zone monitoring: An early warning system, *Ground Water Monitoring Rev.*, 8, 49, 1988.

Kitching, R., Edmunds, W. M., Shearer, T. R., Walton, N. R. G., and Jacovides, J., Assessment of recharge to aquifers, *Hydrol. Sci. Bull.*, Blackwell Scientific Publications, 25, 217–235, 1980.

Klenke, J. M., Flint, A. L., and Nicholson, R. A., A collimated neutron probe for soil-moisture measurements, in *Int. Conf. on Measurement of Soil and Plant Water Status*, Vol. 1, *Soils*, Utah State University, Logan, 1987, 21.

Klute, A., Water retention: laboratory methods, in *Methods of Soil Analysis, Part 1*, 2nd ed., Klute, A., Ed., Agron. Monogr. 9, ASA, CSSA, and SSSA, Madison, WI, 1986, 635.

Klute, A., and Dirksen, C., Hydraulic conductivity and diffusivity: laboratory methods, in *Methods of Soil Analysis, Part 1*, 2nd ed., Klute, A., Ed., Agron. Monogr. 9, ASA, CSSA, and SSSA, Madison, WI, 1986, 687.

Knight, R. B., and Haile, J. P., Construction of the Key Lake tailings facility, in *Proc. Intl. Conf. on Case Histories in Geotech. Eng.*, St. Louis, MO, 1984.

Kmet, P., and Lindorff, D. E., Use of collection lysimeters in monitoring sanitary landfill performance, in *National Water Well Association Conf. on the Characterization and Monitoring of the Vadose (Unsaturated) Zone*, Las Vegas, NV, National Water Well Association, Columbus, OH, 1983.

Knowlton, R. G., Phillips, F. M., and Campbell, A. R., A stable isotope investigation of vapor transport during groundwater recharge in New Mexico, New Mexico Water Resources Research Institute, Tech. Completion Report No. 237, New Mexico State University, Las Cruces, 1989.

Knowlton, R. G., Jr., A stable isotope study of water and chloride movement in natural desert soils, Dissertation, New Mexico Institute of Mining and Technical, Socorro, NM, 1990.

Knowlton, R. G., Parsons, A. M., and Gaither, K. N., Techniques for quantifying the recharge rate through unsaturated soils, in *Current Practices in Ground Water and Vadose Zone Investigations, ASTM STP 1118*, Nielsen, D. M., and Sara, M. N., Eds., ASTM, Philadelphia, 1992.

Knox, R. C., Sabatini, D. A., and Canter, L. W., *Subsurface Transport and Fate Processes*, Lewis Publishers, CRC Press, Boca Raton, FL.

Kool, J. B., Parker, J. C., and van Genuchten, M. T., Determining soil hydraulic properties form one-step outflow experiments by parameter estimation, I. Theory and numerical studies, *Soil Sci. Soc. Am. J.*, 49, 1348, 1985.

Kool, J. B., Parker, J. C., and van Genushten, M. T., ONESTEP: a nonlinear parameter estimation program for evaluating soil hydraulic properties from one-step outflow experiments, Bull. 85-3, Virginia Agricultural Experiment Station, Blacksburg, VA, 1985.

Koorevaar, P., Menelik, G., and Dirksen, C., Developments in soil science 13, *Elements of Soil Physics,* Elsevier, Amsterdam, 1983.

Kramer, J. H., Everett, L. G., and Eccles, L. A., Effects of well construction materials on neutron probe readings with implications for vadose zone monitoring, in *Proc. Fifth Natl. Outdoor Action Conf. on Aquifer Restoration, Ground Water Monitoring, and Geophysical Methods*, NWWA, Dublin, OH, 1991, 1303.

Krupp, H. K., and Elrick, D. E., Miscible displacement in an unsaturated glass bead medium, *Water Resources Res.*, 4, 809, 1968.

Kueper, B. H., and Frind, E. O., Two-phase flow in heterogenous porous media, 2-model application, *Water Resources Res.*, 27(6), 1058–1070, 1991.

Kunze, R. J., and Kirkham, D., Simplified accounting for membrane impedance in capillary conductivity determinations, *Soil Sci. Soc. Am. Proc.*, 26, 421, 1962.

Laase, A. D., A critical review of borehole permeameter solutions, M.S. independent study, Department of Geoscience, New Mexico Institute of Mining and Technology, Socorro, 1989.

Lang, A. R. G., Psychrometric measurement of soil water potential *in situ* under cotton plants, *Soil Sci.,* 106, 460–464, 1967.

Lapalla, E. G., Healy, R. W., and Weeks, E. P., Documentation of computer program VS2D to solve the equations of fluid flow in variably saturated porous media, USGS Water-Resources Investigations Report 83-4099, U.S. Geological Survey, Denver, CO, 1987.

Lapalla, E. G., and Ohland, G. L., Field measurements and modeling applied to estimating recharge rates and potential radionuclide migration; California Low Level Radioactive Waste Disposal Facility, Ward Valley, California, presented at ASA 83rd Annual Mtg., ASA, CSSA, and SSSA, Madison, WI, 1991.

Larson, M. B., and Stephens, D. B., A comparison of methods to characterize unsaturated hydraulic properties of mill tailings, In *Proc. Seventh Symp. on Management of Uranium Mill Tailings, Low-Level Waste and Hazardous Waste*, Colorado State University, Fort Collins, 1985.

Leonard, R. A., Davis, F. M., and Knisel, W. G., Groundwater Loading Effects of Agricultural Management Systems (GLEAMS): A tool to assess soil-climate-management-pesticide interactions, U.S. Department of Agriculture, Tifton, GA, 1989.

Leavitt, M. L., Spatial variability of the hydrologic parameters of soils at Sevilleta National Wildlife Refuge, M.S. thesis, New Mexico Institute of Mining and Technology, Socorro, 1986.

Linden, D. R., and Dixon, R. M., Infiltration and water-table effects of soil air pressure under border irrigation, *Soil Sci. Soc. Am. Proc.*, 37, 94, 1973.

Linsley, R. K., Kohler, M. A., and Paulhus, J. L. H., *Hydrology for Engineers*, McGraw-Hill, New York, 1975, 482.

Linsley, R. K., Kohler, M. A., and Paulhus, J. L. H., *Hydrology for Engineers*, 2nd ed., McGraw-Hill, New York, 1982.

Lohman, S. W., Groundwater hydraulics, U.S. Geological Survey Professional Paper 708, U.S. Government Printing Office, Washington, DC, 1972.

Lowry, W. E., and Narbutovskih, S. M., High resolution gas permeability measurements with the SEAMIST™ system, in *Proc. Fifth Natl. Outdoor Action Conf. on Aquifer Restoration, Ground Watering Monitoring, and Geophysical Methods*, NWWA, Dublin, OH, 1991, 685.

Machette, M. N., Geologic map of the San Acacia Quadrangle, Socorro Co., New Mexico, Geologic Quadrangle Map GQ-1415, U.S. Geological Survey, Washington, DC, 1978.

MacMahon, J. A., *The Audubon Society Nature Guides, Deserts*, 4th ed., Chanticleer Press, New York, 1985.

Mallick, A. U., and Rahman, A. A., Soil water repellency in regularly burned Canadian heartlands: comparison of three measuring techniques, *J. Environ. Manage.*, 20, 207, 1985.

Mallon, B., Martins, S. A., Houpis, J. L., Lowry, W., and Cremer, C. D., SEAMIST™ soil sampling for tritiated water: first year's results, *Proc. Sixth Natl. Outdoor Action Conf. on Aquifer Restoration, Ground Water Monitoring, and Geophysical Methods*, NGWA, Dublin, OH, 1992, 161.

Mann, J. F., Waste water in the vadose zone of arid regions: hydrologic interactions, *Ground Water*, 14, 367, 1976.

Marrin, D. L., and Thompson, G. M., Gaseous behavior of TCE overlying a contaminated aquifer, *Groundwater*, 25, 21, 1987.

Marrin, D. L., and Kerfoot, H. B., Soil-gas surveying techniques, *Environ. Sci. Technol.*, 22, 740, 1988.

Marshall, T. J., and Holmes, J. W., *Soil Physics*, Cambridge University Press, London, 1979 and 1992.

Marthaler, H. P., Vogelsanger, W., Richard, F., and Wierenga, P. J., A pressure transducer for field tensiometers, *Soil Sci. Soc. Am. Proc.*, 47(4), 624–627, 1983.

Massman, J., and Farrier, D. F., Effects of atmospheric pressures on gas transport in the vadose zone, *Water Resources Res.*, 28(3), 777, 1992.

Maxey, G. B., and Eakin, T. E., Groundwater in Railroad, Hot Creek, Reveille, Kawich and Penoyer Valleys, Nye, Lincoln and White Pine Counties, Nevada, in *Contributions to the Hydrology of Eastern Nevada*, State of Nevada, Office of the State Engineer, Water Resources Bulletin No. 12, 1951, 127.

Mazor, E., Applied Chemical and Isotopic Groundwater Hydrology, Halsted Press, New York, 1991.

McCleary-Hanagan, K., and Duffy, C. J., Implications of haline convection for waste disposal in closed basins, in *Geology and Hydrology of Hazardous-waste, Mining-waste, Wastewater, and Repository Sites in Utah*, Utah Geological Associates Publ. 17, 1989, 29.

McCord, J. T., Topographic controls on ground-water recharge at a sandy, arid site, M.S. independent study, New Mexico Institute of Mining and Technology, Socorro, NM, 1985.

McCord, J. T., and Stephens, D. B., Lateral moisture flow beneath a sandy hillslope without an apparent impeding layer, *Hydrol. Proc.*, 1, 225, 1987.

McCord, J. T., and Stephens, D. B., Comment on "Effect of groundwater recharge on configuration of the water-table beneath sand dunes" by T. C. Winter, *J. Hydrol.*, 95, 365, 1987.

McCord, J. T., and Stephens, D. B., Analysis of soil-water movement on a sandy hillslope, New Mexico Water Resources Research Institute, Tech. Completion Report No. 245, New Mexico State University, Las Cruces, 1989.

McCord, J. T., Stephens, D. B., and Wilson, J. L., Hysteresis and state-dependent anisotropy in modelling unsaturated hillslope processes, *Water Resources Res.*, 27, 1501, 1991.

McCuen, R. H., Rawls, W. H., and Brakensisk, D. L., Statistical analysis of the Brooks-Corey and the Green-Ampt parameters across soil textures, *Water Resources Res.*, 17(4), 1005–1013, 1981.

McElroy, D. L., The scale dependence of dispersivity in unsaturated mill tailings, unpublished M.S. independent study paper, New Mexico Institute of Mining and Technology, Socorro, 1987.

McKim,, H. L., Berg, R. L., McGaw, R. W., Atkins, R. T., and Ingersoll, J., Development of a remote reading tensiometer/transducer system for use in subfreezing temperatures, in *Proc. 2nd Conf. on Soil-Water Problems in Cold Regions*, AGU, Edmonton, Alberta, Canada, 1976.

McNeill, J. D., Electromagnetic terrain conductivity measurement at low induction numbers, Tech. Note TN-6, Geonics Limited, Mississauga, Ontario, Canada, 1980.

McWhorter, D. B., and Sunada, D. K., *Ground-Water Hydrology and Hydraulics*, Water Resources Publ., Ft. Collins, CO, 1977.

McWhorter, D. B., and Nelson, J. D., Unsaturated flow beneath tailings impoundments, *J. Geotech. Div., ASCE*, 105, GT 11, 1317, 1979.

McWhorter, D. B., Seepage in the unsaturated zone: a review, in *Proc. Spring Conv., Session on Seepage and Leakage from Dams and Impoundments*, Volpe, R. L., and Kelly, W. E., Eds., ASCE, New York, 1985.

Merrill, S. D., and Rawlins, S. L., Field measurement of soil water potential with thermocouple psychrometers, *Soil Sci.*, 113, 102, 1972.

Meyboom, P., Estimating groundwater recharge from stream hydrographs, *J. Geophys. Res.*, 66, 1203, 1961.

Meyboom, P., Estimate of Groundwater Recharge on the Prairies, Dolman, C. E., Ed., University of Toronto Press, Toronto, 1967, 128.

Mikel, C. D., Field and numerical studies of borehole infiltration tests in heterogeneous soils, M.S. independent study, Department of Geoscience, New Mexico Institute of Mining and Technology, Socorro, 1986

Miller, D. E., and Gardner, W. H., Water infiltration into stratified soil, *Soil Sci. Soc. Am. Proc.*, 26, 115, 1962.

Miller, C. T., Poirier-McNeil, M. M., and Mayer, A. S., Dissolution of trapped nonaqueous phase liquids: mass transfer constraints, *Water Resources Res.*, 26(11), 2783–2796, 1990.

Millington, R. J., Gas diffusion on porous media, *Science,* V-130, 100, 1959.

Moench, A. F., and Kiesel, C. C., Application of the convolution relation to estimating recharge from an ephemeral stream, *Water Resources Res.,* 6, 1087, 1970.

Montazer, P., Weeks, E. P., Thamir, F., Hammermeister, D., Yard, S. N., and Hofrichter, P. B., Monitoring the vadose zone in fractured tuff, *Ground Water Monitoring Rev.,* 8, 72, 1988.

Montazer, P., and Wilson, W. E., Conceptual Hydrologic Model of Flow in the Unsaturated Zone, Yucca Mountain, Nevada, Water Resources Investigations Report 84-4345, U.S. Geological Survey, Lakewood, CO, 1984.

Monteith, J. L., Gas exchange in plant communities, in *Environmental Control of Plant Growth*, Evans, L. T., Ed., Academic Press, New York, 1963, 95.

Montieth, J. L. and Unsworth, M. H., *Principles of Environmental Physics*, 2nd ed., Hodder & Stoughton, London, 1990.

Mook, W. G., Carbon-14 in hydrogeological studies, in *Handbook of Environmental Isotope Geochemistry,* Vol. 1, Fritz, P. and Fontes, J. C., Eds., Amsterdam, Elsevier, 1980, 49–74.

Morel-Seytoux, H. J., and Khanji, J., Derivation of an equation of infiltration, *Water Resources Res.,* 10, 795, 1974.

Moridis, G. J., and McFarland, M. J., Modeling soil water extraction from grain sorghum, in *Proc. Winter Mtg.,* ASCE, St. Joseph, MI, 1982.

Morrison, F. A., Stability of flow from a nuclear cavity, Report UCRL-13799, Lawrence Livermore National Laboratory, Livermore, CA, 1977.

Morrison, R., and Szecsody, J., Sleeve and casing lysimeters for soil pore water sampling, unpublished manuscript, 1983.

Mualem, Y., A new model for predicting the hydraulic conductivity of unsaturated porous media, *Water Resources Res.,* 12, 513, 1976.

Mualem, Y., Hydraulic conductivity of unsaturated porous media: generalized macroscopic approach, *Water Resources Res.,* 14, 325, 1978.

Mualem, Y., Hydraulic conductivity of unsaturated soils: prediction and formulas, in *Methods of Soil Analysis, Part 1: Physical and Mineralogical Methods,* Klute, A., Ed., ASA-SSSA, Madison, WI, 1986, 799.

Muehlberger, E. W., Smith, S. A., Garland, R. B., and Lauerman, B. C., Use of neutron logs to estimate the position of the water-table in mud rotary drillholes, in *Proc. Sixth Natl. Outdoor Action Conf. on Action Restoration, Ground Water Monitoring, and Geophysical Methods,* NGWA, Dublin, OH, 1992, 729.

Mullins, C. E., Matric potential, in *Soil Analysis: Physical Methods,* Smith, K. A., and Mullins, C. E., Eds., Marcel Dekker, New York, 1991, 75.

Mullins, J. A., Carsel, R. F., Sarbrough, J. E., and Ivery, A. M., *PRZM-2, A Model for Predicting Pesticide Fate in the Crop Root and Unsaturated Soil Zones,* User's Manual for Release Version 2.0, U.S. Evironmental Protection Agency, Environmental Research Lab, Athens, GA, 1993.

Narasimhan, T. N., and Witherspoon, P.A., Numerical model for saturated-unsaturated flow in deformable porous media, part I, theory, *Water Resources Res.,* 13(3), 657, 1977.

Narasimhan, T. N., The significance of the storage parameter in saturated-unsaturated ground-water flow, *Water Resources Res.,* 15(3), 569, 1979.

Narasimhan, T. N., and Dreiss, S. J., A numerical technique for modeling transient flow of water to a soil water sampler, *Soil Sci.,* 141, 230, 1986.

Narasimhan, T. N., TRUST: A computer program for transient and steady state fluid flow in multidimensional variably saturated deformable media under isothermal conditions, Report LBL-28927, Lawrence Berkeley Laboratory, Berkeley, CA, 1990.

Nasberg, V. M., The Problem of Flow in an Unsaturated Soil or Injection under Pressure, Izvestja Akademia Nauk, SSSR odt tekh Nauk. No. 9, translated by Mr. Reliant, 1973, B.R.G.M., France, 1951.

Neuman, S. P., and Witherspoon, P., Analysis of nonsteady flow with a free surface using the finite element method, *Water Resources Res.*, 7, 611, 1971.

Neuman, S. P, Saturated-unsaturated seepage by finite elements, *J. Hydraul. Div. Am. Soc. Civil Eng.*, 99, 2233, 1973.

Neuman, S. P., Feddes, R. A., and Bresler, E., Finite element simulation of flow in saturated-unsaturated soils considering water uptake by plants, Third Annual Report, Project AL0-SWC-77, Hydrodynamics and Hydraulic Engineering Lab, Technion, Haifa, Israel, 1974.

Neuman, S. P., Wetting front pressure head in the infiltration model of Green and Ampt, *Water Resources Res.*, 12, 564, 1976.

Neumann, H. H., and Thurtell, G. W., A Peltier cooled thermocouple dewpoint hygrometer for *in situ* measurement of water potentials, in *Psychrometry in Water Relations Research*, Brown, R. W., and van Haveren, B. P., Eds., Utah State University, Logan, 1972, 103.

New Mexico Environment Department, Solid waste management regulations, 1992.

New Mexico Water Resources Research Institute, Quantification of Groundwater Recharge Rates in New Mexico Using Bomb 36CL, Bomb-3H, and Chloride as Soil-Water Tracers, WRRI Report No. 220, Technical Completion Report, Project Nos. 1423638 and 1423654, New Mexico State University, Albuquerque, March 1987.

Nichols, W. D., Geohydrology of the unsaturated zone at the burial site for low-level radioactive waste near Beatty, Nye County, Nevada, USGS Water-Supply Paper 2312, U.S. Government Printing Office, Denver, CO, 1987.

Nielsen, D. R., Biggar, J. W., and Erh, K. T., Spatial variability of field-measured soil-water properties, *Hilgardia*, 42, 215–259, 1973.

Nilson, R. H., Lagus, P. L., McKinnis, W. B., Hearst, J. R., Burkhard, N. R., and Smith, C. F., Field measurements of tracer gas transport induced by barometric pumping, in *Proc. Third Int. Conf., Las Vegas, Nevada, April 12–16, 1992*, Vol. 1, American Nuclear Society, La Grange Park, IL, American Society of Civil Engineers, New York, 1992.

Nimmo, J. R., Rubin, J., and Hammermeister, D. P., Unsaturated flow in a centrifugal field: measurement of hydraulic conductivity and testing of Darcy's law, *Water Resources Res.*, 12, 513, 1987.

Nixon, P. R., and Lawless, G. P., Translocation of moisture with time in unsaturated soil profiles, *J. Geophys. Res.*, 65, 655, 1960.

Nnyamah, J. U., and Black, T. A., Rates and patterns of water uptake in a Douglas-fir forest, *Soil Sci. Soc. Am. J.*, 41: 972, 1977.

Nyhan, J., and Drennon, B., The measurement of soil water tension in a hydrologic study of waste disposal site design, LA-11460-MS (UC-721), Los Alamos National Laboratory, Los Alamos, NM, 1989.

Odeh, A. S., Comparison of solutions to a three-dimensional black-oil reservoir simulation problem, *J. Perol. Technol.*, 13, 1981.

Olhoeft, G. R., Direct detection of hydrocarbon and organic chemicals with ground penetrating radar and complex resistivity, in *Proc. Natl. Conf. on Petroleum Hydrocarbons and Organic Chemicals in Groundwater — Prevention, Detection, and Restoration*, NWWA, Dublin, OH, 1986, 284.

Olsen, R. L., and Davis, A., Predicting the fate and transport of organic compounds in groundwater, part I, *HMC*, 3, 39, 1990.

Olsen, H. W., Willden, A. T., Kiusalaas, N. J., Nelson, K. R., and Poeter, E. P., Volume-controlled hydrologic property measurements in triaxial systems, in *Hydraulic Conductivity and Waste Contaminant Transport in Soils, ASTM STP 1142*, Daniel, D. E., and Trautwein, S. J., Eds., ASTM, Philadelphia, 1993.

Ong, S. K., and Lion, L. W., Mechanisms for trichlorethylene vapor sorption onto soil minerals, *J. Environ. Qual.*, 20(1), 180–188, January–March 1991.

Organization for Economic Co-operation and Development (OECD), *The International HYDROCOIN Project, Level 2: Model Validation*, OECD, Paris, 1990.

Oster, C. A., Review of groundwater flow and transport models in the unsaturated zone, U.S. Nuclear Regulatory Commission Report NUREG/CR-2917, U.S. Gov. Print. Office, Washington, DC, 1982. (Also, Report PNL-4427, Battele/Pacific Northwest Laboratory, Richland, WA.)

Othman, M. A., Benson, C. H., Chamberlain, E. J., and Zimmie, T. F., Laboratory testing to evaluate changes in hydraulic conductivity of compacted clays caused by freeze-thaw: state-of-the-art, in *Hydraulic Conductivity and Waste Contaminant Transport in Soils, ASTM STP 1142*, Daniel, D. E., and Trautwein, S. J., Eds., ASTM, Philadelphia, 1993.

Parizek, R. R., and Lane, B. E., Soil-water sampling using pan and deep pressure-vacuum lysimeters, *J. Hydrol.*, 11, 1, 1970.

Parlange, J.-Y., and Hill, D. E., Theoretical analysis of wetting front instability in soils, *Soil Sci.*, 122, 236, 1976.

Parsons, A. M., Field simulation of waste impoundment seepage in the vadose zone: Site characterization and one-dimensional analytical modeling, M.S. thesis, Department of Geoscience, New Mexico Institute of Mining and Technology, Socorro, 1988.

Passioura, J. E., Determining soil water diffusivities from one-sided outflow experiments, *Aust. J. Soil Res.*, 15, 1, 1976.

Peck, A. J., and Williamson, D. R., Effects of forest clearing on groundwater, *J. Hydrol.*, 94, 47, 1987.

Penman, H. L., Natural evapotranspiration from open water, bare soil and grass, *Proc. R. Soc. London*, 193A, 120, 1948.

Perroux, K. M., and White, I., Designs for disc permeameters, *Soil Sci. Soc. Am. J.*, 52, 1205, 1988.

Peterson, D. M., and Wilson, J. L., Variably saturated flow between streams and aquifers, Tech. Completion Report, New Mexico Water Resources Res. Institute, Socorro, 1988.

Petsonk, A., The BAT method for *in situ* measurement of hydraulic conductivity in saturated soils, thesis in hydrologeology, University of Uppsala, Sweden, 1984.

Petsonk, A. M. and Torstensson, B. A., A multi-function vadose zone probe, 1984.

Phene, C. J., Hoffman, J., and Rawlins, S. L., Measuring soil matric potential *in situ* by sensing heat dissipation within a porous body: I. Theory and sensor construction, *Soil Sci. Soc. Am. Proc.*, 35, 27, 1971.

Philip, J. R., Numerical solution of equations of the diffusion type with diffusivity concentration dependent, *Trans. Faraday Soc.*, 51, 885, 1955.

Philip, J. R. The theory of infiltration. 1. The infiltration equation and its solution, *Soil Sci.*, 83, 345, 1957.

Philip, J. R., Approximate analysis of the borehole permeameter in unsaturated soil, *Water Resources Res.*, 21, 1025, 1985.

Philip, J. R., Approximate analysis of falling-head lined borehole permeameter, *Water Resources Res.*, 29, 3763, 1993.

Phillips, F. M., personal communication, 1995.

Phillips, F. M., Bentley, H. W., Davis, S. N., Elmore, D., and Swanick, G., Chlorine 36 dating of very old groundwater. II., Milk River Aquifer, Alberta, *Water Resources Res.*, 22(13) 2003–2016, 1986.

Phillips, F. M., Mattick, J. L., Duval, T. A., Elmore, D., and Kubik, P. W., Chlorine 36 and tritium from nuclear weapons fallout as tracers for long-term liquid and vapor movement in desert soils, *Water Resources Res.*, 24, 1877, 1988.

Phillips, F. M., Environmental tracers for water movement in desert soils: a regional assessment for the American Southwest, *Soil Sci. Soc. Am. J.*, 58(1), 15–24, 1994.

Phillips, S. P., Hamlin, S. N., and Yates, E. B., Geohydrology, water quality, and estimation of groundwater recharge in San Francisco, California, 1987–92, USGS Water-Resources Investigations Report 93-4019, Sacramento, CA, 1993.

Pignatello, J. J., Sorption dynamics of organic compounds in soils and sediments, *Reactions and Movement of Organic Chemicals in Soils*, SSSA Special Publication Number 22, Madison, WI, 45–80, 1989.

Plummer, L. N., Busenberg, E., and Michel, R. L., Dating of shallow groundwater: Comparison of the transient tracers ^{3}H/^{3}He, chlorofluorocarbons, and ^{85}Kr, *Water Resources Res.*, 30(6), 1693–1708 (Figure 2, p. 1695), June 1994.

Prunty, L., and Alessi, R. S., Prospects for fiber optic sensing in soil, in *Int. Conf. on Measurement of Soil and Plant Water Status*, Vol. 1, *Soils,* Utah State University, Logan, 1987, 261.

Purtymun, W. D., Koopman, F. C., Barr, S., Clements, W. E., Air volume and energy transfer through test holes and atmospheric pressure effects on the main aquifer, Informal Rept. LA-5725-MS, Los Alamos National Laboratory, Los Alamos, NM, 1974.

Raats, P. A. C., Unstable wetting fronts in uniform and non-uniform soils, *Soil Sci. Soc. Am. Proc.*, 37, 681, 1973.

Rabold, R. R., The results of borehole infiltration test with a shallow water-table, M.S. independent study, Department of Geoscience, New Mexico Institute of Mining and Technology, Socorro, 1984.

Ramirez, A., Daily, W., LaBrecque, D., Owen, E., and Chesnut, D., Monitoring an underground steam injection process using electrical resistance tomography, *Water Resources Res.*, 29, 73, 1993.

Rasmussen, V. P., and Campbell, R. H., A simple microwave method for the measurement of soil moisture, in *Proc. Int. Conf. Measurement of Soil and Plant Water Status*, Utah State University, Logan, 1, Soils, 1987, 275.

Rasmussen, T. C., Evans, D. D., Sheets, P. J., and Blanford, J. H., Permeability of Apache Leap Tuff: Borehole and core measurements using water and air, *Water Resources Res.*, 29, 1997, 1993.

Rawlins, S. L., and Campbell, G. S., Water potential: thermocouple psychrometry, in *Methods of Soil Analysis, Part 1*, 2nd ed., Klute, A., Ed., Agron. Monogr. 9, ASA, CSSA, and SSSA, Madison, WI, 1986, 597.

Reaber, D. W., and Stein, T. L., Design and installation of a detection monitoring network at a Class I facility in an arid environment, in *Proc. Fourth Natl. Outdoor Action Conf. on Aquifer Restoration, Ground Water Monitoring and Geophysical Methods*, NGWA, Dublin, OH, 1990, 299.

Redman, J. D., Kueper, B. H., and Annan, A. P., Dielectric stratigraphy of a CNAPL spill and implications for detection with ground penetrating radar, in *Proc. Fifth Natl. Outdoor Action Conf. on Aquifer Restoration, Ground Water Monitoring, and Geophysical Methods*, NWWA, Dublin, OH, 1991, 1017.

Reed, J. E., Bedinger, M. S., and Terry, J. E., Simulation procedure for modeling transient water-table and artesian stress and response, USGS Open-file Report 76-792, U.S. Government Printing Office, Washington, DC, 1976.

Reeve, M. J., and Carter, A. D., Water release characteristic, in *Soil Analysis: Physical Methods*, Smith, K. A., and Mullins, C. E., Eds., Marcel Dekker, New York, 1991, 111.

Reeve, R. C., A method for determining the stability of soil structure based upon air and water permeability measurements, *Soil Sci. Soc. Am. Proc.*, 17, 324, 1953.

Reeves, M., and Duguid, J. O., Water movement through saturated-unsaturated porous media: a finite element Galerkin model, Report ORNL-S4927, Oak Ridge National Laboratory, Oak Ridge, TN, 1975

Renard, K. G., *The Hydrology of Semiarid Rangeland Watersheds*, U.S. Department of Agriculture Agricultural Research Service, ARS 41-162, 25, 1970.

Reynolds, W. D., Elrick, D. E., and Topp, G. C., A reexamination of the constant head well permeameter method for measuring saturated hydraulics conductivity above the water table, *Soil Sci.*, 136, 250, 1983.

Reynolds, W. D., Elrick, D. E., and Clothier, B. E., The constant head well permeameter: effect of unsaturated flow, *Soil Sci.*, 139, 172, 1985.

Reynolds, W. D., and Elrick, D. E., In situ measurement of field-saturated hydraulic conductivity, sorptivity and the α-parameter using the Guelph permeameter, *Soil Sci.*, 140, 292, 1985.

Reynolds, W. D., and Elrick, D. E., A method for simultaneous *in situ* measurement in the vadose zone of field saturated hydraulic conductivity, sorptivity, and the conductivity-pressure head relationship, *Ground Water Monitoring Rev.*, 6, 84, 1986.

Rhoades, J. D., Monitoring soil salinity: A review of methods, in *Establishment of Water Quality Monitoring Programs*, American Water Resources Association, Minneapolis, MN, 1978, 150.

Richards, L. A., The usefulness of capillary potential to soil-moisture and plant investigators, *J. Agric. Res.*, 37, 719, 1928.

Richards, L. A., Capillary conduction of liquids in porous mediums, *Physics*, 1, 318, 1931.

Richards, S. J., and Weeks, L. V., Capillary conductivity values from moisture yield and tension measurements on soil columns, *Soil Sci. Soc. Am. Proc.*, 17, 206, 1953.

Riesenauer, A. E., Methods for solving problems of multidimensional, partially saturated steady flow in soils, *J. Geophys. Res.*, 68, 5725, 1963.

Ripple, C. D., Rubin, J., and Hylckama, T. E. A., Estimating steady-state evaporation rates from bare soils under conditions of high water-table, in *Techniques in Unsaturated-Zone Hydrology*, Water-Supply Paper 2019-A, USGS, Menlo Park, CA, 1972, A1, 39.

Robbins, G. A., and Gemmel, M. M., Factors requiring resolution in installing vadose zone monitoring systems, in *Proc. Fifth Natl. Symp. Expo. on Aquifer Restoration and Ground Water Monitoring*, NWWA, Worthington, OH, 1985, 184.

Robinson, A. R., and Rohwer, C., Measuring seepage from irrigation channels, USDA Tech. Bull. 1203, 1959.

Rojstocyer, S. A., Moisture movement through layered soils of highly contracting texture, M.S. dissertation, University of Illinois, Urbana, 1981.

Rorabaugh, M. I., Estimating changes in bank storage and groundwater contribution to streamflow, *Int. Assoc. Scientific Hydrology Publication,* 63, 432–441, 1964.

Rose, C. W., Evapotranspiration — some growth areas, in *Water Shed Management on Range and Forest Lands*, Utah Water Research Lab, Heady et al., Eds., Utah State University, Logan, 1976, 83.

Rosenberg, N. J., Blad, B. L., and Verma, S. B., *Microclimate: The Biological Environment*, 2nd ed., Wiley-Interscience, John Wiley & Sons, New York, 1983.

Rosenbloom, J., Mock, P., Turin, H. J., Lawson, P., and Brown, J., Look at groundwater quality to set VOC cleanup levels, *Remediation*, 2, 399, 1992.

Ross, B., A conceptual model of deep unsaturated zones with negligible recharge, *Water Resources Res.*, 20, 1627, 1984.

Rubin, J., Numerical method for analyzing hysteresis-affected, post-infiltration redistribution of soil moisture, *Soil Sci. Soc. Am. Proc.*, 31, 13, 1967.

Russell, H., Instrumentation and monitoring of excavations, *Bull. Assoc. Eng. Geol.*, 18, 91, 1981.

Rutledge, A. T., and Daniel, C. C., III, Testing an automated method to estimate groundwater recharge from streamflow records, *Ground Water J.*, 32(2), 180–189, 1994.

Salverda, A. P., and Dane, J. H., An examination of the Guelph permeameter for measuring the soil's hydraulic properties, *Geoderma*, 57, 405, 1993, Elsevier, Amsterdam.

Sammis, T. W., Evans, D. D., and Warrick, A. W., comparison of methods to estimate deep percolation rates, *Water Resources Bull.*, 18(3), 465–470, 1982.

Sammis, T. W. and Weeks, D. L., Hydrology and Water Resources in Arizona and the Southwest, *Proceedings* of the 1977 meetings of the Arizona Section, American Water Resources Assn. and the Hydrology Section, Arizona Academy of Science, Las Vegas, NV, April 15, 1977, 7, 235–240.

Scanlon, B. R., Unsaturated flow along arroyos and fissures in the Hueco Bolson, Texas, Final contract report to Texas Low-level Radioactive Waste Disposal Authority, Bureau of Economics and Geology, University of Texas, Austin, 1990.

Scanlon, B. R., Wang, F. P., and Richter, B. C., Field studies and numerical modeling of unsaturated flow in the Chihuahuan Desert, Texas Rep. Invest. 199. Bur. of Econ. Geol., University of Texas, Austin, 1991.

Scanlon, B. R., Evaluation of liquid and vapor water flow in desert soils based on chlorine 36 and tritium tracers and nonisothermal flow simulation, *Water Resources Res.*, 28, 285, 1992.

Schroeder, P. R., Gibson, A. C., and Smolen, M. G., The hydrologic evaluation of landfill performance (HELP) model, USEPA Doc. No. EPA/530-SW-84-010, U.S. Environmental Protection Agency, Office of Solid Waste and Emergency Response, Washington, DC, 1984.

Schwarzenbach R. P. and Westall, J., Transport of nonpolar organic compounds from surface to groundwater, laboratory sorption studies, *Env. Sci. Tech.*, 15(11), 1360–1366, 1981.

Sebenik, P. G., and Thomas, J. L., Water consumption by phreatophytes, *Prog. Agric. Ariz.*, 19, 10, 1967.

Severson, R. C., and Grigal, D. F., Soil solution concentrations: effect of extraction time using porous ceramic cups under constant tension, *Water Resources Bull.*, 12, 1161, 1976.

Shackelford, C. D., Waste-soil interactions that alter hydraulic conductivity, in *Hydraulic Conductivity and Waste Contaminant Transport, ASTM STP 1142*, Daniel, D. E., and Trautwein, S. J., Eds., ASTM, Philadelphia, 1994.

Sharma, M. L., Barron, R. J. W., and Williamson, D. R., Soil water dynamics of lateritic catchments as affected by forest clearing for pastures, *J. Hydrol.*, 94, 29, 1987.

Sharma, M. L., Recharge estimation from the depth-distribution of environmental chloride in the unsaturated zone — Western Australian examples, in *Estimation of Natural Groundwater Recharge*, Simmers, I., Ed., D. Reidel Publishing Co., Norwell, MA, 1988, 159.

Sharma, M. L., Barron, R. J. W., and Craig, A. B., Influence of land use on natural groundwater recharge in the unconfined aquifers of the Swan Coastal Plain, Western Australia, CSIRO Division of Water Resources, Wembley, Western Australia, 1988.

Sharma, M. L., and Craig, A. B., Comparative recharge rates beneath bambria woodland and two pine plantations on the Gnangara mound, Western Australia, in *Proc. Symp. on Groundwater Recharge*, Mandurah, Balkmas, Rotterdam, 1989, 11.

Shaw, B., and Baver, L. D., An electrothermal method for following moisture changes of the soil *in situ*, *Soil Sci. Soc. Am. Proc.*, 4, 78, 1939.

Shoemaker, C. A., Culver, T. B., Lion, L. W., and Peterson, M. G., Analytical models of the impact of two-phase sorption on subsurface transport of volatile chemicals, *Water Resources Res.*, 26, 745, 1990.

Siegrist, R. L., and Jenssen, P. C., Evaluation of sampling method effects on volatile organic compound measurements in contaminated soils, *Environ. Sci. Technol.*, 24, 1387, 1990.

Silka, L. R. and Jordan, D. L., Vapor analysis/extraction, in *Geotechnical Practice for Waste Disposal*, David, D. E., Ed., Chapman & Hall, London, 1993, 379–429.

Simmers, I., Natural groundwater recharge estimation in (semi-)arid zones: some state-of-the-art observations, in *Proc. Sahel Forum, State-of-the-Art of Hydrology and Hydrogeology in the Arid and Semi-Arid Areas of Africa, Ouagadougou, Burkina Faso*, Stout, G. E., and Demissie, M., Eds., UNESCO, New York, 1989, 374.

Simpson, E. S., and Duckstein, L., Finite-state mixing-cell models, in *Karst Hydrology and Water Resources,* Vol. 2, Yevjevich, V., Ed., Water Resource Publications, Ft. Collins, CO, 489–508, 1976.

Smith, W. O., Infiltration in sands and the relation to groundwater recharge, *Water Resources Res.*, 5, 539, 1967.

Smith, J. A., Chiou, C. T., Kammer, J. A., and Kile, D. E., Effect of soil moisture on the sorption of trichloroethene vapor to vadose zone soil at Picatinny Arsenal, New Jersey, *Environ. Sci. Technol,*, 24(5), 676–683, 1990.

Solomon, D. K., and Sudicky, E. A., Tritium and helium 3 isotope ratios for direct estimation of spatial variations in groundwater recharge, *Water Resources Res.,* 27(9), 2309–2319, September 1991.

Sophocleous, M., Analysis of water and heat flow in unsaturated-saturated porous media, *Water Resources Res.*, 15(5), 1195–1206, October 1979.

Sophocleous, M., Combining the soilwater balance and water-level fluctuation methods to estimate natural groundwater recharge: practical aspects, *J. Hydrol.,* 124, 229–241, 1991.

Sophocleous, M., Groundwater recharge estimation and regionalization: the Great Bend Prairie of central Kansas and its recharge statistics, *J. Hydrol.*, 137, 113–140, 1992.

Sophocleous, M., and Perry, C. A., Experimental studies in natural groundwater-recharge dynamics: the analysis of observed recharge events, *J. Hydrol.*, 81, 297–332, 1985.

Sosebee, R. E., Hydrology: the state of the science evapotranspiration, in *Water Shed Management in Range and Forest Lands*, Heady, Ed., Utah Water Resource Lab, Utah State University, Logan, 1976, 95.

Spanner, D. C., The Peltier effect and its use in the measurement of suction pressure, *J. Exp. Bot.*, 11, 145, 1951.

Srivastava, R., and Yeh, T.-C.J., Analytical solutions for one-dimensional, transient infiltration toward the water-table in homogeneous and layered soils, *Water Resources Res.*, 27(5), 753, 1991.

Stannard, D. I., Use of a hemispherical chamber for measurement of evapotranspiration, Open-File Report 88-452, U.S. Geological Survey, Denver, CO, 1990.

Stannard, D. I., *Tensiometers — Theory, Construction, and Use, Ground Water and Vadose Zone Monitoring, ASTM STP 1053*, Nielsen, D. M., and Johnson, A.I., Eds., ASTM, Philadelphia, 1990, 34–51.

Stark, A. M., Field simulation of waste impoundment seepage in the vadose zone: Late stage infiltration and drainage in a heterogeneous, partially saturated soil profile, M.S. thesis, Department of Geoscience, New Mexico Institute of Mining and Technology, Socorro, 1992.

Stein, T. L., Seasonal and spatial variations in soil-moisture flow at a semi-arid field site, M.S. independent study, Department of Geoscience, New Mexico Institute of Mining and Technology, Socorro, 1990.

Stephens, D. B., Analysis of constant head borehole infiltration tests in the vadose zone, Ph.D. dissertation, University of Arizona, Tucson, 1979.

Stephens, D. B., and Neuman, S. P., Analysis of borehole infiltration tests above the water table, Tech. Report on National Resource System 35, University of Arizona, Tucson, 1980.

Stephens, D. B., and Neuman, S. P., Vadose zone permeability tests: summary, *J. Hydraul. Eng.*, 108, 623, 1982.

Stephens, D. B., and Neuman, S. P., Vadose zone permeability tests: steady state, *J. Hydraul. Eng.*, 108, 640, 1982.

Stephens, D. B., and Neuman, S. P., Vadose zone permeability tests: unsteady flow, *J. Hydraul. Eng.*, 108, 660, 1982.

Stephens, D. B., and Neuman, S. P., Free surface and saturated-unsaturated analysis of borehole infiltration tests above the water-table, *Adv. Water Res.*, 5, 111, 1982.

Stephens, D. B., Groundwater flow and implications for groundwater contamination north of Prewitt, New Mexico, U.S.A., *J. Hydrol.*, 61, 391, 1983.

Stephens, D. B., Neuman, S. P., Tyler, S., Lambert, K., Watson, D., Rabold, R., Knowlton, R., Byers, E., and Yates, S., *In situ* determination of hydraulic conductivity in the vadose zone using borehole infiltration tests, Tech. Report 180, Water Resources Research Institute, New Mexico State University, Las Cruces, 1983.

Stephens, D. B., Tyler, S., Lambert, K., and Yates, S., Field experiments to determine saturated hydraulic conductivity in the vadose zone, in *Role of the Unsaturated Zone in Radioactive and Hazardous Waste Disposal*, Mercer, J., Ed., Butterworths, Stoneham, 1983, 113.

Stephens, D. B., Lambert, K., and Watson, D., Influence of entrapped air on field determinations of hydraulic properties in the vadose zone, in *Proc. Conf. on Vadose Zone Characterization and Monitoring of the Vadose (Unsaturated) Zone*, NWWA, Columbus, OH, 1983.

Stephens, D. B., Significance of natural groundwater recharge in site selection for mill tailings disposal, *AIME Trans. Soc. Mining Eng.*, 280, 2062, 1985.

Stephens, D. B., A field method to determine unsaturated hydraulic conductivity using flow nets, *Water Resources Res.*, 21, 45, 1985.

Stephens, D. B., and Rehfeldt, K. R., Evaluation of closed-form analytical models to calculate conductivity in a fine sand, *Soil Sci. Soc. Am. J.*, 49, 12, 1985.

Stephens, D. B., Knowlton, R. G., Jr., Stanfill, M., and Hirtz, E. M., Field study to quantify seepage from a fluid impoundment, in *Proc. Conf. on Characterization and Monitoring of the Vadose (Unsaturated) Zone*, NWWA, Dublin, OH, 1985, 283.

Stephens, D. B., and Knowlton, R., Jr., Soil water movement and recharge through sand at a semi-arid site in New Mexico, *Water Resources Res.*, 22, 881, 1986.

Stephens, D. B., Knowlton, R. G., Jr., McCord, J., and Cox, W., Field study of natural groundwater recharge in a semi arid lowland, Technical Completion Report, Project No. 1345679, New Mexico Water Resources Research Institute, WRRI Report No. 177, New Mexico State University, Las Cruces, 1986.

Stephens, D. B., Lambert, K., and Watson, D., Regression models for hydraulic conductivity and field test of the borehole permeameter, *Water Resources Res.*, 23, 2207, 1987.

Stephens, D. B. and Heermann, S. E., Dependence of anisotropy on saturation in a stratified sand, *Water Resources Res.*, 24, 770, 1988.

Stephens, D. B., Unruh, M., Havlena, J., Knowlton, R., Jr., Mattson, E., and Cox, W., Vadose zone characterization of low-permeability sediments using field permeameters, *Ground Water Monitoring Rev.*, 8, 59, 1988.

Stephens, D. B., Cox, W., and Havlena, J., Field study of ephemeral stream infiltration and recharge, New Mexico Water Resources Research Institute, Tech. Completion Report No. 228, New Mexico State University, Las Cruces, 1988.

Stephens, D. B., Parsons, A. M., Mattson, E. D., Black, K., Flanigan, K., Bowman, R. S., and Cox, W. B., A field experiment of three-dimensional flow and transport in a stratified soil, in *Proc. Int. Conf. and Workshop on the Validation of Flow and Transport Models for the Unsaturated Zone*, New Mexico State University, Socorro, 1988, 401.

Stephens, D. B., Hicks, E., and Stein, T., Field analysis on the role of three-dimensional moisture flow in groundwater recharge and evapotranspiration, New Mexico Water Resources Research Institute, Tech. Completion Report No. 260, New Mexico State University, Las Cruces, 1991.

Stephens, D. B., Application of the borehole permeameter, in *Advances in Measurements of Soil Physical Properties: Bringing Theory into Practice*, SSSA Special Publication No. 30, Topp, G. C., Reynolds, W. D., and Green, R. E., Eds., SSSA, Madison, WI, 1992, 43.

Stephens, D. B., A comparison of calculated and measured unsaturated hydraulic conductivity of two uniform soils in New Mexico, in *Proc. Int. Workshop on Indirect Methods for Estimating the Hydraulic Properties of Unsaturated Soils*, van Genuchten, M. T., Ed., University of California, Riverside, 1992, 249.

Stephens, D. B., Hydraulic conductivity assessment of unsaturated soils, in *Hydraulic Conductivity and Waste Contaminant Transport in Soils, ASTM STP 1142*, Daniel, D. E., and Trautwein, S. J., Eds., ASTM, Philadelphia, 1993.

Stephens, D. B., A perspective on diffuse natural recharge mechanisms in areas of low precipitation, *Soil Sci. Soc. Am. J.*, 58(1), 40–48, 1994.

Stevenson, C. D., Simple apparatus for monitoring land disposal systems by sampling percolating soil waters, *Environ. Sci. Eng.*, 12, 329, 1978.

Stockton, J. G., and Warrick, A. W., Spatial variability of unsaturated hydraulic conductivity, *Soil Sci. Soc. Am. Proc.*, 35, 847, 1971.

Stone, W. J., Natural recharge in southwestern landscape — examples from New Mexico, in Southwestern ground water issues, *Proc. Natl. Water Well Assoc.*, Tempe, AZ, October 20–22, 1986, NWWA, Dublin, OH, pp. 595–602.

Stutman, M., A novel passive sorbent collection apparatus for site screening of semivolatile compounds, in *Proc. 3rd Int. Conf. on Field Screening Methods for Hazardous Waste and Toxic Chemicals*, 1993.

Su, N., A formula for computation of time-varying recharge of groundwater, *J. Hydrol.*, 160, 123–135, 1994.

Swinbank, W. C., A measurement of vertical transfer of heat and water vapour and momentum in the lower atmosphere with some results, *J. Meteorol.*, 8, 135, 1951.

Taniguchi, M., and Sharma, M.L., Determination of groundwater recharge using the change in soil temperature, *J. Hydrol.*, 148, 219–229, 1993.

Tanner, C. B., Energy balance approach to evapotranspiration from crops, *Soil Sci. Soc. Am. Proc.*, 24, 1, 1960.

Tanner, B. D., Tanner, M. S., Dugas, W. A., Campbell, E. C., and Bland, B. L., Evaluation of an operational eddy correlation system for evapotranspiration measurements, in *Advances in Evapotranspiration, Proceedings of the National Conference on Advances in Evapotranspiration*, December 16–17, 1985, Chicago, American Society of Agricultural Engineers, St. Joseph, MI, 1985, 87–99.

Taylor, S. A., and Ashcroft, G. L., *Physical Edaphology: The Physics of Irrigated and Nonirrigated Soils*, W. H. Freeman, San Francisco, 1972.

Taylor and Stewart, 1960.

Terletskaya, N. M., Determination of permeability in dry soils, Hydroelectric Waterworks No. 2, February, Moscow, 1954.

Theis, C. V., Amount of groundwater recharge in the southern high plains, *Am. Geophys. Union Trans.*, 18, 564, 1937.

Theis, C. V., The source of water derived from wells, *Civil Eng.*, 10(5), 277, 1940.

Theis, C. V., The effect of a well on the flow of a nearby stream, *Am. Geophys. Union Trans.*, 22, 734, 1941.

Thomas, A. M., *In situ* measurement of moisture in soil and similar substances by "fringe" capacitance, *J. Sci. Instrum.*, 79, 1699, 1966.

Thomas, G. W., and Phillips, R. E., Consequences of water movement in macropores, *J. Environ. Qual.*, 8, 149, 1979.

Thornthwaite, C. W., and Holzman, B., Measurement of evaporation from land and water surface, *USDA Tech. Bull.*, 817, 1, 1942.

Thornthwaite, C. W., An approach toward a rational classification of climate, *Geogr. Rev.*, 38, 55, 1948.

Tillman, N., Meyer, T. J., and Ranlet, K., Use of soil gas surveys to enhance monitoring well placement and data interpretations, in *Proc. Third Nat. Outdoor Action Conf. on Aquifer Restoration, Ground Water Monitoring, and Geophysical Methods*, NGWA, Dublin, OH, 1989, 239.

Tollner, E. W., Cheshire, J. M., Jr., and Verma, B. P., X-ray computed tomography and nuclear magnetic resonance for studying soil systems, in *Int. Conf. on Measurement of Soil and Plant Water Status*, Vol. 1, *Soils*, Utah State University, Logan, 1987, 247.

Tolman, K. J., and Thompson, G. M., Mobile soil gas sampling and analysis: applications and limitations, in *Proc. Third Nat. Outdoor Action Conf. on Aquifer Restoration, Ground Water Monitoring, and Geophysical Methods*, NGWA, Dublin, OH, 1989, 251.

Toorman, A. F., Wierenga, P. J., and Hills, R. B., Estimation of soil-water parameters from transient laboratory experiments, in *Agronomy Abstracts*, ASA, CSSA, and SSSA, Madison, WI, 1990, 220.

Topp, G. C., Davis, J. L., and Annan, A. P., Electromagnetic determination of soil water content: Measurement in coaxial transmission lines, *Water Resources Res.*, 16, 579, 1980.

Topp et al. 1983.

Topp and Davis 1985.

Topp, G. C., The application of time-domain reflectometry (TDR) to soil water content measurement, in *International Conference on Measurement of Soil and Plant Water Status*, Vol. 1, *Soils*, Utah State University, Logan, 1987, 85.

Torstensson, B. A., A new system for groundwater monitoring, *Ground Water Monitoring Rev.*, 4, 131, 1984.

Travis, B. J., TRACR3D: a model of flow and transport in porous/fractured media, LA-9667-MS, UC-32 and UC-70, Issued: May 1984, Los Alamos National Laboratory, Los Alamos, NM, 1984.

Tyler, S. W., Moisture monitoring in large diameter boreholes, in *Proceedings from Conference on Characterization and Monitoring of the Vadose (Unsaturated) Zone*, NWWA, Dublin, OH, 1985, 97.

Tyler, S. W. and Walker, G. R., Root zone effects on tracer migration in arid zones, *SSSAJ*, 58(1), 25–30, 1994.

U.S. Bureau of Reclamation, *Earth Manual*, 2nd ed., U.S. Government Printing Office, Washington, DC, 1974.

U.S. Department of Agriculture, U.S. Salinity laboratory diagnosis and improvement of saline and alkaline soils, *Agriculture Handbook No. 60*, 1954 and 1969.

U.S. Environmental Protection Agency, Hazardous waste management system: Standards for owners and operators of hazardous waste treatment storage and disposal facilities, *Fed. Reg.*, 45, 33191, 33206, 33207, 1980.

U.S. Environmental Protection Agency, Hazardous waste management system: permitting requirements for land disposal facilities, *Fed. Reg.*, 47, 32328, 1982.

U.S. Environmental Protection Agency, *Permit Guidance Manual on Unsaturated Zone Monitoring for Hazardous Waste Land Treatment Units*, EPA 530-SW-86-040, U.S. Environmental Protection Agency, Washington, DC, 1986.

U.S. Environmental Protection Agency, *Test Methods for Evaluating Solid Waste, Volume 1B: Laboratory Manual, Physical/Chemical Methods*, PB88-239223, Part 2 of 4, U.S. Government Printing Office, Washington, DC, 1986.

Unruh, M. E., Corey, C., and Robertson, J. M., Vadose zone monitoring by fast neutron thermalization (neutron probe) — a 2-year case study, in *Proceedings on the Fourth National Outdoor Action Conference on Aquifer Restoration, Ground Water Monitoring, and Geophysical Methods*, NGWA, Dublin, OH, 1990, 431.

Vachaud, G., Vauclin, M., and Khanji, D., and Wakil, M., Effects of air pressure on water flow in an unsaturated stratified vertical column of sand, *Water Resources Res.*, 9(1), 160, 1973.

Valsaraj, K. T. and Thibodeaux, L. J., Equilibrium absorption of chemical vapors on surface soils, landfills and landfarms — a review, *J. Haz. Mat.*, 19, 79–99, 1988.

van Genuchten, M. T. and Alves, W. J., Analytical solutions of the one-dimensional convective-dispersive solute transport equation, *Technical Bulletin*, U.S. Department of Agriculture, 1661, 1982, 149 p.

van Genuchten, M. T., and Wierenga, P. J., Mass transfer studies in sorbing porous media, I. Analytical solutions, *Soil Sci. Soc. Am. J.*, 40, 473, 1976.

van Genuchten, M. T., Calculating the unsaturated hydraulic conductivity with a new closed-form analytical model, Res. Report 78-WR-08, Princeton University, Princeton, NJ, 1978.

van Genuchten, M. T., Numerical solutions of the one-dimensional saturated-unsaturated flow equation, Report 78-WR-09, Princeton University, Princeton, NJ, 1978.

van Genuchten, M. T., A closed-form equation for predicting the hydraulic conductivity of unsaturated soils, *Soil Sci. Soc. Am. J.*, 44, 892, 1980.

Veihmeyer, F. J., and Hendricksen, A. H., The moisture equivalent as a measure of the field capacity of soils, *Soil Sci. Soc. Am. Proc.*, 35, 631–634, 1931.

Verge, M. J., A three-dimensional saturated-unsaturated groundwater flow model for practical application, Ph.D. dissertation, University of Waterloo, Waterloo, Ontario, Canada, 1976.

Vomocil, J. A., Porosity, in *Methods of Soil Analysis, Part 1, Agron. Monogr. 9*, Black, C. A., Evans, D. D., White, J. L., Ensminger, L. E., and Clark, F. E., Eds., ASA, Madison, WI, 1965, 299.

Voss, C. I., Saturated-Unsaturated Transport (SUTRA): A finite-element simulation model for saturated-unsaturated, fluid-density-dependent groundwater flow with energy transport or chemically-reactive single-species solute transport, U.S. Geological Survey Report, Scientific Publications, Washington, DC, 1984.

Wagenet, R. J., Principles of Salt Movement in Soils, in *Chemical Mobility and Reactivity in Soil Systems, SSSA Special Publication Number 11*, Proceedings of a Symposium Sponsored by Divisions S-1, S-2, and A-5 of the ASA and SSSA in Atlanta, GA, November 29 to December 3, 1981, pub. SSSA/ASA, Madison, WI, 1983, 122.

Ward, A., Wells, L. G., and Phillips, R. E., Characterizing unsaturated hydraulic conductivity of western Kentucky surface mine spoils and soils, *Soil Sci. Soc. Am. J.*, 47, 847–854, 1983.

Warrick, A., Biggar, J. W., and Nielsen, D. R., Simultaneous solute and water transfer for unsaturated soil, *Water Resources Res.*, 7, 1216, 1971.

Warrick, A. W., Mullen, G. J., and Nielsen, D. R., Predictions of the soil water flux based upon field-measured soil-water properties, *Soil Sci. Soc. Am. J.*, 41(1), 14–19, 1977.

Warrick, A. W., and Amoozegar-Fard, A., Soil water regimes near porous cup water samplers, *Water Resources Res.*, 13, 203, 1977.

Warrick, A. W., and Yeh, T.-C. J., One-dimensional, steady vertical flow in a layered soil profile, *Adv. Water Resources*, 13, 207, 1990.

Warrick, A. W., Lomen, D. O., and Islas, A., An analytical solution to Richards' equation for a draining soil profile, *Water Resources Res.*, 26, 253, 1990.

Watson, K. K., An instantaneous profile method for determining the hydraulic conductivity of unsaturated porous materials, *Water Resources Res.*, 2, 709, 1966.

Watson, P., Sinclair, P., and Waggoner, R., Quantitative evaluation of a method for estimating recharge to the desert basins of Nevada, *J. Hydrol.*, 31(3/4), 335, 1976.

Weast, R. C., Ed., *Handbook of Chemistry and Physics*, CRC Press, Boca Raton, FL, 1974.

Weeks, E. P., and Sorey, M. L., Use of finite-difference arrays of observation wells to estimate evapotranspiration from groundwater in the Arkansas River Valley, Colorado, Water Supply Paper 2029-C, U.S. Geological Survey, Washington, DC, 1973.

Weeks, E. P., Field determination of vertical permeability to air in the unsaturated zone, U.S. Geological Survey Prof. Paper No. 1051, U.S. Government Printing Office, Washington, DC, 1978.

Weeks, E. P., Does the wind blow through Yucca Mountain?, in *Proc. Workshop V: Flow and Transport Through Unsaturated Fractured Rock — Related to High-Level Radioactive Waste Disposal*, Evans, D. D., and Nicholson, T. J., Eds., NUREG-CP-0040, U.S. Nuclear Regulatory Commission, U.S. Government Printing Office, Washington, DC, 1991.

Weeks, E. P., Does the wind blow through Yucca Mountain?, in *Proc. Workshop V: Flow and Transport Through Unsaturated Fractured Rock — Related to High-Level Radioactive Waste Disposal*, January 7–10, 1991, Tucson, AZ, Evans, D. D., and Nicholson, T. J., Eds., NUREG-CP-0040, U.S. Nuclear Regulatory Commission, U.S. Government Printing Office, Washington, DC, 43, 1993.

Wellings, S. R., Recharge of the upper chalk aquifer at a site in Hampshire, England, 1. Water balance and unsaturated flow, *J. Hydrol.*, 69, 259–273, 1984.

Weppernig, R., and Stute, M., Dating of shallow groundwater: comparison of the transient tracers ^{3}H/^{3}He, chlorofluorocarbons, and ^{85}Kr, *Water Resources Res.*, 30(6), 1693–1708, (Figure 2, p. 1695), June 1994.

Wierenga, P. J., Hagan, R. M., and Nielsen, D. R., Soil temperature profiles during infiltration and redistribution of cool and warm irrigation water, *Water Resources Res.*, 6, 230–238, 1970.

Wierenga, P. J., Solute distribution profiles computed with steady-state and transient water movement models, *Soil Sci. Soc. Am. J.*, 41, 1050, 1977.

Willis, W. O., Evaporation from layered soils in the presence of a water-table, *Soil Sci. Soc. Am. Proc.*, 24, 239, 1960.

Wilson, J. L., Contrad, S. H., Mason, W. R., Peplinski, W., and Hagen, E., *Laboratory Investigation of Residual Liquid Organics*, USEPA/600/6-90/004, Robert S. Kerr Environmental Research Laboratory, Ada, OK, 1990.

Wilson, L. G., DeCook, K. J., and Neuman, S. P., Regional research for southwest alluvial basins: Final report to the U.S. Geological Survey, Water Resources Research Center, Department of Hydrology and Water Resources, University of Arizona, Tucson, 1980.

Wilson, L. G., Methods for sampling fluids in the vadose zone, in *Ground Water and Vadose Zone Monitoring, ASTM STP 1053*, Nielsen, D. M., and Johnson, A. I., Eds., ASTM, Philadelphia, 1989, 7.

Wilson, J. L. and Gelhar, L. W., Dispersive mixing in a partially saturated porous medium, Water Resources and Hydrodynamics, Department of Civil Engineering, Massachusetts Institute of Technology, Boston, MA, Report No. 191, 1974.

Winger R. J., Jr., In-place permeability tests used for subsurface drainage investigation, U.S. Bureau of Reclamation, Division of Drainage and Groundwater Engineering, Denver, CO, 1965.

Winter, T. C., Effect of groundwater recharge on configuration of the water table beneath sand dunes and on seepage in lakes in the sandhills of Nebraska, USA, *J. Hydrol.*, 86, 221, 1986.

Wood, W. W. and Petraitis, M. J., Origin and distribution of carbon dioxide in the unsaturated zone of the Southern High Plains of Texas, *Water Resources Res.*, 20(9), 1193–1208, 1984.

Wooding, R.A., Steady infiltration from a shallow circular pond, *Water Resources Res.*, 4, 1259, 1968.

Yao, T.-M. and Hendrickx, J. M. H., Stability of wetting fronts in dry homogeneous soil under low infiltration rates, *Soil Sci. Soc. Am. J.,* in press.

Yao, T.-M. and Hendrickx, J. M. H., personal communication, 1993.

Yeh, T.-C. J., One-dimensional steady infiltration in heterogeneous soils, *Water Resources Res.*, 25, 2149, 1989.

Yeh, T.-C. J., Guzman, A., Srivastava, R., and Gagnard, P. E., Numerical simulation of the wicking effect in liner systems, *Ground Water J.*, 32(1), 2, 1994.

Yeh, T.-C. J., and Gelhar, L. W., Unsaturated flow in heterogeneous soils, in *Role of the Unsaturated Zone in Radioactive and Hazardous Waste Disposal*, Mercer, J.W. et al., Eds., Ann Arbor Science, Ann Arbor, MI, 1983, 71.

Yeh, T.-C. J., Gelhar, L., and Gutjahr, A. L., Stochastic analysis of unsaturated flow in heterogeneous soils, 1. Statistically isotropic media, *Water Resources Res.*, 21, 447, 1985.

Yeh, G. T. and Ward, D. S., FEMWATER: A Finite-Element Model of Water Flow through Saturated-Unsaturated Porous Media, ORNL-5567, Oak Ridge National Laboratory, Oak Ridge, TN, 1980.

Yeh, T.-C. J., Gelhar, L., and Gutjahr, A. L., Stochastic analysis of unsaturated flow in heterogeneous soils, 2. Stochastically anisotropic media with variable α, *Water Resources Res.*, 21, 457, 1985.

Yeh, T.-C. J., Gelhar, L., and Gutjahr, A. L., Stochastic analysis of unsaturated flow in heterogeneous soils, 3. Observations and applications, *Water Resources Res.*, 21, 471, 1985.

Yong, R. N., and Hoppe, E. J., Application of electric polarization to contaminant detection in soils, Geotechnical Research Centre, McGill University, Montreal, Quebec, Canada, 1989.

Youngs, E. G., An infiltration method of measuring the hydraulic conductivity of unsaturated porous materials, *Soil Sci.*, 109, 307, 1964.

Zanger, C. N., Theory and problems of water percolation, United States Bureau of Reclamations Engineering Monograph No. 8, 1953.

Zaslavsky, D., Theory of unsaturated flow into a non-uniform soil profile, *Soil Sci.*, 97, 400, 1964.

Zaslavsky, D., and Sinai, G., Surface hydrology: I — Explanation of phenomena, *ASCE J. Hydrol. Div.*, 107, 1, 1981.

Zaslavsky, D., and Sinai, G., Surface hydrology: III — Causes of lateral flow, *ASCE J. Hydrol. Div.*, 107, 37, 1981.

Zaslavsky, D., and Sinai, G., Surface hydrology: IV — Flow in sloping, layered soil, *ASCE J. Hydrol. Div.*, 107, 53, 1981.

Zaslavsky, D., and Sinai, G., Surface hydrology: V — In-surface transient flow, *ASCE J. Hydrol. Div.*, 107, 65, 1981.

Zegelin, S. J., and White, I., Design for a field sprinkler infiltrometer, *Soil Sci. Soc. Am. J.*, 46, 1129, 1982.

Zegelin, S. J., and White, I., Improved field probes for soil water content and electrical conductivity measurement using time domain reflectometry, *Water Resources Res.*, 25(11), 2367–2376, 1989.

Zimmerman, C. F., Price, M. T., and Montgomery, J. R., A comparison of ceramic and teflon *in situ* samplers for nutrient pore water determinations, in *Estuarine and Coastal Marine Science*, Academic Press, London, 1978, 93.

Zohdy, A. A. R., Eaton, G. P., and Mabey, D. R., Application of surface geophysics to groundwater investigations, *Techniques of Water-Resources Investigations of the U.S. Geological Survey*, 2, D1, 1974.

Index

A

Absorption, horizontal infiltration, 25–26, 73
Accelerator/mass spectrometer, 64
Acoustic velocity logs, 229
Active soil gas detector, 235
Adsorption
  breakthrough and, 34
  onto colloidal surface, 39
  linear, 36
  nonlinear, 36
  of organic compounds, 36
  soil gas sampling and, 242
Adsorption-desorption reactions, 35
Adsorptive forces, and matric potential, 3–4
Advection, 241
  defined, 31
  wind-induced, 104
Advection-dispersion equation, 34, 39
Air
  entrapped. See Entrapped air
  molar mass of, 31
  temperature of, 51, 56
  in three-phase flow, 27
Air and gas permeameters, 153–156
  borehole type, 155–156
  natural air pressure method, 156
  vertical cylinder type, 154–155
Air compression, 78
Air entrapment. See Entrapped air
Air-entry permeameter, 142–144, 159–160, 167
Air-entry pressure head, 10–13
  calculation of, 144
  in hanging water column method, 188–189
  for selected porous materials, 246
  in steady-state flow methods, 171
Air permeability, 300
  factors affecting, 153–154
  testing within the SEAMIST™ system, 236
Air pressure. See Atmospheric pressure
Air-water interface
  force imbalance at the, 4
  as semipermeable membrane, 6
Alice Springs, Australia, 110, 112
Alluvial aquifers, 131
Alluvium, 113, 122
  fans, 108
  surficial, 285
Alundum™, 246, 252

Ambrosia deltoidea, 101–102
American Society for Testing and Materials, methods
  D-1950, 4
  D-2434–68, 139
  D-3385–88, 151
  D-5084–90, 139
  D-5126, 141, 144, 151, 156
  D-5220–92, 221
  D-5298–92, 215
  Guide D-4700, 243
Americium, 220, 222
Analytical models. *See* Numerical simulation
Anemometer, 49
Anion exclusion, 34–35
Anisotropy of hydraulic conductivity, 23–24, 84, 124–125, 129, 157
Antecedent soil moisture, 74
  displacement of, 82–84
  effect on ponded infiltration rate, 77
  limitations of measurement by electrical resistivity, 231
Apache Leap Tuff site, Arizona, 156
Apparent electrical conductivity, 231
Apparent resistivity of the soil, 230, 232
Aquiclude, 2
AQUIFEM-N, 60
Aquifers
  alluvial, 131
  Darcy flux in, 57–58
  hydraulic gradient in, *vs.* vadose zone, 17
  perched. *See* Perched aquifer
  response to recharge, 109
  Richards equation modification for, 28–29
  in San Juan Basin, 127–133
  as source of soil gas, 241
  stream-aquifer system, 117–118
  water-level fluctuations, 58–59
Aquitard, 2, 17, 108
Arkansas River Valley, Colorado, evapotranspiration in, 50
Arroyos (wadis), 116, 124
Artifical recharge, 126
Atmosphere, chlorofluorocarbon concentration, 69
Atmospheric pressure
  effect on soil gas sampling, 241–242
  gaseous-phase transport and, 104
  in natural air pressure permeameters, 156
  and soil-water potential, 1–2, 8